Texts in Philosophy

Volume 11

# PhiMSAMP

Philosophy of Mathematics: Sociological
Aspects and Mathematical Practice

Texts in Philosophy Series Editors
Vincent F. Hendriks                          vincent@hum.ku.dk
John Symons                                  jsymons@utep.edu
Dov Gabbay                                   dov.gabbay@kcl.ac.uk

# PhiMSAMP

## Philosophy of Mathematics: Sociological Aspects and Mathematical Practice

edited by

**Benedikt Löwe**

and

**Thomas Müller**

ISBN 978-1-904987-95-6


College Publications
Scientific Director: Dov Gabbay
Managing Director: Jane Spurr
Department of Computer Science
King's College London, Strand, London WC2R 2LS, UK

http://www.collegepublications.co.uk


Original cover design by orchid creative          www.orchidcreative.co.uk
Printed by Lightning Source, Milton Keynes, UK

---

# Contents

# Preface

Mathematics is a lively science with its own social norms, rules of behaviour, and a special technique for generating new results: mathematical proof. Proof is often contrasted with the epistemic techniques of all other sciences, stressing its deductive nature in comparison to induction and abduction in other sciences. To practitioners of mathematics, mathematical knowledge feels more objective than other knowledge, and authors going back to Plato have paid tribute to this intuition of higher objectivity by granting mathematical knowledge a special status.

In an attempt to explain this undeniable feeling of exceptional objectivity, philosophy of mathematics has emphasized the deductive nature of mathematics, and the debates on foundations of mathematics in the early 20th century tied philosophy of mathematics closely to the developing field of mathematical logic. Research areas like proof theory suggested that there is a direct correspondence between a mathematical proof (which is done by human beings on a piece of paper or the blackboard) and its formal representation (which is a sequence of symbols in a formal language). The idea that all of mathematical activity can in principle be represented by sequences of formal statements in some adequate system of logic obstructed the view towards what mathematicians are really doing. In fact, sociology of science mostly ignored mathematics presumably under the assumption that the human component of mathematical research is negligible.[1]

We strongly believe that a philosophy of mathematics needs to be adequate for actual mathematical practice and provide an explanation for the procedures of the mathematical community. Empirical and conceptual work needs to be brought together in order to reach this goal. In recent years, more philosophers of mathematics have started to take mathematical practice seriously, as is witnessed by the conferences entitled *Perspectives on Mathematical Practices* (Brussels, 2002 & 2007) and the formation of *Association for the Philosophy of Mathematical Practice* (APMP) in 2009.

But the solution to the puzzle of the objectivity of mathematical knowledge cannot be solved by philosophers alone. Involvement with mathematical practice means that other disciplines, such as the history of science, the fields of science education, sociology of science, cognitive science, and possibly psychology hold parts of the answer to our questions. Interdisciplinary exchange of ideas is a necessity in our attempts to understand the special nature of mathematics. The purpose of the research network PhiMSAMP ("Philosophy of Mathematics: Sociological Aspects and Mathematical Prac-

---

[1]Cf. (Heintz, 2000, p. 9): "[d]ie Soziologie [begegnet] der Mathematik mit einer eigentümlichen Mischung aus Devotion und Desinteresse" (sociology meets mathematics with a queer mixture of devotion and lack of interest).

tice") was to catalyze this interdisciplinary exchange and to create a basis of communication between the involved research areas.

### The network PhiMSAMP

The network PhiMSAMP was a *Wissenschaftliches Netzwerk* funded by the *Deutsche Forschungsgemeinschaft* (DFG) (MU 1816/5-1) and was coordinated by the two editors of this volume. The ideas for PhiMSAMP go back to 2004, and we had a preliminary workshop (PhiMSAMP-0) in Bonn on 8 May 2005, during which the future members of the network presented their research interests to each other.

The network started its activities in 2006. The first event was a satellite workshop to the conference GAP.6 in Berlin with the title *Towards a new epistemology of mathematics* (14–16 September 2006; PhiMSAMP-1) with an open call for papers. We selected 15 papers to be presented at PhiMSAMP-1 and edited a special issue of the journal *Erkenntnis* with selected papers of the workshop afterward (Buldt et al., 2008a). The editors of this special issue wrote an introductory paper (Buldt et al., 2008b) that can serve as a manifesto for the PhiMSAMP network, explaining how empirical and historical studies can serve as background for a practice-based philosophy of mathematics.

The declared main goal of the network was the production of the present volume. To that end, the network members formed groups of several authors from different background in order to shed light on particular (philosophical) aspects of mathematical practice. Three of our six workshops were dedicated to this internal exchange between network members. At PhiMSAMP-2 in Utrecht (19–21 October 2007), we encountered researchers from sociology and cognitive science (Vanessa Dirksen, Konstanz; Keith Stenning, Edinburgh; Vincent Buskens, Utrecht; and Hansjörg Neth, then Troy NY) who gave tutorials for the philosophical members of the network. The paper by De Cruz, Neth, and Schlimm in this volume is an immediate product of the resulting collaboration. The workshops PhiMSAMP-4 (Brussels, 24–25 October 2008) and PhiMSAMP-5 (Hatfield, 29–30 June 2009) were purely internal workshops during which network members worked on the paper projects started earlier.

The conference PhiMSAMP-3, held in Vienna, 16–18 May 2008, was a well attended international event. It was organized by Stefan Götz and Esther Ramharter and funded by the *Institut für Philosophie der Universität Wien*, the *Fakultät für Mathematik der Universität Wien*, the *Bundesministerium für Wissenschaft und Forschung*, and the *Magistratsabteilung 7 der Stadt Wien*. In total, 32 talks were presented to a large number of participants of which we list the speakers and the titles of the talks on pp. xii–xiii. After the conference, we issued a call for papers asking the speakers to submit paper versions of their presentations for publication in this volume (see below).

The final workshop of the network will take place after this book is published: PhiMSAMP-6 will be held in Amsterdam and Utrecht on 22 and 23 April 2010 in cooperation with the Dutch *Vereniging voor Logica en Wijsbegeerte der Exacte Wetenschappen* in order to celebrate the publication of this volume.

The members of PhiMSAMP were involved in a number of other conferences and workshops related to PhiMSAMP activities as well. In March 2007, the conference PMP 2007 ("Perspectives on Mathematical Practices") was held in Brussels, co-organized by François and Van Kerkhove (and with Löwe and Van Kerkhove in the programme committee). Eva Müller-Hill was one of the organizers of the 18th *Novembertagung* (November 2007) held in Bonn: the *Novembertagung* is a meeting of doctoral students and junior researchers in philosophy and history of mathematics; in 2007, it was expanded—under the influence of the PhiMSAMP spirit—to include mathematics education in its scope.

François, Löwe, Müller, and Van Kerkhove organized the conference *Foundations of the Formal Sciences VII: Bringing together philosophy and sociology of science* in Brussels (21–24 October 2008). This conference had a slightly wider scope, including other sciences than mathematics; it hosted the internal workshop PhiMSAMP-4 as a satellite event, and the organizers will publish a proceedings volume with selected papers (François et al., 2010). The conference *Two Streams in the Philosophy of Mathematics: Rival Conceptions of Mathematical Proof* at the University of Hertfordshire in Hatfield was also closely related to PhiMSAMP topics and was taken as an opportunity for our network to hold the internal meeting PhiMSAMP-5; that internal meeting was also used for discussions about the future developments of the PhiMSAMP platform.

During its existence, the network PhiMSAMP had nodes in Amsterdam (The Netherlands), Bonn (Germany), Bremen (Germany), Brussels (Belgium), Darmstadt (Germany), Dortmund (Germany), Fort Wayne IN (United States of America), Konstanz (Germany) Montréal QC (Canada), Siegen (Germany), Utrecht (The Netherlands), and Vienna (Austria). Due to the mobility of network members, changing jobs between different universities and countries, some of the nodes were replaced by others during the life of the network. The following researchers were members of the network for at least some time during the existence of PhiMSAMP (in alphabetic order): Joachim Bromand, Bernd Buldt, Stephan Cursiefen, Helen De Cruz, Johannes Emrich, Karen François, Hannes Leitgeb, Katja Lengnink, Nikola Leufer, Benedikt Löwe, Thomas Müller, Eva Müller-Hill, Henrik Nordmark, Esther Ramharter, Dirk Schlimm, Marius Thomann, Susanne Prediger, and Bart Van Kerkhove.

Of course, there are some activities of the network that are not properly represented in this volume. As a representative for all these, we would like to mention Eva Müller-Hill (2010) who wrote her PhD dissertation on *socio-empirically informed epistemology of mathematics* during the duration of PhiMSAMP (her time as a PhD student coincided almost exactly with the duration of the DFG network). Her work (cf. also Wilhelmus, 2008; Löwe et al., 2010; Müller-Hill, 2009) is closely related to the network and exemplifies the empirical side of PhiMSAMP.

**This volume**

The seventeen papers in this volume fall into two categories. Eight of the papers have been written as part of the network activities. These papers have been conceived during the internal meetings of the network, and—in most cases—written in collaboration among network members. In the final phase, these internal papers were read by the other network members in a system of open comment and criticism. These eight papers are the papers by Buldt / Schlimm, De Cruz / Neth / Schlimm, François / Van Kerkhove, Geist / Löwe / Van Kerkhove, Lengnink / Leufer, Lengnink / Schlimm, Löwe / Müller, and Ramharter.

The remaining nine papers represent written versions of talks given at the conference PhiMSAMP-3, *Is Mathematics Special?* After the conference, all speakers were invited to submit paper versions of their talks. All submissions underwent a thorough refereeing process according to the standards of leading journals of our field, involving 21 referees. Finally, nine of these papers were accepted for publication: Borovik, Bremer, François / Van Bendegem, Johansen, Larvor, Nickel, Schiemer, Sørensen, and Wagner. Not all of these papers are research papers in philosophy: in order to tighten the link between mathematical practice and philosophy of mathematics, we also included two first-hand accounts of thoughts about mathematical practice by research mathematicians (Borovik and Nickel).

During the production of the volume, Edgar Andrade (Amsterdam) helped us with typesetting and bibliographic work. The production of the volume was generously funded by the *Deutsche Forschungsgemeinschaft* (DFG, MU 1816/5-1). Certainly, this book would not exist without the help and support of the numerous referees involved and the members of the PhiMSAMP network.

Amsterdam & Utrecht, March 2010                                    B. L.  T. M.

# Bibliography

Buldt, B., Löwe, B., and Müller, T., editors (2008a). *Towards a new epistemology of mathematics*. Special issue of the journal *Erkenntnis*; Volume 68, Issue 3.

Buldt, B., Löwe, B., and Müller, T. (2008b). Towards a new epistemology of mathematics. *Erkenntnis*, 68(3):309–329.

François, K., Löwe, B., Müller, T., and Van Kerkhove, B., editors (2010). *Foundations of the Formal Sciences VII. Bringing together philosophy and sociology of science*. Studies in Logic. College Publications, London. To appear.

Heintz, B. (2000). *Die Innenwelt der Mathematik. Zur Kultur und Praxis einer beweisenden Disziplin*. Springer, Vienna.

Löwe, B., Müller, T., and Müller-Hill, E. (2010). Mathematical knowledge: A case study in empirical philosophy of mathematics. In Van Kerkhove, B., De Vuyst, J., and Van Bendegem, J. P., editors, *Philosophical Perspectives on Mathematical Practice*, volume 12 of *Texts in Philosophy*, pages 185–203. College Publications, London.

Müller-Hill, E. (2009). Formalizability and knowledge ascriptions in mathematical practice. *Philosophia Scientiae*, 13(2):21–43.

Müller-Hill, E. (2010). *Die epistemische Rolle formalisierbarer mathematischer Beweise. Formalisierbarkeitsbasierte Konzeptionen mathematischen Wissens und mathematischer Rechtfertigung innerhalb einer sozio-empirisch informierten Erkenntnistheorie der Mathematik*. PhD thesis, Rheinische Friedrich-Wilhelms-Universität Bonn.

Wilhelmus, E. (2008). Socio-empirical epistemology of mathematics. *The Reasoner*, 2(2):3–4.

# List of presentations at PhiMSAMP-3

# Is mathematics special?

## Alexandre V. Borovik*

School of Mathematics, University of Manchester, Alan Turing Building, Oxford Road, Manchester M13 9PL, United Kingdom

E-mail: borovik@manchester.ac.uk

## 1 Introduction

> *Toutes les grandes personnes ont d'abord été des enfants*
> *(Mais peu d'entre elles s'en souviennent.)*
> Antoine de Saint-Exupéry, *Le Petit Prince.*

This paper is an attempt to launch a new and somewhat unusual research programme aimed at gaining a better understanding of the specific nature of mathematical practice. I propose to systematically record and analyse logical difficulties experienced—and occasionally overcome—by children in their early learning of mathematics. Quite naturally, this limits my study to analysis of recollections of my fellow mathematicians and to interviews with so-called "mathematically able" children—only they posses an adequate language which allows them to describe their personal experiences. My approach is justified by the success of Vadim Krutetskii's (1976) classical study of mathematically able children. His book provides remarkable insights into mathematical thinking, and adult professional mathematicians instantly recognise themselves in Krutetskii's young subjects.

I am a mathematician and restrict myself to describing hidden structures of elementary mathematics which may intrigue and—like shadows in the

*I am grateful to my correspondents AB, BB, BC, VČ, ŞUE, AG, EHK and LW for sharing with me their childhood memories, to parents of DW for allowing me to write about the boy, to David Pierce for many comments and for a permission to use his paper (Pierce, 2009), to Mikael Johansson who brought my attention to the cohomological nature of carries, and to Eren Mehmet Kıral and Sevan Nişanyan for help with Turkish numerals.

Anonymous referees have helped me to understand what my paper is about and provided some very useful insights. This text would not appear without kind invitations to give a talk at the conference "Is Mathematics Special?" in Vienna in May 2008 and to give a lecture course "Elementary mathematics from the point of view of 'higher' mathematics" at Nesin Mathematics Village in Şirince, Turkey, in July 2008. The final version was prepared during my visit to the Bilgi University, Istanbul, in Spring 2009. I am grateful to all my colleagues at Bilgi and especially to fantastically hospitable staff of Santral-residence. My work on this paper was partially supported by a grant from the John Templeton Foundation. The opinions expressed in this paper are those of the author and do not necessarily reflect the views of the John Templeton Foundation.

Finally, my thanks go to the blogging community—I have picked in the blogosphere some ideas and quite a number of references—and especially to numerous (and mostly anonymous) commentators on my blog, especially to my old friend who prefers to be known only as Owl.

Benedikt Löwe, Thomas Müller (*eds.*). *PhiMSAMP. Philosophy of Mathematics: Sociological Aspects and Mathematical Practice.* College Publications, London, 2010. Texts in Philosophy 11; pp. 1–27.

*Received by the editors:* 1 September 2008; 16 April 2009.
*Accepted for publication:* 12 May 2009.

night—sometimes scare an inquisitive child. I hope that my notes could be useful to specialists in mathematical education and in psychology of education. But I refrain from making any recommendations on mathematics teaching. For me, the primary aim of my project is to understand the nature of hardcore mainstream "research" mathematics.

I hope that my proposal provides an answer to the question in the title of the conference: "Is mathematics special?" Of course it is! The emphasis on child's experiences makes my programme akin to linguistic and cognitive science. However, when a linguist studies formation of speech in a child, he studies language, not the structure of linguistic as a scientific discipline. When I propose to study formation of mathematical concepts in a child, I wish to get insights into the interplay of mathematical structures in *mathematics*. Mathematics has an astonishing power of reflection, and a self-referential study of mathematics by mathematical means plays an increasingly important role within mathematical culture. I simply suggest to make a step further (or step aside, or step back in life) and take a look back in time, in one's child years.

Some very incisive comments from anonymous referees of this paper helped me to better specify its scope, and I reiterate: (**1**) I am neither a philosopher nor a psychologist. (**2**) This paper is not about philosophy of mathematics, it is about mathematics. (**3**) This paper is not about psychology of mathematics, it is about mathematics. (**4**) This paper is not about mathematical education, it is about mathematics. (**5**) But this paper has a secondary purpose: it is an attempt to trigger the chain of memories in my readers. Every, even the most minute, recollection of difficulties and paradoxes of their early mathematical experiences are most welcome.

## 2   Adding one by one

My colleague EHK told me about a difficulty she experienced in her first encounter with arithmetic, aged 6.[1] She could easily solve "put a number in the box" problems of the type

$$7 + \square = 12,$$

by counting how many 1's she had to add to 7 in order to get 12 but struggled with

$$\square + 6 = 11,$$

because she did not know where to start. Worse, she felt that she could not communicate her difficulty to adults. Her teacher forgot to explain to her that addition was commutative.

---

[1] EHK is female, English, has a PhD in Mathematics, teaches mathematics at a highly selective secondary school.

FIGURE 1. *L'Evangelista Matteo e l'Angelo.* Guido Reni, 1630–1640. Pina-coteca Vaticana. Source: *Wikipedia Commons.* Public domain. Guido Reni was one of the first artists in history of visual arts who paid attention to psychology of children. Notice how the little angel counts on his fingers the points he is sent to communicate to St. Matthew.

Another one of my colleagues, AB,[2] told me how afraid she was of sub-traction. She could easily visualise subtraction of 4 from 100, say, as a stack of 100 objects; after removing 4 objects from the top, 96 are left. But what happens if we remove 4 objects from the bottom of the stack?

A brief look at axioms introduced by Dedekind (but commonly called Peano axioms) provides some insight in EHK's and AB's difficulties.

Recall that the Peano axioms describe the properties of natural numbers $\mathbb{N}$ in terms of a "successor" function $S(n)$. (There is no canonical notation for the successor function, in various books it is denoted $s(n)$, $\sigma(n)$, $n'$, or even $n++$, as in popular computer languages C and C++.)

**Axiom 1** 1 is a natural number.

**Axiom 2** For every natural number $n$, $S(n)$ is a natural number.

---

[2] AB is female, Turkish, has a PhD in Mathematics, teaches mathematics in a research-led university. She was 6 years old at the time of that story.

Axioms 1 and 2 define a unary representation of the natural numbers: the number 2 is $S(1)$, and, in general, any natural number $n$ is

$$S^{n-1}(1) = S(S(\cdots S(1) \cdots)) \quad (n-1 \text{ times}).$$

As we shall soon see, the next two axioms deserve to be treated separately; they define the properties of this representation.

**Axiom 3** For every natural number $n$ other than 1, $S(n) \neq 1$. That is, there is no natural number whose successor is 1.

**Axiom 4** For all natural numbers $m$ and $n$, if $S(m) = S(n)$, then $m = n$. That is, $S$ is an injection.

The final axiom (Axiom of Induction) has a very different nature and is best understood as a method of reasoning about all natural numbers.

**Axiom 5** If $K$ is a set such that 1 is in $K$, and for every natural number $n$, if $n$ is in $K$, then $S(n)$ is in $K$, then $K$ contains every natural number.

Thus, Peano arithmetic is a formalisation of that very counting by one that EHK did, and addition is defined in precisely the same way as EHK learned to do it: by a recursion

$$m + 1 = S(m); \quad m + S(n) = S(m+n).$$

Commutativity of addition is a non-trivial (although still accessible to a beginner) theorem. To force you to feel some sympathy to poor little EHK and to poor little AB, I reproduce *verbatim* its proof from Edmund Landau's famous book *Grundlagen der Analysis* (1930).

## 3   Landau's definition of addition

I use notation from Landau's (1966) book:

$$S(n) = n'.$$

Landau's proof consists of two self-contained Theorems, 4 and 6, at the very beginning of his book. Although this is not emphasised by him, the two theorems are not using Axioms 3 and 4. Landau starts by *defining* addition:

> **Theorem 4** (and at the same time Definition 1) To every pair of numbers $x, y$, we may assign in exactly one way a natural number, called $x + y$, such that
>
> (1) $x + 1 = x'$ for every $x$,

(2) $x + y' = (x + y)'$ for every $x$ and every $y$.

*Proof.* (A) First we will show that for each fixed $x$ there is at most one possibility of defining $x + y$ for all $y$ in such a way that $x + 1 = x'$ and $x + y' = (x + y)'$ for every $y$.

Let $a_y$ and $b_y$ be defined for all $y$ and be such that

$$a_1 = x', \quad b_1 = x', \quad a_{y'} = (a_y)', \quad b_{y'} = (b_y)' \text{ for every } y.$$

Let $\mathcal{M}$ be the set of all $y$ for which

$$a_y = b_y.$$

(I) $a_1 = x' = b_1$; hence 1 belongs to $\mathcal{M}$.

(II) If $y$ belongs to $\mathcal{M}$, then $a_y = b_y$, hence by Axiom 2,

$$(a_y)' = (b_y)',$$

therefore

$$a_{y'} = (a_y)' = (b_y)' = b_{y'},$$

so that $y'$ belongs to $\mathcal{M}$.

Hence $\mathcal{M}$ is the set of all natural numbers; i.e., for every $y$ we have $a_y = b_y$.

(B) Now we will show that for each $x$ it is actually possible to define $x + y$ for all $y$ in such a way that

$$x + 1 = x' \text{ and } x + y' = (x + y)' \text{ for every } y.$$

Let $\mathcal{M}$ be the set of all $x$ for which this is possible (in exactly one way, by (A)).

(I) For $x = 1$, the number $x + y = y'$ is as required, since
$x + 1 = 1' = x'$,
$x + y' = (y')' = (x + y)'$.
Hence 1 belongs to $\mathcal{M}$.

(II) Let $x$ belong to $\mathcal{M}$, so that there exists an $x + y$ for all $y$. Then the number $x' + y = (x + y)'$ is the required number for $x'$, since

$$x' + 1 = (x + 1)' = (x')'$$

and

$$x' + y' = (x + y')' = ((x + y)')' = (x' + y)'.$$

Hence $x'$ belongs to $\mathcal{M}$. Therefore $\mathcal{M}$ contains all $x$.  Q.E.D.

It is time to pause and ask a natural question: why is the proof of *consistency* of the inductive definition of addition is so difficult? David Pierce (2009) published a timely reminder about this conundrum of the foundations of arithmetic:

> A set with an initial element and a successor-operation may admit proof by induction without admitting inductive or rather *recursive* definition of functions.

Historically, this observation was made explicit by Dedekind (1888, Remark 130) but overlooked by Peano (1889). Landau himself (1966, Preface for the Teacher) confesses to committing an error—detected by his teaching assistant—in the first version of his lectures. In a more algebraic language, the issue is clarified by Henkin (1960):

> If we consider a *unary algebra*, that is, an algebraic structure $\mathcal{N} = \langle N; 1, S \rangle$ consisting of a ground set $N$ together with a constant symbol 1 and a unary function $S$ (it will automatically satisfy Axioms 1 and 2), then
>
> - $\mathcal{N}$ satisfies the Axiom of Induction (Axiom 5) if and only if $\mathcal{N}$ is generated by 1;
>
> - $\mathcal{N}$ satisfies the Axiom of Induction (Axiom 5) together with Axioms 3 and 4 if and only if $\mathcal{N}$ is a free generated unary algebra freely generated by element 1.

Landau implicitly (and Henkin explicitly) shows that addition and —later in the book—multiplication can be defined by induction alone. But, as we have just seen, the argument takes some work.

David Pierce makes an incisive comment:

> Indeed, if one thinks that the recursive definitions of addition and multiplication—

$$
\begin{aligned}
n + 0 &= n, \\
n + (k + 1) &= (n + k) + 1; \\
n \cdot 0 &= 0, \\
n \cdot (k + 1) &= n \cdot k + n
\end{aligned}
$$

> —are *obviously* justified by induction alone, then one may think the same for exponentiation, with

$$
\begin{aligned}
n^0 &= 1 \\
n^{k+1} &= n^k \cdot n.
\end{aligned}
$$

However, while addition and multiplication are well-defined on $\mathbb{Z}/n\mathbb{Z}$ (which admits induction), exponentiation is not; rather, we have

$$(x, y) \quad \longmapsto \quad x^y$$
$$(\mathbb{Z}/n\mathbb{Z})^* \times \mathbb{Z}/\varphi(n)\mathbb{Z} \quad \longrightarrow \quad \mathbb{Z}/n\mathbb{Z},$$

where $(\mathbb{Z}/n\mathbb{Z})^*$, as usual, denotes the group of invertible elements of the residue ring $\mathbb{Z}/n\mathbb{Z}$. Indeed, the recursive definition of exponentiation fails in $\mathbb{Z}/3\mathbb{Z}$,

| $n$ | $n^2$ | $n^3$ | $n^3 \times n$ | $n^4$ |
|---|---|---|---|---|
| 2 | 1 | 2 | 1 | 2, |

but holds in $\mathbb{Z}/6\mathbb{Z}$:

| $n$ | $n^2$ | $n^3$ | $n^4$ | $n^5$ | $n^6$ | $n^7$ |
|---|---|---|---|---|---|---|
| 1 | 1 | 1 | 1 | 1 | 1 | 1 |
| 2 | 4 | 2 | 4 | 2 | 4 | 2 |
| 3 | 3 | 3 | 3 | 3 | 3 | 3 |
| 4 | 4 | 4 | 4 | 4 | 4 | 4 |
| 5 | 1 | 5 | 1 | 5 | 1 | 5 |
| 6 | 6 | 6 | 6 | 6 | 6 | 6 |

The former is an exception rather than rule, as clarified by Don Zagier's theorem.

**Theorem 3.1.** (Don Zagier, 1996) The identities

$$a^1 = a, \qquad a^{b+1} = a^b \times a \tag{1}$$

hold on $\mathbb{Z}/n\mathbb{Z}$ if and only if $n \in \{0, 1, 2, 6, 42, 1806\}$.

I share David Pierce's (2009) indignation at the state of affairs:

> Yet the confusion continues to be made, even in textbooks intended for students of mathematics and computer science who ought to be able to understand the distinction. Textbooks also perpetuate related confusions, such as suggestions that induction and 'strong' induction (or else the 'well-ordering principle') are logically equivalent, and that either one is sufficient to axiomatize the natural numbers. [...]

> This is one example to suggest that getting things straight may make a pedagogical difference.

But I have to admit that I shared the widespread ignorance until David Pierce brought my attention to the issue—despite the fact that, in a calculus course that I took in the first year of my university studies, the lecturer (Gleb Pavlovich Akilov) explicitly proved the existence of a function of natural argument defined by a recursive scheme (see Akilov and Dyatlov, 1979).

## 4   Induction and recursion

From a pedagogical point of view, recursion could be simpler than induction, and for two different reasons.

Firstly, recursion goes back, to smaller numbers and simpler cases. Secondly, recursion is a calculation, which is psychologically easier for children to handle than a proof. A childhood story from BB raises yet another point:[3] recursion could be more concrete than induction.

> In our math circle we covered induction (domino analogy, proofs of summation formulae such as
>
> $$1 + \cdots + n = \binom{n+1}{2},$$
>
> and varied other examples). I did passably well on the problems, but still I did not understand what the induction is really for, until the end-of-year competition. I failed to solve a single problem: arrange all binary strings of length 10 around the circle so that two adjacent differ in precisely one position (it is known as cyclic Gray code of size 10). It was when I was told the solution that I felt that I finally understood the induction. The missing element was probably the fact that I did not realize that the statement proved by induction is an honest mathematical statement that pertains to concrete numbers like 10, and not only to $x$, $y$, $n$, $m$ and 1996, among which only the latter is a number, but so big and arbitrary that it could as well be denoted by $n$.

One of many recursive rules (the so-called "binary-reflexive" algorithm) for construction of a cyclic Gray code is obvious from the examples of codes of sizes 1, 2, 3, 4 given in Figure 2. This particular recursive algorithm is remarkable indeed for being "self-proving" (in the sense of Barry Mazur's (2007) concept of "theorems that prove themselves").

## 5   Landau's proof of commutativity of addition

Now we return to Landau and commutativity of addition.

> **Theorem 6** (Commutative Law of Addition)
>
> $$x + y = y + x.$$
>
> *Proof.* Fix $y$, and let $\mathcal{M}$ be the set of all $x$ for which the assertion holds.

---

[3] BB was 11 or 12 years old at the time of the story. He is male, Russian, currently a PhD student in pure mathematics.

| 0 | 00 | 000 | 0000 |
|---|----|-----|------|
| 1 | 01 | 001 | 0001 |
|   | 11 | 011 | 0011 |
|   | 10 | 010 | 0010 |
|   |    | 110 | 0110 |
|   |    | 111 | 0111 |
|   |    | 101 | 0101 |
|   |    | 100 | 0100 |
|   |    |     | 1100 |
|   |    |     | 1101 |
|   |    |     | 1111 |
|   |    |     | 1110 |
|   |    |     | 1010 |
|   |    |     | 1011 |
|   |    |     | 1001 |
|   |    |     | 1000 |

FIGURE 2.

(I) We have $y + 1 = y'$ and furthermore, by the construction in the proof of Theorem 4, $1 + y = y'$, so that

$$1 + y = y + 1$$

and 1 belongs to $\mathcal{M}$.

(II) If $x$ belongs to $\mathcal{M}$, then $x + y = y + x$, therefore

$$(x + y)' = (y + x)' = y + x'.$$

By the construction in the proof of Theorem 4, we have

$$x' + y = (x + y)'$$

hence

$$x' + y = y + x',$$

so that $x'$ belongs to $\mathcal{M}$. The assertion therefore holds for all $x$.

Q.E.D.

Notice that it follows from Landau's proof that addition is defined and commutative on any 1-generated unary algebra, in particular, on the algebra shown on Figure 3.

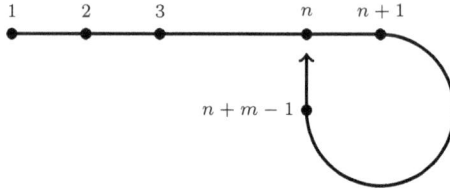

FIGURE 3. A unary algebra with commutative addition (with thanks to David Pierce).

Landau's book is characterised by a specific austere beauty of entirely formal axiomatic development, dry, cut to the bone, streamlined. Not surprisingly, it is claimed that logical austerity and precision were Landau's characteristic personal traits.[4]

*Grundlagen der Analysis* opens with two prefaces, one intended for the student and the other for the teacher; we already quoted *Preface for the Teacher*, it is a remarkable pedagogical document. The preface for the student is very short and begins thus:

1. Please don't read the preface for the teacher.

2. I will ask of you only the ability to read English and to think logically—no high school mathematics, and certainly no higher mathematics. [...]

3. Please forget everything you have learned in school; for you haven't learned it.

   Please keep in mind at all times the corresponding portions of your school curriculum; for you haven't actually forgotten them.

4. The multiplication table will not occur in this book, not even the theorem,
$$2 \times 2 = 4,$$
   but I would recommend, as an exercise for Chap. I, section 4, that you define
   $$2 = 1+1,$$
   $$4 = (((1+1)+1)+1),$$

and then prove the theorem.

---

[4]Asked for a testimony to the effect that Emmy Noether was a great woman mathematician, Landau famously said: "I can testify that she is a great mathematician, but that she is a woman, I cannot swear."

Also, I would like to offer as an exercise for the reader to prove Theorem 5 from Landau's book: associativity of addition of natural numbers. It is easy. By the way, will the addition on the unary algebra of Figure 3 be associative?

Perhaps the reader would agree that little EHK and little AB had good reasons to be confused.

## 6   Children may take their thoughts very seriously

And here comes a really precious point communicated to me, in different words, by the two referees. I quote one of them:

> That children find start unknown problems harder than result unknown problems is well established (e.g., Riley et al., 1983) but attributing it to a lack of understanding of commutativity does not seem to be the answer: children show a grasp of commutativity in some form before they typically succeed on these problems (Canobi, 2004).

What follows is based on my earlier conversations with EHK and AB:

EHK, in her own words, did know what commutativity was; she understood that 3 apples and 2 apples was the same as 2 apples and 3 apples. She simply was not told that the same principle was applicable in her problem, because her problem was not about real life apples, it was about some entities obtained by counting one by one; an educated adult would say that the problem was about abstract objects.

AB's visualisation of subtraction as removal of objects, one by one, from the top of a stack was not about a real life stack of books, say (to start with, there were 100 of them—she distinctively remembers the concrete number in an arithmetic problem). It was a visualisation of an abstract concept of a stack (and exactly of the same stack as used in the computer science as a formalisation of push-down, "first come, first go" storage). Hers was an example of "inverse vision" in terminology of Bill Thurston (1994), visual representation of an abstract construction. She wanted to try to remove the books from the bottom of the stack, but was not sure whether she had to apply to them normal expectations of behaviour of real time objects: it did not matter for real life books where to be removed from, from the top or from the bottom, but in the second case the stack would collapse, and the stalk was important, it represented the essence of the counting algorithm.

Importantly, both EHK and AB emphasised a feeling of frustration they had because of their inability to communicate their difficulty to adults. It could be suggested, at least at a metaphorical level, that their problems were rooted in them taking the first mental structures of mathematics growing in their minds very seriously. And they do not make a representative sample— after all, both got a PhD in mathematics later in their lives.

## 7   "Named" numbers

I take the liberty to tell a story from my own life;[5] I believe it is relevant for the principal theme of the paper.

When, as a child, I was told by my teacher that I had to be careful with *"named"* numbers and not to *add* apples and people, I remember asking her why in that case we can *divide* apples by people:

$$10 \text{ apples } : 5 \text{ people} = 2 \text{ apples}. \tag{2}$$

Even worse: when we distribute 10 apples giving 2 apples to a person, we have

$$10 \text{ apples } : 2 \text{ apples} = 5 \text{ people} \tag{3}$$

Where do "people" on the right hand side of the equation come from? Why do "people" appear and not, say, "kids"? There were no "people" on the left hand side of the operation! How do numbers on the left hand side know the name of the number on the right hand side?

There were much deeper reasons for my discomfort. I had no bad feelings about dividing 10 apples among 5 people, but I somehow felt that the problem of deciding how many people would get apples if each was given 2 apples from the total of 10, was completely different. (My childhood experience is confirmed by experimental studies, cf. (Squire and Bryant, 2002). I suggest to call operation (2) *sharing* and (3) *dispensing* or *distribution*.)

In the first problem you have a fixed data set: 10 apples and 5 people, and you can easily visualize giving apples to the people, in rounds, one apple to a person at a time, until no apples were left. But an attempt to visualize the second problem in a similar way, as an orderly distribution of apples to a queue of people, two apples to each person, necessitated dealing with a potentially unlimited number of recipients. In horror I saw an endless line of poor wretches, each stretching out his hand, begging for his two apples. This was visualization gone astray. I was not in control of the queue! But reciting numbers, like chants, while counting *pairs* of apples, had a soothing, comforting influence on me and restored my shattered confidence in arithmetic.[6]

---

[5]Call me AVB; I am male, Russian, have a PhD in Mathematics, teach mathematics at a research-led university in Britain.

[6]To scare the reader into acceptance of the intrinsic difficulty of division, I refer to the paper *Division by three* (2006) by Peter Doyle and John Conway. I quote their abstract: "We prove without appeal to the Axiom of Choice that for any sets $A$ and $B$, if there is a one-to-one correspondence between $3 \times A$ and $3 \times B$ then there is a one-to-one correspondence between $A$ and $B$. The first such proof, due to Lindenbaum, was announced by Lindenbaum and Tarski in 1926, and subsequently 'lost'; Tarski published an alternative proof in 1949. We argue that the proof presented here follows Lindenbaum's original." Here, of course, 3 is a set of 3 elements, say, $\{0, 1, 2\}$. An exercise for the reader: prove this in a naïve set theory *with* the Axiom of Choice.

FIGURE 4. The First Law of Arithmetic: you do not add fruit and people. Giuseppe Arcimboldo, *Autumn*. 1573. Musée du Louvre, Paris.

I did not get a satisfactory answer from my teacher and only much later did I realize that the correct naming of the numbers should be

$$10 \text{ apples} : 5 \text{ people} = 2 \frac{\text{apples}}{\text{people}}, \ 10 \text{ apples} : 2 \frac{\text{apples}}{\text{people}} = 5 \text{ people}. \quad (4)$$

It is a commonplace wisdom that the development of mathematical skills in a student goes alongside the gradual expansion of the realm of numbers with which he or she works, from natural numbers to integers, then to rational, real, complex numbers:

$$\mathbb{N} \subset \mathbb{Z} \subset \mathbb{Q} \subset \mathbb{R} \subset \mathbb{C}.$$

What is missing from this natural hierarchy is that already at the level of elementary school arithmetic children are working in a much more sophisticated structure, a graded ring

$$\mathbb{Q}[x_1, x_1^{-1}, \ldots, x_n, x_n^{-1}].$$

of Laurent polynomials in $n$ variables over $\mathbb{Q}$, where symbols $x_1, \ldots, x_n$ stand for the names of objects involved in the calculation: apples, persons, etc. This explains why educational psychologists confidently claim that the operations (1) and (2) have little in common (Squire and Bryant, 2002)— indeed, operation (2) involves operands of much more complex nature.

Usually, only Laurent monomials are interpreted as having physical (or real life) meaning. But the addition of heterogeneous quantities still makes sense and is done componentwise: if you have a lunch bag with (2 apples + 1 orange), and another bag, with (1 apple +1 orange), together they make

(2 apples +1 orange) + (1 apple +1 orange) = (3 apples +2 oranges).

Notice that this gives a very intuitive and straightforward approach to vectors.[7] Of course, there is no need to teach Laurent polynomials to kids; but it would not harm to teach them to teachers. I have an ally in François Viéte who in 1591 wrote in his *Introduction to the Analytic Art* that

> If one magnitude is divided by another, [the quotient] is heterogeneous to the former [...] Much of the fogginess and obscurity of the old analysts is due to their not paying attention to these [rules].

## 8   Digression into Turkish grammar

A logical difference between operations of sharing and dispensing is reflected in the grammar of the Turkish language by presence of a special form of numerals, *distributive numerals*.

What follows are accounts of David Pierce, Eren Mehmet Kıral and Sevan Nişanyan. David Pierce reports:

> Turkish has several systems of numerals, all based on the cardinals; as well as a few numerical peculiarities. The cardinals begin "bir, iki, üç, dört, beş, altı, ..." (one, two, three, ...) These answer the question "Kaç?" (How many?) The ordinals take the suffix -inci, adjusted for vowel harmony: "birinci, ikinci, üçüncü, dördüncü, beşinci, altıncı, ..." (first, second, third, ...). These answer the

---

[7]By the way, this "lunch bag" approach to vectors allows a natural introduction of duality and tensors: the total cost of a purchase of amounts $g_1, g_2, g_3$ of some goods at prices $p^1, p^2, p^3$ is a "scalar product"-type expression $\sum g_i p^i$. We see that the quantities $g_i$ and $p^i$ could be of completely different nature. The standard treatment of scalar (dot) product in undergraduate linear algebra usually conceals the fact that dot product is a manifestation of duality of vector spaces and creates immense difficulties in the subsequent study of tensor algebra.

question "Kaçıncı?". The distributives take the suffix -(ş)er: "birer, ikişer, üçr, ... " Used singly, these mean "one each, two each" and so on, as in "I want two fruits from each of these baskets"; they answer the question "Kaçar?".

Eren Mehmet Kıral continues:

> When somebody is distributing some goods s/he might say "Beşer beşer alın" (each one of you take five) or "İ kişer elma alın" (take two apples each). I do not know if it is a grammatical rule (or if it is important) but when the name of the object being distributed is not mentioned then the distributive numeral is repeated as in the first example.
>
> The numeral may also be used in a non distributive problem. If somebody is asking students (or soldiers) to make rows consisting of 7 people each then s/he might say "Yedişer yedişer dizilin." (Get into rows of seven)
>
> I believe the repetition in these numerals exists to give the feeling of distribution. Since in any distribution event you give more than one time you say *beşer beşer* and indicate how many objects you are giving to the first two people, and the rest is assumed to be the same.

In that context, a story told to me by one of my colleagues, ŞUE is very interesting.[8] His experience of arithmetic in his (Turkish) elementary school, when he was about 8 or 9 years old, had a peculiar trouble spot: he could factorise numbers up to 100 before he learnt the times table, so he could instantly say that 42 factors as $6 \times 7$, but if asked, on a different occasion, what is $6 \times 7$, he could not answer. Also, he could not accept the concept of division with remainder: if a teacher asked him how many 3s go into 19 (expecting an answer: 6, and 1 is left over), little ŞUE was very uncomfortable—he knew that 3 did not go into 19. ŞUE added:

> But I did not pay attention to 19 being prime. I had the same problem when I was asked how many 3s go into 16. It is the same thing: no 3s go in 16. Simply because 3 is not a factor of 16. This is perhaps because of distributive numerals I some how built up an intuition of factorising, but perhaps for the same reason (because of the intuition that distributive gave) I could not understand division with remainder.

ŞUE came in peace with remainder only at the first year at university, when the process of division with remainder was introduced as a formal technique.

As we can see, ŞUE does not dismiss the suggestion that distributive numerals of his mother tongue could have made it easier for him to form

---

[8]ŞUE is Turkish, male, undergraduate mathematics student.

concept of divisibility and prime numbers (although he did not know the term "prime number") before he learned multiplication.

## 9 Fractions and inductive limits

In response to my call for personal stories about difficulties in studying (early) mathematics AG sent me the following e-mail:[9]

> When I was about 9 years old, I've first learned at school about fractions, and understood them quite well, but I had difficulties in understanding the concept of fractions that were bigger than 1, because you see we were thought that fractions are part of something, so I could understand the concept of, for example 1/3 (you a take a piece of something you divided in 3 equal pieces and you take one), but I couldn't understand what meant 4/3 (how can you take 4 pieces when there are only 3?). Of course I get it in several days, but I remember that I was baffled at first.

I am surprised to see how frequently such memories are related to subtle play of hidden mathematical structures, like dance of shadows in a moonlit garden; these shadows can both fascinate and scare an imaginative child. As a child, I myself was puzzled by expressions like 5/4, but my worries were resolved by pedagogical guidance: I was taught to think about fractions as named numbers of special kind: quarter apples. Fractions like 5/4 are not result of dividing 5 apples between 4 people, since this operation of division is not yet defined; they come from making sufficient number of material objects of new kind, "quarter apples" and then counting five "quarter apples". AG was less fortunate:

> The word for fraction in Romanian is *fractie*, and that was the terminology used by my teacher and in textbooks. The word is used frequently in common language to express a part of something bigger, like fraction is used in English (I think . . . ), so I think that it's very likely that my difficulty could be of linguistic nature, I can't eliminate also the fact that actually that was how fractions were introduce to us—pupils (like parts of an object) and only later the notion was extended, and so maybe I had problems accommodating to the new notion. Or maybe both reasons . . .

I was luckier: in Russian the word for "fraction" is reserved for arithmetic and is not used in the everyday language.[10] Also, I was clearly told to think about simple fractions $1/n$ as "named numbers" of special kind. In effect, when dealing with quarter apples, we are working in the additive semigroup $\frac{1}{4}\mathbb{N}$ generated by $\frac{1}{4}$.

---

[9]AG is male, Romanian, a student of Computer Science.
[10]Unless you are a hunter and use the same word for lead shot pellets.

What happens next is much more interesting and sophisticated: we have
to learn how to add half apples with quarter apples. This is done, of course,
by dividing each half in two quarters, which amounts to constructing a
homomorphism

$$\frac{1}{2}\mathbb{N} \longrightarrow \frac{1}{4}\mathbb{N}.$$

Since both $\frac{1}{2}\mathbb{N}$ and $\frac{1}{4}\mathbb{N}$ are canonically isomorphic to $\mathbb{N}$, we, being adults
now, can make a shortcut in notation and write this homomorhism simply
as

$$\mathbb{N} \longrightarrow \mathbb{N}, \quad z \mapsto 2 \times z.$$

In effect, we have a direct system

$$\mathbb{N} \xrightarrow{k\times} \mathbb{N}, \quad k = 2, 3, 4, \dots$$

—or, if you prefer less abstract notation—

$$\frac{1}{n}\mathbb{N} \xrightarrow{\text{Id}} \frac{1}{kn}\mathbb{N}.$$

Then we do something outrageous: we take its inductive (or direct, which
is an equivalent term) limit. In the primary school, of course, taking the
inductive limit is called bringing fractions to a common denominator.

Here are formal definition of a direct set, direct system, direct (inductive) limit: A *directed set* is a nonempty set $A$ together with a reflexive
and transitive binary relation $\leq$ (that is, a *preorder*), with the additional
property that every pair of elements has an upper bound.

Notice that the directed set in our definition is the set $\mathbb{N}$ with the divisibility relation. It is not linearly ordered and has a pretty sophisticated
structure by itself!

Let $(I, \leq)$ be a directed set. Let $\{A_i \mid i \in I\}$ be a family of objects
indexed by $I$ and suppose we have a family of homomorphisms $f_{ij} : A_i \to
A_j$ for all $i \leq j$ with the following properties: $f_{ii}$ is the identity in $A_i$,
$f_{ik} = f_{jk} \circ f_{ij}$ for all $i \leq j \leq k$. Then the pair $(A_i, f_{ij})$ is called a *direct
system* over $I$.

In our case, the direct system is formed by semigroups $\frac{1}{k}\mathbb{N}$, $k \in \mathbb{N}$ with
natural embeddings

$$\frac{1}{n}\mathbb{N} \to \frac{1}{m}\mathbb{N} \text{ if } n \text{ divides } m.$$

The underlying set of the direct (inductive) limit, $A$, of the direct system $(A_i, f_{ij})$ is defined as the disjoint union of the $A_i$'s modulo a certain
equivalence relation $\sim$:

$$A = \bigsqcup A_i / \sim$$

Here, if $x_i$ is in $A_i$ and $x_j$ is in $A_j$, $x_i \sim x_j$ if there is some $k$ in $I$ such that $f_{ik}(x_i) = f_{jk}(x_j)$.

The sum of $x_i \in A_i$ and $x_j \in A_j$ is defined as the equivalence class of $f_{ik}(x_i) + f_{jk}(x_j)$ for some $k$ such that $i \leq k$ and $j \leq k$.

The result, of course, is the additive semigroup of positive rational numbers $\mathbb{Q}^+$. Only then we define multiplication on $\mathbb{Q}^+$. I leave a category theoretical construction of multiplication as an exercise to the reader. It is much helped by the fact that endomorphisms of the additive semigroup $\mathbb{Q}^+$ are invertible and form a group (the multiplicative group of positive rational numbers $\mathbb{Q}^\times$) which acts on $\mathbb{Q}^+$ simply transitively: for every non-zero $r, s \in \mathbb{Q}^+$ there exists a unique $q \in \mathbb{Q}^\times$ such that $s = q \cdot r$. And here is a story from BC which illustrates this point:[11]

> I could not understand the "invert and multiply" rule for dividing fractions. I could obey the rule, but *why* was multiplying by 4/3 the same as dividing by 3/4?
>
> My teachers could not explain, but I was used to that. I couldn't work it out for myself either, which was less usual.
>
> Finally I asked my father, who was an accountant. He said: if you divide everything into halves, you have twice as many things. Suddenly not just fractions but the whole of algebra made sense for the first time.

## 10   Palindromic decimals and palindromic polynomials

My next case study is based on conversations with an 8 year old boy, DW, in May 2007. DW's parents sent me a file of DW's book. It included the following paragraphs, reproduced here *verbatim*:

> What's weird about 1, 11, 111, 1111 etc when you square then?
>
> $1^2 = 1$. $11^2 = 121$. Keep on doing this with the other numbers. (If necessary use a calculator).
>
> Solutions see page it counts up e.g. $1111111^2 = 1234567654321$
>
> But when you have $1, 111, 111, 111^2$ the answer is different. Figuring out (or using the calculator) when are the next square numbers in the pattern after $1111111111^2$?
>
> Solutions see page 1234567900987654321, 123456790120987654321, 123456790123430987654321, 1234567901234320987654321 and 123456790123454320987654321!
>
> Do you notice a pattern?

---

[11]BC was 10 years old. He is male, a non-English Western European.

I wrote to DW:

> Indeed, there is something weird. I believe you have figured out that

$$
\begin{aligned}
1 \times 1 &= 1 \\
11 \times 11 &= 121 \\
111 \times 111 &= 12321 \\
1111 \times 1111 &= 1234321 \\
11111 \times 11111 &= 123454321 \\
111111 \times 111111 &= 12345654321 \\
1111111 \times 1111111 &= 1234567654321 \\
11111111 \times 11111111 &= 123456787654321 \\
111111111 \times 111111111 &= 12345678987654321
\end{aligned}
$$

> There is a wonderful palindromic pattern in the results. But mathematics is interested not so much in beautiful patterns but in reasons why the patterns cannot be extended without loss of their beauty. In our case, the pattern breaks at the next step (judging by your book, you have already noticed that):

> $$1111111111 \times 1111111111 = 1234567900987654321$$

> The result is no longer symmetric. Why? What is the difference from the previous 9 squares? Can you give any suggestions?

I had some brief e-mail exchanges with DW which suggested that he might have an explanation, but could not clearly express himself. Our discussion continued when he visited me (with his mother) in Manchester on 8 May 2007.

I wrote on the whiteboard in my office (the following images are photographs of actual writing on the board)

$$1^2 = 1$$
$$11^2 = 121$$
$$111^2 = 12321$$
$$1111^2 = 1234321$$

and asked DW whether the symmetric pattern of results continued indefinitely. DW instantly answered "No" and also instantly wrote on the board, apparently from his memory:

$$111111111\stackrel{?}{=}1234567\,0098765432\,1$$

"Good," said I, "but let us try to figure out why this is happening", and wrote on the board:

$$\times\ \begin{array}{cccc}1&1&1&1\\1&1&1&1\end{array}$$

"Yes," said DW, "this is column multiplication"—"And what are the sums of columns'?" —"1, 2, 3, 4, 3, 2, 1," dictated DW to me, and I wrote down the result. "Will the symmetric pattern continue indefinitely?", I asked. "No," was DW's answer, "when there are 10 1's in a column, 1 is added on the left and there is no symmetry."—"Yes," said I, "carries break the symmetry. But let us look at another example," and I wrote:

$$1^2 = 1$$
$$(1+x)^2 = 1 + 2x + x^2$$
$$(1+x+x^2)^2 = 1 + 2x + 3x^2 + 2x^3 + x^4$$

DW was intrigued and made a couple of experiments (and it appeared from his behaviour that he was using mostly mental arithmetic, writing down the result, term by term, with pauses):

$$(1+x+x^3)(1+x+x^3) = 1+x+x^2+x+x^2+x^4+x^3+x^2+x^3+x^4 =$$
$$x^4+2x^3+3x^2+2x+1$$

$$(1+X+X^2+X^3)^2 = 1+2x+3x^2+4x^3+3x^4+2x^5+x^6$$

Finally, he said with obvious enthusiasm: "Yes, it is the same pattern!"—
"Wonderful, " I answered, "let us see why this is happening. I'll give you
a hint: multiplication of polynomials can be written as column multiplication", and started to write:

$$X^4+X^3+X^2+X+1$$
$$X^4+X^3+X^2+X+1$$
$$\overline{\qquad X^4 \quad X^3 \quad X^2 \quad X \quad 1}$$
$$X^5 \quad X^4 \quad X^3 \quad X^2 \quad X$$
$$X^6 \quad X^5 \quad X^4 \quad X^3 \quad X^2$$
$$X$$

DW did not let me finish, grabbed the marker from my hand and insisted
on doing it himself:

$$X^{10}+X^9+X^8+X^7+X^6+X^5+X^4+X^3+X^2+X+1$$
$$X^{10}+X^9+X^8+X^7+X^6+X^5+X^4+X^3+X^2+X+1$$
$$\overline{X^{10}+X^9+X^8+X^7+X^6+X^5 \quad X \, X^4+X^3+X^2+X+1}$$
$$X^2 \quad X$$

He stopped after he barely started the second line and said very firmly:
"Yes, it is like with numbers"—"Well," I said, "but will the pattern break

down or will continue forever?" That was the first time when DW fell
in deep thought (and I was a bit uncomfortable about the degree of his
concentration and retraction from the real world). This was also the first
time when his response was not instantaneous—perhaps, whole 20 seconds
passed in silence. Then he suddenly smiled happily and answered: "No,
it will not break down!"—"Why?" I inquired. "Because when you add
polynomials, the coefficients just add up, there are no carries." At that
point I decided to stop the session on the pretext that it was late and the
boy was perhaps tired, but, to round up the discussion, made a general
comment: "You know, in mathematics polynomials are sometimes used
to explain what is happening with numbers". The last word, however,
belonged to DW: "Yes, 10 is $x$."

## 11   DW: a discussion

### 11.1   The so-called "able" children.

DW is a classical example of what is usually called a "mathematically able
child". He mastered, more or less on his own, some mathematical routines—
multiplication of decimals and polynomials—which are normally taught to
children at much later age. He also showed instinctive interest in detecting beautiful patterns in behaviour of numbers, and, which is even more
important, in limits of applicability of patterns, in their breaking points.

DW understands what generalisation is and, moreover, loves making
quick, I would say recklessly quick generalisations. Krutetskii (1976) lists
this trait among characteristic traits of "mathematically able" children:
very frequently, they are children, who, after solving just one problem, already know how to solve *any* problem of the same type.

But let us return to the principal theme of the present paper: hidden structures of elementary mathematics. In our conversation, DW was
shown—I emphasise, for the first time in his life—a beautiful but hidden
connection between decimals and polynomials—*and was able to see it*!

In our little exercise, DW advanced (a tiny step) in *conceptual* understanding of mathematics: he had seen an example of how one mathematical
structure (polynomials) may hide inside another mathematical structure
(decimals).

My final comment is that although DW made a small, but important step
towards deeper understanding of mathematics, this step is not necessarily
visible in the standard mathematics education framework. It is unlikely that
a school assignment will detect him making this small step. Procedurally,
in this small exercise DW learned next to nothing—he multiplied numbers
and polynomials before, he will multiply them with the same speed after.

One should not think, however, that the "procedural" aspect of mathematics is of no importance. DW's ability to do this tiny bit of "concep-

tual" mathematics would be impossible without him mastering the standard routines (in this case, column multiplication of decimals and addition and multiplication of polynomials).

## 11.2 Decimals and polynomials: an epiphany

DW's words "10 is $x$" are a formulation of analogy between decimals and polynomials which is not frequently emphasised in schools but, when discovered by children on their own, is experienced as an epiphany. These expression is taken from another childhood story, by LW.[12]

> When I was in the fourth grade (about 9 years old), we learned long division. I had enormous difficulty learning the method, though I could divide 3- and 4-digit numbers by 1- to 2-digit numbers in my head. I don't recall exactly how I did the divisions in my head, though I suspect that the method was similar to long division. I recall that it was broadly based on "seeing how much I needed to add to the result to move on." However, I couldn't seem to remember long division, despite being able to follow a list of instructions on homework. My homework on long division took hours to finish. I think the issue was that my teacher and parents never explained *why* long division actually worked, so it seemed like a disconnected list of steps that had little relation to one another. On a quiz, my teacher thought I had cheated, since I had written no work on any of the problems.
>
> I only became able to do long division in high school when I learned how to do long division with polynomials. At that time, the teacher went to great pains to explain why it worked. While doing a homework, I had an epiphany: long division for polynomials was a close cousin of long division for numbers. Suddenly I could do long division of numbers.
>
> I now take pains when teaching calculus to try and explain why formulas are true whenever possible. This has lead to mixed reactions on my evaluations, with some students saying that they enjoy knowing why the formulas are true and others saying that they have enough trouble learning formulas in the first place without having to *also* learn why they're true. For me, the two are deeply connected. If I don't know why something is true (at least in broad outline), I find it significantly harder to remember.

LW's story is an evidence in support of an observation made by Krutetskii (1976): "mathematically inclined children" may appear to be slow because they are frequently trying to solve a more general problem than the

---

[12]LW was 9 years at the time of his story, he is male, learned mathematics in American English (his native tongue), currently is a second year student in a PhD program in pure mathematics.

one given and understand the underlying reasons for a rule they are told to apply unquestionably.

## 11.3   Cohomology

But what was the mathematical object that DW dealt with? It is, of course, carry in the addition of decimals, one of the extreme cases when the deceptive simplicity of elementary school arithmetic hides a very sophisticated structure.

In Molière's *Le Bourgeois Gentilhomme*, Monsieur Jourdain was surprised to learn that he had been speaking prose all his life. I was recently reminded that, starting from my elementary school and then all my life, I was calculating 2-cocycles.

Indeed, a carry in elementary arithmetic, a digit that is transferred from one column of digits to another column of more significant digits during addition of two decimals, is defined by the rule

$$c(a, b) = \begin{cases} 1 & \text{if } a + b > 9 \\ 0 & \text{otherwise} \end{cases}.$$

One can easily check that this is a 2-cocycle from $\mathbb{Z}/10\mathbb{Z}$ to $\mathbb{Z}$ and is responsible for the extension of additive groups

$$0 \longrightarrow 10\mathbb{Z} \longrightarrow \mathbb{Z} \longrightarrow \mathbb{Z}/10\mathbb{Z} \longrightarrow 0.$$

DW discovered (without knowing the words "2-cocycle" and "cohomology") that carry is doing what cocycles frequently do: they are responsible for break of symmetry.

## 12   Conclusions

I can easily continue a list of my case studies. Many of them can be found in my forthcoming book (Borovik, 2010). All my examples lead to the same conclusions:

### 12.1   Even basic elementary mathematics is immensely rich.

I hope that this thesis is self-evident from the examples considered in the paper. For example, elementary school mathematics actually uses inductive limits and cohomology.

Although I am trying to escape from giving any advice on methods and approaches of mathematics instruction, I cannot avoid making a few general remarks.

First of all, glossing over difficulties presented by hidden structures may seriously imperil students' progress.

Next, it is desirable (but perhaps not always realistic) that teachers are aware about the hidden structures and are able to guide pupils around dangerous spots—perhaps without needlessly alerting them every time.

## 12.2   Hidden structures: obstacles or opportunities?

Another preliminary conclusion of this study is that the sensitivity to the presence of "hidden" structures appears to be an important component of mathematical ability in so-called "mathematically gifted" children. Special forms of this observation can be found already in the classical study by Vadim Krutetskii (1976) who pinpoints numerous instances of children using "symmetry" considerations or intentionally resorting to more general arguments.

And this is point where a reminder about non-representativity of the sample of respondents is necessary. One of the respondents (BB, quoted in Section 4) formulated this caveat very explicitly:

> I would like to point out that the stories are provided by the people who are [...] atypical in terms of mathematical thinking and learning. Most people who have responded have not only gone to deal with unusually abstract concepts in their career, but actually do mathematics. So, the examples here might represent not so much the major difficulties that need to be overcome (as in finding the correct way of thinking of division apples by apples) before an understanding can be reached, but the signs that the understanding has already been reached, and difficulty is purely semantic, i.e., how to express it.

Perhaps, stories told in this paper represented difficulties for the teacher and missed opportunities for the child.

## 12.3   Intrinsic logical structures of human languages

We had a chance to see in several stories that children's perception of mathematics can be affected by logical and mathematical structures (such as systems of numerals) of the language of mathematics instruction and their mother tongue. Intrinsic structures of natural languages are engaged in a delicate interplay with hidden structures of mathematics.

For me personally, this is a serious practical issue. Every autumn, I teach a foundation year (that is, zero level) mathematics course to a large class of students which includes 70 foreign students from countries ranging from Afghanistan to Zambia. Students in the course come from a wide variety of socioeconomic, cultural, educational and linguistic backgrounds. But what matters in the context of the present paper are invisible differences in the logical structure of my students' mother tongues which may have huge impact on perception of mathematics. For example, the connective "or" is strictly exclusive in Chinese: "one or another but not both", while in English "or" is mostly inclusive: "one or another or perhaps both". Meanwhile, in mathematics "or" is always inclusive and corresponds to the expression "and/or" of bureaucratic slang. In Croatian, there are two connectives "and": one *parallel*, to link verbs for actions executed simultaneously, and another *consecutive*.[13]

But it is as soon as you approach definite and indefinite articles that you get in a real linguistic quagmire. In words of my correspondent VČ:[14]

> [In Croatian, there are] no articles. There are many words that can "serve" as indefinite articles (*neki*=some, for example), but not no particularly suitable word to serve as definite article (except adjective *određeni* = definite, I guess). Many times when speaking mathematics, I (in desperation) used English articles to convey meaning (eg. *Misliš da si našao **a** metodu ili **the** metodu za rješavanje problema tog tipa? = You mean you found **a** method or **the** method for solving problems of that type?*

Psycholinguistic aspects of learning mathematics are a deserving area of further study.

## Bibliography

Akilov, G. P. and Dyatlov, V. N. (1979). *Elements of Functional Analysis.* Novosibirsk State University, Novosibirsk. In Russian.

Beman, W. W., editor (1901). *Richard Dedekind. Essays on the Theory of Numbers.* Open Court Publishing, Chicago IL.

Borovik, A. V. (2010). *Mathematics under the Microscope.* American Mathematical Society, Providence, RI.

Canobi, K. H. (2004). Individual differences in children's addition and subtraction knowledge. *Cognitive Development*, 19:81–93.

Dedekind, R. (1888). *Was sind und was sollen die Zahlen?* Vieweg, Braunschweig. Page numbers refer to the English translation in (Beman, 1901).

Doyle, P. G. and Conway, J. H. (2006). Division by three. Unpublished.

Henkin, L. (1960). On mathematical induction. *The American Mathematical Monthly*, 67(4):323–338.

Krutetskii, V. A. (1976). *The Psychology of Mathematical Abilities in School children.* The University of Chicago Press, Chicago IL. Edited by Jeremy Kilpatrick and Izaak Wirszup, translated by Joan Teller.

---

[13] Rudiments of "consecutive and" can be found in my native Russian and traced to the same ancient Slavic origins.

[14] VČ is male, Croatian, a lecturer in mathematical logic and computer science.

Landau, E. (1930). *Grundlagen der Analysis. Das Rechnen mit ganzen, rationalen, irrationalen, komplexen Zahlen.* Akademische Verlagsgesellschaft, Leipzig. Quoted after the English edition in Landau (1966).

Landau, E. (1966). *Foundations of Analysis. The arithmetic of whole, rational, irrational and complex numbers.* Chelsea, New York NY. Translated from the German by F. Steinhardt.

Mazur, B. (2007). How did Theaetetus prove his theorem? In Kalkavage, P. and Salem, E., editors, *The Envisoned Life: Essays in honor of Eva Brann*, pages 230–253. Paul Dry Books, Philadelphia PA.

Peano, G. (1889). *Arithmetices principia, nova methodo exposita.* Bocca, Torino. Quoted after the English translation in (van Heijenoort, 1967, pp. 83–97).

Pierce, D. (2009). Induction and recursion. *Bulletin of Symbolic Logic*, 15(1):126–127.

Riley, M. S., Greeno, J. G., and Heller, J. I. (1983). Development of children's problem-solving ability in arithmetic. In Ginsburg, H. P., editor, *The development of mathematical thinking*, pages 153–196. Academic Press, New York NY.

Squire, S. and Bryant, P. (2002). The influence of sharing on children's initial concept of division. *Journal of Experimental Child Psychology*, 81(1):1–43.

Thurston, W. P. (1994). On proof and progress in mathematics. *Bulletin of the American Mathematical Society*, 30(2):161–177.

van Heijenoort, J., editor (1967). *From Frege to Gödel. A Source Book in Mathematical Logic, 1879–1931.* Harvard University Press, Cambridge MA.

Viète, F. (1591). *The Analytic Art.* Mettayer, Tours. Quoted from the edition in Witmer (1983).

Witmer, T., editor (1983). *François Viète: The Analytic Art.* The Kent State University Press, Kent OH.

Zagier, D. (1996). Problems posed at the St. Andrews Colloquium. Unpublished.

# Contradictions in mathematics

## Manuel Bremer

Philosophisches Institut, Heinrich-Heine-Universität Düsseldorf, Universitätsstraße 1,
40225 Düsseldorf, Germany
E-mail: bremer@mbph.de

Contradictions are typically seen as anathema to mathematics. As formalism sees consistency as the only condition to consider some mathematical structure as an object of study, inconsistency becomes the single excluding criterion. For mathematical realists accepting inconsistencies comes down to accepting inconsistent objects, which just seems bizarre.

In this paper I consider two inconsistency friendly approaches in (the philosophy of) mathematics. In a recent study *How Mathematicians Think*, William Byers (2007) argues that one way of mathematical progress is by way of contradiction. The first paragraph outlines Byers' thesis, but it turns out that contradictions play a role only *ex negativo*. In contrast to that the approach of inconsistent mathematics claims contradictions to be real. Especially in inconsistent arithmetic contradictions are said to play a vital role. They turn out to provide a framework for a finitist position which endorses inconsistent numbers.

## 1 Byers' creative use of contradictions

William Byers (2007) claims mathematics to be at core a creative activity. Mathematical reasoning, according to Byers, is not primarily algorithmic or based on proof systems, but is based on using (great) "ideas" to shed new light on mathematical objects and structures. These ideas not only are placed at the centre of mathematical understanding, which Byers calls "turning on the light", but also propel mathematical progress. Byers presents a couple of examples in which a crucial step forward in the development of mathematics depended on the presence of two at first sight unrelated or even barely compatible perspectives on some mathematical structure. He starts with the discovery of the irrational numbers (like $\sqrt{2}$), where $\sqrt{2}$ is clearly present as a geometric object (the length of the hypothenuse of the right angled triangle with unit length sides) but is not allowed for by (early Greek) arithmetic. The real numbers "provide a context" (p. 38) in which the two perspectives are unified. Another famous example is the *Fundamental Theorem* of the calculus, which says "that there is in fact one process in calculus that is integration when it is looked at in one way and differentiation when it is looked at in another" (p. 50). The core of mathematics, according to Byers, is finding such situations and being able to understand them by providing a more comprehensive view. This process is creative and

Benedikt Löwe, Thomas Müller (*eds.*). *PhiMSAMP. Philosophy of Mathematics: Sociological Aspects and Mathematical Practice.* College Publications, London, 2010. Texts in Philosophy 11; pp. 29–37.

*Received by the editors:* 9 October 2008; 11 December 2008.
*Accepted for publication:* 20 April 2009.

not algorithmic. Proofs only sum up the discovery and preserve the results in text books. Mechanical proofs Byers sees as "trivial" (p. 373) whereas "deep" proofs are framed in expressing some (great) "idea" (like re-ordering infinite series makes it obvious to see a sum formula). Some proofs (like diagonalization with Cantor's original insights) are part of discovery, but these are rather the exception than the rule.

Good mathematicians are, therefore, those who hit on "ideas" (like Cantor hitting on diagonalization and the continuum hypotheses). Even more revolutionary are "great ideas". An example of a great idea is formalism. Formalism provided a unifying perspective on the whole of mathematics. When Hilbert started with formalizing Euclid's geometry *"formalism* was born and, in the process, the whole notion of truth was radically transformed" (p. 291). A great idea is then inflated (like in Hilbert's claims on behalf of formalism) and then again delimited in a wider perspective (like when Gödel's Theorems hit formalism). As ideas are outbursts of creativity "the answer to the question of whether a computer could ever do mathematics is clearly 'No!' " (p. 369). Byers finally relates his view to the question of how mathematics is to be taught, namely by getting students understand the ideas to "turn on the light".

One of the central methodological concepts – besides ambiguity – Byers uses in analysing the examples he presents is 'contradiction'. The very subtitle of his book reads 'Using Ambiguity, Contradiction, and Paradox to Create Mathematics'.

'Contradiction' is understood by Byers in two ways. On the one hand we have two seemingly contradictory perspectives in some of the mathematical problems he presents. For example, one may see $\sqrt{2}$ as a decimal, "an 'infinite' indefinite object" (p. 97), but also as a finite geometric object. One can see "$2 + 3 = 5$" both as expressing a fact of identity (i.e., something static) as well as expressing the process of adding (i.e., something dynamic). But of course the fact can be established by going through the process of adding, the two perspectives are finally compatible and not inconsistent. The paradox of zero (as something that is nothing) vanishes with axiomatization.

On the other hand some seminal proofs work by using contradictions, or so Byers claims. For a simple example, one can argue for the proposition that a straight line falling on parallel straight lines makes the alternate angles equal to one another in the following fashion:

> Suppose on the contrary, that angle $\alpha$ is not equal to angle $\beta$, for example, angle $\alpha$ is smaller than angle $\beta$. Then by adding angle $\gamma$ we end up with
>
> angle $\alpha$ + angle $\gamma$ < angle $\beta$ + angle $\gamma$ = two right angles.

Thus the interior angles on the same side are less than two right angles. The parallel postulate tells us that in this situation the lines must meet, contradicting the assumption that they are parallel. (p. 95)

Other famous examples are Cantor's use of diagonalization or Gödel's Theorems.

Now, if we look at these examples it becomes obvious that none of the mathematicians in question endorses any of the contradictions. Quite the opposite. In these indirect proofs contradictions are used as a threat to establish the opposite result. Just for reduction some innocent looking assumption is made which turns out to be contradictory and thus untenable. What we really see here is not a creative use of contradictions, but the creative use of indirect proof methods. Mathematics still avoids the contradictory. Even supposedly incompatible perspectives on one and the same structure have to be kept distinct from contradictions. The paradoxical calls for resolution. The perceived incompatibility is the very reason to look for another solution. At last Byers admits "we cannot leave it at that—things *must* be reconciled" (p. 111). "New stages of mathematical development arise out of a 'resolution' of a set of paradoxes" (p. 186).

## 2   Inconsistent mathematics and finitism

To have *an inconsistent number theory* means at least that within the theorems of number theory there is some sentence $A$ with $A$ being a theorem and $\neg A$ being a theorem at the same time. Supposedly this contradiction corresponds to some object/number $a$ being an inconsistent object. So inconsistent mathematics is connected to inconsistent ontology. Its underlying logic has to be paraconsistent (cf. Bremer, 2005). Changing the basic logic used in mathematics to a paraconsistent logic makes mathematics in a weak sense paraconsistent: If there were to turn up some inconsistency in mathematics, it would not explode. Explosion would happen in standard logic as one can derive any sentence whatsoever from a contradictions, as it implies everything; this move typically is blocked in paraconsistent logics.

The problems with having $F(a)$ and $\neg F(a)$ for some object $a$ seem less pressing if $a$ is some mathematical object than if $a$ is a physical object. Mathematical objects are either non-existent—*mere* theory, taken instrumentally—or they are in some elusive Platonic realm where strange things may well happen. If on the other hand one is a reductionist realist about mathematics (mathematics being about structures of reality or mathematical entities rather being concrete entities dealt with by mereology) then inconsistent mathematics is as problematic as your cat being (wholly) black and not being (wholly) black at the same time.

Philosophers, when concerned with mathematics, focus on number theory, since the ontological questions of mathematics ("What and where are mathematical objects?", "Are there infinite sets?", ...) and the epistemological questions of mathematics ("How do we know of numbers?", "Is mathematics merely conventional?", ...) do arise already with number theory.

Taking set and model theory as part of logic anyway, logicians are also mainly concerned with number, since a lot of meta-logical theorems make us of the device of arithmetization.

The same goes for the general theory of automata and computability. I follow this focus here and so this section of the paper concerns itself mostly with arithmetic. This may not be enough for a mathematician trying to assess the power of inconsistent mathematics. She looks for inconsistent theories at least of the power of the calculus. There are actually such theories, e.g., presented by Chris Mortensen (1995).

One of the most fundamental mathematical theories is arithmetic (as given for instance by the Dedekind/Peano axioms). We are here concerned mainly with first order representations. In distinction to an axiomatic arithmetic theory like Peano Arithmetic there is the arithmetic $\mathsf{N}$ (being the set of true first order arithmetic sentences in the standard interpretation). $\mathsf{N}$ is negation complete (either $A$ or $\neg A$ is in $\mathsf{N}$), not axiomatisable, not decidable, and, of course, infinitely large.

Given its first order representation there are a lot of well-known theorems about arithmetic (e.g., Peano Arithmetic being negation- and $\omega$-incomplete).

Using the compactness theorem for first order logic, one can prove that there are non-standard models of Peano Arithmetic, which contain additional numbers over and above the natural numbers. These additional numbers behave consistently, however. Consistency provides them in the first place. Inconsistent arithmetic may concern itself with the opposite deviance: Having arithmetics where there are less numbers than in standard arithmetic.

This is of utmost philosophical interest, since the infinite is a really problematic concept leading to the ever larger cardinalities of "Cantor's paradise", and finitism (in the sense of the assumption that there are only *finitely many* objects, even of mathematics) is therefore an option worth exploring and pursuing.

Robert Meyer (1976) was the first to give a non-triviality proof of a Relevant (paraconsistent) arithmetic. The system $\mathbf{R}^{\#}$ is an extension of the first order version of Relevant logic $\mathbf{R}$ with axioms mirroring those of Peano Arithmetic save that the "$\supset$" in them has been replaced by the Relevant "$\rightarrow$". Induction is present as a rule. $\mathbf{R}^{\#}$ is non-trivial in that

$0 = 1$ is not provable. This non-triviality can be established by finitistic methods.

Inconsistent arithmetics that are finite (in their models) may have any finite size you like. They contain one largest number. A largest number is a very special number indeed: usually it is understood as verifying a statement as to the identity of that number to its (supposed) successor. One may object that this is difficult to conceive. But as often with mathematical or scientific claims our capacities for (visual) imagery may not have the last word here. We can characterise a largest number as that item which fulfils certain axioms or theorems. Seen thus a largest number is an entity postulated by a (mathematical) theory like, say, a large cardinal.

Since we do not know which number really is the largest we may assume that one of these finitistic arithmetics is true, although we don't know which. Which one it is is not that important, since all these arithmetics have common properties:

We can define a sequence $N_n$ of finite, inconsistent arithmetics with the following properties (cf. Priest, 1994a,b, 1997):

**(i)** For every $n \in \mathbb{N}$, we have $\mathsf{N} \subset \mathsf{N}_n$.

**(ii)** For every $n \in \mathbb{N}$, $\mathsf{N}_n$ is inconsistent.

**(iii)** For every $n \in \mathbb{N}$, if $A$ is a (negated) equation for numbers $< n$, then $A \in \mathsf{N}$ if and only if $A \in \mathsf{N}_n$.

**(iv)** Every $\mathsf{N}_n$ is decidable.

**(v)** For every $n \in \mathbb{N}$, $\mathsf{N}_n$ is representable in $\mathsf{N}_n$ (i.e., we have a truth predicate for $\mathsf{N}_n$).

**(vi)** For the proof predicate B of $\mathsf{N}_n$, every instance of $B(\ulcorner A \urcorner) \supset A$ is in $\mathsf{N}_n$.

**(vii)** If $A$ is not a theorem of $\mathsf{N}_n$, then $\neg B(\ulcorner A \urcorner)$ is in $\mathsf{N}_n$.

**(viii)** If $G_n$ is the Gödel sentence for $\mathsf{N}_n$, then both $G_n$ and $\neg G_n$ are in $\mathsf{N}_n$.

These inconsistent arithmetics $\mathsf{N}_n$ thus have quite remarkable properties: The theories $\mathsf{N}_n$ are negation complete (by (i)) and inconsistent (by (ii) and (viii)). By (iv), they have all the nice properties that $\mathsf{N}$ does not have, though the $\mathsf{N}_n$ are complete. By (v), we can define a truth predicate in the language of arithmetic for the same language, and by (vi), $\mathsf{N}_n$ has an ordinary proof predicate, using a standard Gödel numbering if enough numbers are available. Finally, by by (vii) in conjunction with (iii), we have not only that $N_n$ is non-trivial (by excluding some the equations that are excluded by $N$), but that this non-triviality can be established within $N_n$ itself.

Let us give a few comments about the proof of the properties of the theories $N_n$: First of all, a theory with fewer numbers has fewer counterexamples to a given arithmetic sentence. Thus, more sentences are true (in general, this is called the "collapsing lemma"). Since $N$ is negation complete, any properly stronger theory will have to add a sentence whose negation is already in $N$. Thus, for at least one $A$, the resulting theory must contain both $A$ and $\neg A$. This means that the logic of these arithmetic theories has to be a paraconsistent logic.

Representability of truth is a consequence of (iv) and (i). The same holds for the representability of the proof predicate, (vi). Once the proof predicate is representable in the decidable theory $N_n$, we can represent non-provability, and thus have (vii) and finally (viii).

The most interesting property is (iii) which results from the way the domain of a corresponding model is constructed.

A model of a theory $N_n$ is constructed as a filtering of an ordinary arithmetic model. In general one can reduce the cardinality of some domain by substituting for the objects equivalence classes given some equivalence relation. The equivalence classes provide then the substitute objects. Since the objects within the equivalence class are equivalent in the sense of interest in the given context the predicates still apply (now to the substitute object). The trick in case of $N_n$ is to chose the filtering which puts every number $< n$ into its equivalence class, and nothing else; and puts all numbers $\geq n$ into $n$'s equivalence class. As a result of this for $x < n$ the standard equations are true (of $[x]$), while in case of $y \geq n$, *everything* that could be said of such a $y$ is true of $[n]$. So we have immediately $n = n$ (by identity) and $n = n + 1$ (since for $y = n + 1$ in $N$ this is true). So, the domain of a theory $N_n$ is of cardinality $n$. The number $n$ becomes an inconsistent object of $N_n$. Drawing the successor function by arrows, the structure of a model of $N_n$ looks like this:

$$0 \longrightarrow 1 \longrightarrow 2 \longrightarrow \cdots \longrightarrow n\,\circlearrowright$$

These models are called "heap models". The logic of $N_n$ has to be paraconsistent. And it has to have restrictions on standard first order reasoning as well.

Mortensen chooses **RM3#** as basic system and finitizes it by substituting for a number $n$ the number $n$ modulo some $m$. Thus the domain becomes $\{0, 1, 2, \ldots, m - 1\}$. The models then are no longer heap models but *circular* of size $m$. The resulting arithmetic **RM3**$^m$ is negation complete, non-trivial and decidable. **RM3**$^m$ is *axiomatisable* by adding to

**RM3#** the axioms:
$$\vdash 0 = m$$
and all instances of the following axiom scheme for $n \in \{0, 1, \ldots, m - 1\}$:

$$\vdash (0 = n \leftrightarrow 0 = 1).$$

The approach "modulo some $m$" has at least the same deviant results than the heap models mentioned before: In **RM3**[5] we have $4 + 2 = 6$ and $4 \times 6 = 4$; this approach gets deviant sentences for some *known* numbers!

Arithmetic is constructed as a finite theory. One can generalize the steps of this procedure to apply it to other mathematical theories. Van Bendegem (1993) distinguishes the following steps:

(i) Take any first-order theory $T$ with finitely many predicates. Let $M$ be a model of $T$.

(ii) Reformulate the semantics of $T$ in a paraconsistent fashion (i.e., the mapping to truth values and overlapping extensions of $P^+$ and $P^-$).

(iii) If the models of $M$ are infinite, define an equivalence relation $R$ over the domain $D$ of $M$ such that $D/R$ is finite.

(iv) The model $M/R$ is a finite paraconsistent model of the given first-order theory $T$ such that validity is at least preserved.

The restriction to theories with finitely many predicates is no real restriction in any field of applied mathematics or formal linguistics, since no physical device (be it human or machine) can store a non-enumerable list of basic predicates. Van Bendegem hints at finite version of the theory of integers and the theory of rational numbers. Mortensen (1995) considers some inconsistent version of the calculus.

The *Löwenheim/Skolem-Theorem* is one of the limitative or negative meta-theorems of standard arithmetic and first-order logic. It says that any theory presented in first-order logic has a *denumerable* model. This is strange, since there are first order representations not only of real number theory (the real numbers being presented there as uncountable/non-denumerable), but of set theory itself. Thus the denumerable models are deviant models (usually Herbrand models of self-representation), but they cannot be excluded. Given the general procedure to finitize an existing mathematical first order theory using paraconsistent semantics, there is a paraconsistent strengthened version of the *Löwenheim/Skolem-Theorem*:

**Theorem 2.1.** Any mathematical theory presented in first order logic has a *finite* paraconsistent model.

## 3   The benefits of inconsistency

A mathematics that does not commit us to the infinite is a nice thing for anyone with reductionist and/or realist leanings. As far as we know the universe is finite, and if space-time is (quantum) discrete there isn't even an infinity of space-time points. The largest number may be indefinitely large. So we never get to it (e.g., given our limited resources to produce numerals by writing strokes). If there is a largest number n there is the corresponding inconsistent arithmetic $N_n$. We can presuppose $N_n$ being our arithmetic. Since N and $N_n$ agree on all finite and computational mathematics it is hard to see whether we lose anything important at all by switching to $N_n$. If we have paraconsistency anyway for other reasons, we get this finitism for free, it seems. So why not take it? In as much as $N_n$ is correct no correct reasoning transcends the finite. Hilbert wouldn't have rejoiced, probably, since $N_n$ of course is inconsistent itself. The drawback of all this is, of course, the problem of an ontology of inconsistent entities—at least if you are a realist.

If there are inconsistent versions of more elaborated mathematical fields like the calculus one may draw some general philosophical conclusions: Firstly, if there are corresponding inconsistent versions of these mathematical theories with comparable strength to the original theories then consistency is not the fundamental mathematical concept, but functionality (of the respective basic concepts) may well be.

Moreover, if the justification of mathematics depends on its applicability and the inconsistent versions are of comparable applicability then they are justified not just as mathematical theories, but even in the wider perspective of grasping fundamental structures of reality; there no longer will be available the argument from mathematical describability to the consistency of the world.

One final worry may be, which contradictions to accept. If we (mis-)understood inconsistent arithmetic as allowing for any old contradictions, this would be the end of all serious study, wouldn't it? Roughly the inconsistent mathematician accepts only those contradictions that are forced on her by one of the following two commitments:

The first is a commitment to some principles which are intuitively valid and superior to their circumscribed rivals, like Naïve Comprehension in set theory. Going into the details here would lead us to the wider debate surrounding dialetheism and paraconsistency, but dialetheist typically argue that the supposed solutions to the antinomies like a semantic or set theoretic hierarchy are even more mysterious than dialetheism and additionally violate our intuitive conceptions, of sets for example (cf. Bremer, 2005, pp. 20–31). The accepted contradictions are then these which follow (more or less immediately) from the presence of the accepted principles like Naïve Comprehension (e.g., the Russell Set being a member of itself and not being a member of itself). These contradictions are isolated, however, as a paraconsistent logic bars the spreading of contradictions.

The second is the commitment to a finitist universe. The broader picture of the universe one might like to endorse may be that of a finite universe without additional abstract entities. This is a highly controversial metaphysical agenda, but one that because of its naturalness should be taken as a serious rival to the more ontologically promiscuous standard picture of infinite Platonism (cf. Bremer, 2007). The accepted contradictions are then those which immediately concern the largest, and therefore inconsistent, number(s). Again the presence of a paraconsistent logic prevents these contradictions from spreading.

## Bibliography

Bremer, M. (2005). *An Introduction to Paraconsistent Logics*. Peter Lang, Frankfurt a.M.

Bremer, M. (2007). Varieties of finitism. *Metaphysica*, 8(2):131–148.

Byers, W. (2007). *How Mathematicians Think: Using ambiguity, contradiction, and paradox to create mathematics*. Princeton University Press, Princeton NJ.

Meyer, R. K. (1976). Relevant arithmetic. *Bulletin of the Section of Logic of the Polish Academy of Science*, 5:133–137.

Mortensen, C. (1995). *Inconsistent mathematics*. Kluwer Academic, Dordrecht.

Priest, G. (1994a). Is arithmetic consistent? *Mind*, 103(411):337–349.

Priest, G. (1994b). What could the least inconsistent number be? *Logique et Analyse*, 37:3–12.

Priest, G. (1997). Inconsistent models of arithmetic part I: Finite models. *Journal of Philosophical Logic*, 26(2):223–235.

Van Bendegem, J. P. (1993). Strict, yet rich finitism. In Wolkowski, Z., editor, *First International Symposium on Gödel's Theorems*, pages 61–79. World Scientific Press, Singapore.

# Loss of vision: How mathematics turned blind while it learned to see more clearly

Bernd Buldt[1], Dirk Schlimm[2,*]

[1] Department of Philosophy, Indiana University–Purdue University Fort Wayne (IPFW), 2101 East Coliseum Boulevard, Fort Wayne, IN 46805, United States of America

[2] Department of Philosophy, McGill University, 855 Sherbrooke Street West, Montreal QC, H3A 2T7, Canada

E-mail: buldtb@ipfw.edu; dirk.schlimm@mcgill.ca

## 1 Introduction

### 1.1 Overview

The aim of this paper is to provide a framework for the discussion of mathematical ontology that is rooted in actual mathematical practice, i.e., the way in which mathematicians have introduced and dealt with mathematical objects. Using this framework, some general trends in the development of mathematics, in particular the transition to modern abstract mathematics, are formulated and discussed. Our paper consists of four parts: First, we begin with a critical discussion of the notion of *Aristotelian abstraction* that underlies a popular folk ontology and folk semantics of mathematics; second, we present a conceptual framework based on the distinction between *bottom-up* and *top-down* approaches to the introduction of mathematical objects; in the third part we briefly discuss a number of historical episodes in terms of this framework, illustrating a general move towards top-down approaches and resulting in changes of the nature of mathematical objects; finally, the effects of this change with regard to the role of visualization in mathematics are discussed.

That mathematical objects are *abstract* posed a significant problem for philosophers already in ancient Greece. However, it is a commonplace that in the 19th century mathematics became *more* abstract.[1] What this 'more' consists in, we claim, can be explicated as a shift from a traditional notion of abstraction that goes back to Aristotle to a non-Aristotelian conception of abstraction.[2] This is closely related to the trend we identify in the development of 19th century mathematics, which reveals an increased attention to the study of mathematical *relations* as opposed to mathematical objects,

---

*The authors would like to thank the *Wissenschaftliches Netzwerk* PhiMSAMP funded by the *Deutsche Forschungsgemeinschaft* (MU 1816/5-1) for travel support. The authors also wish to thank Brendan Larvor and Rachel Rudolph for numerous helpful remarks on an earlier draft of this paper.

[1]See, e.g., (Ferreirós and Gray, 2006; Gray, 2008).

[2]For a similar point, regarding theories in psychology, see (Lewin, 1931).

Benedikt Löwe, Thomas Müller (*eds.*). *PhiMSAMP. Philosophy of Mathematics: Sociological Aspects and Mathematical Practice.* College Publications, London, 2010. Texts in Philosophy 11; pp. 39–58.

*Received by the editors:* 13 November 2009; 22 February 2010.
*Accepted for publication:* 23 February 2010.

based on the increased emphasis of top-down characterizations as opposed
to bottom-up ones (see Section 3). We argue that these developments are
best understood from a *structuralist* perspective, as opposed to a traditional
Aristotelian view that is based on the notion of *substance*.

## 1.2   Structures

To our mind, the development of mathematics during the 19th century
shows, certainly not in each single move mathematicians of the time took,
but in their overwhelming majority nonetheless, a clear tendency to prepare
and level the ground for 20th century structuralism. The term 'structuralism,' however, is ambiguous and can mean at least two different things.[3]

There is, first, the structuralism of the Bourbaki group, inspired by
advances in set theory. Here one starts with a set of elements, a domain $D$,
and then defines by set-theoretic operations alone a number of relations $R$
over $D$ that obey certain axioms, thus yielding a structure $\mathfrak{S}$ composed of $D$
and $R$.[4] As such it can be seen as a basically bottom-up approach: one starts
out with intuitively given objects conceived of as elements over which first-
order quantifiers can range (like the natural numbers or points in a plane)
and then defines a relational super-structure by set-theoretic constructions
(identifying relations with certain $n$-tuples of the Cartesian product, etc.).
Studying the properties of a structure in this context entails knowing how
it is built up, its *Bauplan*. This reading of Bourbakian structuralism is well-
known in particular among philosophers as it resembles the approach taken
by model-theoretic semantics.

There is, second, the structuralism of those who champion category theory, inspired by advances in algebra.[5] Here one starts with a class of structured objects $A, B, C, \ldots$ (complex objects that are already equipped with
structural features), a class of mappings $f, g, h, \ldots$ among them, such that
those mappings have a number of desirable or 'natural' properties, and then
studies those transformations that preserve the structure of those objects.
As such it is a top-down approach: one starts with objects that usually
are quite complex (like all sets, all topological spaces) and assumes them
as already given. Studying the properties of a structure in this context entails knowing what their structure-preserving transformations are or what
structures admit such transformations, but doesn't require us to know the
structure's *Bauplan*.[6]

---

[3] For a more detailed discussion of different versions of structuralism, see (Reck and
Price, 2000).

[4] See, e.g., (Bourbaki, 1950, §3, esp. p. 225 ff.).

[5] See, e.g., (Awodey, 1996).

[6] And to the extent that a certain structure is characterizable in the language of
category theory—say, a group $(G, \cdot)$ as a category with a single object $\star$ and all elements
$a$ of $G$ as morphisms $a : \star \rightarrow \star$, or a poset as a category in which there is at most one

While in general we lean towards an interpretation of structuralism as inspired by category-theory, the notion of structuralism underlying this paper is more broadly defined and also more vague. For lack of an established term and for reasons to be adduced below we shall call it 'non-Aristotelian structuralism.' We shall be concerned with what many felt (and many still feel) is a development towards a mathematics that is more (sometimes too) 'abstract' and much less intuitive than it should be—a mathematics that features objects not very amenable to visualizations; many think this a sufficient reason to dismiss these developments. We think what causes this uneasiness is a conflict, not well-understood and hence unresolved, between, on one hand, a 'folk ontology and semantics' that starts with concretely given objects and their properties and, on the other hand, a non-Aristotelian structuralism that does not need such objects.

## 2    Folk ontology and folk semantics

We are surrounded in our daily lives by middle-sized concrete objects that have properties conveyed to us through our senses; this, we are inclined to think, captures what the furniture of the world is. This was also the starting point for Plato; but things quickly proved to be much more difficult.[7] When he set out to refute the Sophists and in particular their claim that there is neither truth nor falsehood but that man is the measure of all things, he was faced with two opposing viewpoints that had emerged from Ionian natural philosophy. There was, first, Heraclitus' doctrine that the true nature of things—which love to disguise themselves and trick us into holding mere subjective opinions[8]—is to be in constant flux propelled by never ending opposition. Second, there was Parmenides' doctrine that what truly exists is eternal and immutable, implying that the language of change is deceptive and that everything we can hope to know must therefore be statements that hold without any exceptions. These were serious issues of the time. Plato's teacher Cratylos inferred that, based on Heraclitean doctrines, it is impossible for a language to have a denotational semantics and decided to stop speaking but to point with his finger instead, while Antisthenes, a student of Socrates like Plato, but later following Parmenides' lead, found it only possible to argue for the truth of analytical sentences and limited his utterances to sentences like 'a man is a man.' Plato needed to develop

---

arrow between any two objects—then this doesn't reveal the 'familiar' *Bauplan* of the structure either.

[7]Scholars disagree on what the correct interpretation of Plato in the context of his time is. We cannot hope to settle any of these disputes here; all we can do is to clearly say where we stand, i.e., to acknowledge that our own account of Plato's philosophy of language is heavily indebted to Rehn (1982).

[8]'Nature loves to hide itself' (φύσις κρύπτεσθαι φιλεῖ) is one of his more famous statements; see (Diels and Kranz, 1952, frg. **B** 123) or (Marcovich, 2001), frg. 8.

a theory of language that was able to refute the relativism of the Sophists and to establish that declarative sentences can indeed be true or false; at the same time he needed to accommodate Heraclitean and Parmenidean arguments and find a way to reconcile the idea of permanent objects, assumed to be one, with fleeting properties, which are many (the venerable problem of 'unity *vs* multiplicity'). Plato's solution to this entangled knot of problems was to develop a comprehensive theory of language and then base crucial arguments upon that theory. Due to his efforts—and for his times this was quite an accomplishment—Plato might have very well been the first to clearly identify the grammatical structure of subject and predicate as underlying declarative sentences; a sentence according to Plato always is an artful 'composition' (σύνθεσις) or close 'intertwining' (συμπλοκή) of 'nouns and verbs' (ὀνομάτων καί ῥημάτων).[9]

Aristotle adopted Plato's basic insights about the linguistic functions of nouns and verbs but not his teacher's conclusions (e.g., that knowledge of the physical world—knowledge here understood in its emphatic, Parmenidean meaning of the word—is not possible as the world forms a realm of change, becoming, not of being). Both, however, agree with Parmenides that any sentence is 'about something' (περί τινος), where this 'something' always refers to a 'state of affairs' (πρᾶγμα).[10] For Plato, this was just a necessary condition to ensure the 'matter-of-factness' that characterizes any declarative sentence, while Aristotle extended the 'about something' structure to a 'something about something' structure (τι κατά τινος). This, then, was according to Aristotle the proper structural analysis: A declarative sentence (λόγος ἀποφαντικός) features a subject $S$ (ὑποκείμενον) and a predicate $P$ (κατηγορία), where the 'predicate something' is about the 'subject something,' or, as Aristotle would also formulate it, $P$ is 'predicated of' (κατηγορεῖν) or 'belongs to' (ὑπάρχειν) $S$.

Unlike Plato, who addressed the Sophistic and Ionian challenges mainly in the realm of language, Aristotle took an ontological turn—the solution, he remarked, belongs to another field of investigation[11]—and stipulated that the grammatical subject $S$ always denotes (σημαίνει) an 'ousia' (οὐσία),[12] and it was this concept of ousia (or, the 'what is' (τί εστι) that was meant to shoulder the main bulk of explanatory work.[13]

---

[9]See Plato, *Cratylus* 424e–425a; *Sophistes* 261c–262d.

[10]See Plato, *Sophistes* 263a, resp. *ibid.* 262e; Aristotle, *Topica* I.8, 103b7; English translation can be found in (Cooper, 1997; Barnes, 1984).

[11]Aristotle, *De Interpretatione* V, 17a15.

[12]Defined as what can be predicated but is never predicated of; see Aristotle, *Categoriae* V.

[13]Like in the case of Plato, there is quite some disagreement among scholars on the details of a proper understanding of Aristotle's ontological doctrines, his '*prima philosophia*' (πρώτη φιλοσοφία). As we are only interested in the mainstream views that emerged from it, we feel free to gloss over all these difficulties. We skip in particular Aristotle's quite

We can capture the basics of what we need in the following by modifying the account given by Spade (1985, p. 236ff). An individual substance $s$, denoted by a grammatical subject $S$, does not change and thus allows for knowledge, but also acts as a pincushion for its changing properties $p$, represented by pins that come and go and denoted by predicates $P$. Some of these 'pin-properties,' however, cannot be removed without ripping the cushion apart. For they are properties that are 'essential' to $s$, while all other properties, those whose pins may be added or removed, are 'accidental.' We cannot take away the property of rationality from a human being without creating a freak of nature; but anyone can dye their hair a different color every day without losing their humanity.

According to this approach knowledge is firmly rooted in sense experience and one arrives at an abstract object by zooming in on only certain properties that constitute it. For example, if a basic geometrical object, like a square, is conceived of as a boundary surface of a solid die, then it does not exist independently of the die. The mind, however, can treat it as an abstract object by focusing on just the square's properties and thereby grasping the latter's form.[14] The mental processes of focussing on some aspects but neglecting others that enable the mind to take on the form of an abstract object, i.e., to identify, grasp, and know it, do not necessarily resemble the means we use to give a logical description of what it means to identify or define an abstract object. The logical reconstruction usually employs the language of abstraction. If, in the example above, the composition of the die has a number of properties $p_1, \ldots, p_n$, then, by eliminating many of them (like material, color, weight, etc.), we arrive at a sub-set $p_{i_1}, \ldots, p_{i_k}$ of the die's properties that characterizes an abstract object, or, in Aristotle's language, a 'secondary ousia,' like a square. We shall call this method of arriving at new objects from old ones 'Aristotelian' or 'eliminative abstraction.' In more general terms, if '$o_1[abc]$' denotes some object that has, among others, the properties $a, b, c$ and $o_2[bcd]$ another object with properties $b, c, d$, then an object like $o[bc]$—that is characterized by what objects $o_1$ and $o_2$ have in common—is obtained by eliminating those properties that the two objects do not share, $a$ and $d$, and possibly others. Due to

---

complex theory of forms (or *causae*) and how they contribute to the unity of objects, especially when two or more of them coincide, and ignore the intricate theory of how the soul, as the form of the human body, can come to know something by taking on the form of that something. We follow common practice since Boethius, though, and render 'ousia' as 'substance.'

[14]See Aristotle, *De Anima*, pt. 7. 'The so-called abstract objects the mind thinks just as, if one had thought of the snub-nosed not as snub-nosed but as hollow, one would have thought of an actuality without the flesh in which it is embodied: it is thus that the mind when it is thinking the objects of mathematics thinks as separate elements which do not exist separate. In every case the mind which is actively thinking is the objects which it thinks.'

the denotational power of language, where subjects denote substances and predicates denote properties, the process of abstraction is available in the realm of language as well. The possibility of linguistic abstraction as well as ontological abstraction has led to two different interpretations of Aristotle, but we omit further discussion of this issue.[15]

Aristotle thus established, after a heated debate that lasted for many generations and was fueled by conflicting intuitions about what the furniture of the universe is and how language can refer to it, what would eventually become a linguistic and ontological paradigm for the next two millennia. And the resulting views were not too disquieting: We are surrounded by middle-sized concrete objects whose properties are given by the senses; new objects can be obtained by eliminative abstraction, and all objects and their properties are amenable to human knowledge.[16]

The reason to call the Aristotelian paradigm 'folk ontology and semantics' is that its underlying intuitions strike most people as so natural that it requires a serious effort not think along its lines. Kant even went a step further and turned thinking according to substance and predicate from a psychological propensity into a logical necessity, i.e., made it *a priori*.[17]

Within this paradigm all concept formation is always bottom-up and well-founded; a concept cannot be legitimately formed unless each property $P$ it contains can ultimately be traced back to some concretely given object that instantiates $P$ or its subordinated constituents. It is this that seems to have motivated both the slogan of empiricism that nothing is in the mind that was not in the senses before and Kant's *dictum* that concepts without objects given in intuition must be empty.[18] Unsurprisingly, textbooks in the semantics of natural languages often present concepts arranged in a tree-structured hierarchy very much along the lines of Plato and Aristotle, and similar to the Porphyrian trees that emerged from that tradition.

Mathematicians appear to have embraced this approach as well whenever they proved new mathematical entities to exist by constructing them, in a bottom-up fashion, from already existing mathematical objects; like von Staudt constructed projective points as sets of 'real' points, Dedekind real numbers as sets of rational numbers, Hamilton imaginary numbers as pairs of real numbers, and so forth. (More on this in the next two sections.)

---

[15]See (Mueller, 1970; Lear, 1982).

[16]Recent decades have seen a revival of viable alternatives to an Aristotelian ontology based on the notion of substance, like mereology and process ontology. We shall not, however, explore their prospects in this article.

[17]By turning substance and predicate into pure concepts of the understanding and by basing a synthetic judgement *a priori*, i.e., the first analogy of experience, on the notion of substance; see (Kant, 1781, B 106, and B 244ff, resp.).

[18]See (Cranefield, 1970) on the history of the phrase *nihil est in intellectu quod non prius fuerit in sensu*, and see (Kant, 1781, B 75) for Kant's *dictum*.

The Aristotelian paradigm emerged to accommodate the needs and the language of everyday life and of sciences that hardly scratch on the surface of things; it doesn't seem to be the best choice available when it comes to understanding modern mathematics, whose development picked up incredible speed during the 19th century. We therefore wish to suggest that an approach that leaves behind the Procrustean bed of an Aristotelian ontology and the shackles of his doctrines is better suited to describe modern mathematics.

This proposal is by no means new. In particular Cassirer in his book *Substance and Function* argued for a similar point.[19] We find, however, first, his Neo-Kantian conclusions to be no longer defensible and, second, some recent accounts on structuralism to be so confused that we believe it is worthwhile to revisit the topic.[20]

We shall try to provide the evidence necessary to support our theses by way of example, for two reasons. First, a fuller scrutiny of the historical evidence would require a book-length study, something we cannot hope to accomplish within the confines of an article. Second, and much more importantly, we do not claim that the mathematical community as a whole moves (or has ever moved) like one solid block in just one direction; nothing could be farther from the truth. We would rather compare the historical development of the mathematical community with the movement of a body that various people try to pull in different directions. The vector that describes the actual movement of the body will then be the sum all those individual vectors that represent the various people. The 'vector' that describes the historical movement of the mathematical community as a whole results likewise, we suggest, from adding all individual vectors, and it points clearly, we think, in the direction of a non-Aristotelian structuralism (see Figure 1). We shall therefore be content with highlighting just a selection of those achievements that contributed more than other developments to pull mathematics towards that structuralism and readily admit that it is easy to find examples that suggest otherwise; sometimes even one and the same person can serve as a witness for both sides. While these alleged counterexamples clearly prove how diverse and vibrant the community of mathematicians has been at any given time, we also claim that their 'associated vectors'

---

[19]Cassirer argued for the stronger claim that all of modern science has moved away from an Aristotelian ontology of substances; see (Cassirer, 1910). Although we believe Cassirer to be basically correct about this, we have to limit our attention to mathematics.

Brendan Larvor was kind enough as to point out to the authors the work of Albert Lautman, who provided another account of the shift in cognitive style from 19th to 20th century mathematics. Since this paper is a programmatic outline only, we shall not engage in a detailed discussion here; see, however, (Larvor, 2010).

[20]See, e.g., the controversy between Hellman (2003) and Awodey (2004), or the self-inflicted difficulties Shapiro (2000) runs into when he tries to reconcile structuralism with what we called folk ontology and semantics.

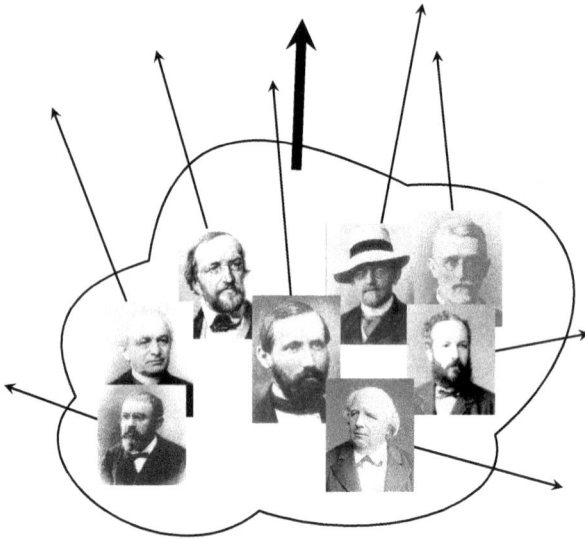

FIGURE 1. How the mathematical community *really* moves ...

have never carried weight enough to pull mainstream mathematics in their direction.

## 3   Towards a new ontology and semantics of mathematics

We have seen in the previous section that the notion of Aristotelian abstraction that underlies the 'folk semantics and ontology' can be interpreted both ontologically and linguistically. These two perspectives allow for the introduction of abstract objects in two different ways: in an ontological 'bottom-up' fashion and in a linguistic 'top-down' fashion.

Since our aim is not to discuss mathematics *per se*, presented in some kind of canonical form, but mathematics as a historical enterprise, its methods are certainly not fixed and have changed over time. In the following, we suggest a framework for discussing some of these developments. In particular, the move away from Aristotelian abstraction and towards 'more' abstract objects is interpreted as a move towards more 'top-down' characterizations of mathematical objects.

The main components of our framework for discussing the historical development of mathematics are illustrated in Figure 2. According to it the

                            Description
         *Top-down*             ↓
                            Structure(s)

                            Complex objects
         *Bottom-up*             ↑
                            Given objects

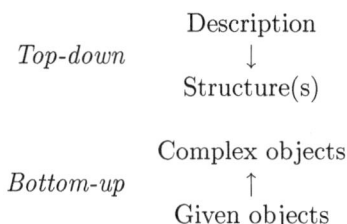

FIGURE 2. 'Bottom-up' and 'top-down' characterizations of mathematical objects.

introduction of mathematical objects can be achieved in two distinct ways:

I. They can be *constructed* by various means from other mathematical objects that are considered to be previously given. Historically, cuts of rational numbers and ideals have been introduced by Dedekind in this way. We refer to this approach as 'bottom-up,' and as will be discussed below, it is closely related to the 'folk ontology and semantics' paradigm mentioned above.

II. Alternatively, mathematical structures can be defined by linguistic descriptions purely in terms of their relational properties. Such definitions can be of various degrees of specificity, with 'implicit definitions' by systems of axioms being the most common ones (e.g., for groups and natural numbers). This 'top-down' approach is characteristic for modern, abstract mathematics.

The distinction between these two modes of introducing mathematical objects or structures reflects Hilbert's distinction between the 'genetic' and the 'axiomatic method' (Hilbert, 1900). As an example of the genetic method Hilbert mentions the extension of the concept of number to include real numbers, through the successive definition of negative numbers and rational numbers as pairs, and the definition of real numbers as cuts of rational numbers. These definitions are all instances of what we call the 'bottom-up' approach, since the new objects are introduced as (set-theoretic) constructions on the basis of the natural numbers, which are taken as given from the outset. Hilbert's example of the top-down, axiomatic method is Euclidean geometry, where

> one customarily begins by assuming the existence of all the elements, i.e., one postulates at the outset three systems of things (namely, the points, lines, and planes) and then [...] brings these elements into relationship with one another by means of certain axioms [...] (Hilbert 1900, p. 180; quoted from Ewald 1996, p. 1092).

$$\textit{Aristotelian abstraction:} \quad \begin{array}{c} \text{Semi-intuitive/quasi-empirical objects} \\ \uparrow \\ \text{Physical objects (world)} \end{array}$$

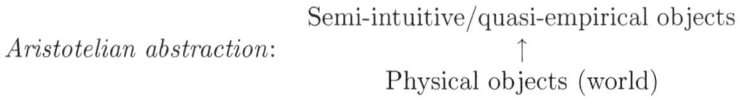

FIGURE 3. Aristotelian abstraction as a special case of bottom-up construction.


The notion of 'construction' employed in our description of the bottom-up approach is very general and should not be confused with the use of constructive, as opposed to classical, methods. In particular, many debates among mathematicians and philosophers, like that between Kronecker and Dedekind, are exactly about what kinds of means should be taken as legitimate for the construction of new objects. As Gray has argued, exactly such disagreements were frequently the source of anxieties that put a strain on discussions of that time (Gray, 2004). Moreover, which means are licensed by the mathematical community changed considerably during the historical development of mathematics: Cantor's and Dedekind's use of set-theoretic definitions, e.g., represented a significant extension of these means.

In general, mathematical constructions take genuine mathematical objects, like numbers, functions, or spaces as their starting point; but there is an important exception to this. A special case of this general notion of construction is *Aristotelian abstraction*, where the given objects are taken to be physical objects and the means of construction involve the deletion of particular properties of these objects (see Figure 3). Thus, according to the 'folk ontology and semantics' this particular kind of bottom-up approach is anchored in perceptible, real world objects. Such a grounding—understood either epistemically or ontologically—as tenuous as it may be, need not exist for mathematical concepts that are defined in a top-down fashion.

The determination of mathematical objects or structures by linguistic means in terms of their relations to others finds its most mature form in the implicit definitions based on systems of axioms. The axiomatic definitions of algebraic structures or the axiomatizations of various geometries are prominent examples. Drobisch's notion of 'abstraction by variation' is another example of this method of introducing mathematical concepts (Drobisch, 1875).

Despite the fact that the the top-down and bottom-up approaches are distinct in nature, in practice they are often employed side by side (see Figure 4). On the one hand, a system of objects that is constructed is often introduced with the explicit aim of satisfying particular axiomatic conditions. Hamilton's quaternions, designed to be an instance of a system in which multiplication is not commutative, and Dedekind's constructions of a simply infinite system and the system of cuts of rational numbers are good

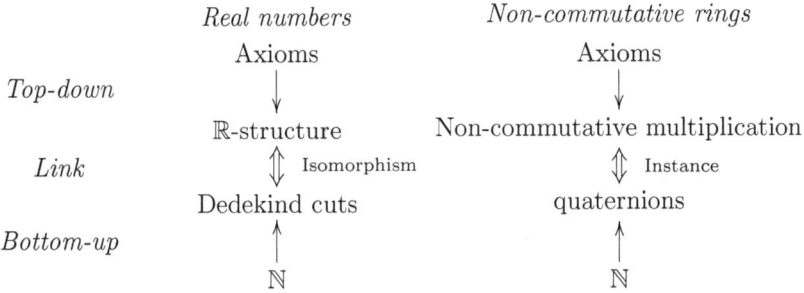

$$
\begin{array}{ccc}
 & \textit{Real numbers} & \textit{Non-commutative rings} \\
 & \text{Axioms} & \text{Axioms} \\
\textit{Top-down} & \downarrow & \downarrow \\
 & \mathbb{R}\text{-structure} & \text{Non-commutative multiplication} \\
\textit{Link} & \updownarrow \ \text{Isomorphism} & \updownarrow \ \text{Instance} \\
 & \text{Dedekind cuts} & \text{quaternions} \\
\textit{Bottom-up} & \uparrow & \uparrow \\
 & \mathbb{N} & \mathbb{N}
\end{array}
$$

FIGURE 4. Examples of connecting bottom-up and top-down approaches.

examples.[21] On the other hand, axioms are often introduced to characterize a system of mathematical objects that has been constructed previously (e. g., the axiomatization of a topological space intended to capture some properties of the real line).[22] In other words, the new objects that are generated in a bottom-up fashion are often intended to instantiate a mathematical concept that has been defined using the top-down method; and vice versa, new objects that are defined in a top-down fashion are meant to be instantiated by objects that were (previously) constructed bottom-up. Through this connection the two approaches are linked and mathematicians often alternate between the two.

In our framework *mathematical work* happens in three places: In the bottom-up constructions of new mathematical objects, in developing appropriate descriptions that are the starting points for the top-down approach, and in establishing possible connections between the structures and the constructed objects (e.g., showing that they satisfy all postulated properties, or that they are even isomorphic). Both top-down and bottom-up approaches involve finding fundamental concepts and fruitful definitions, and working out their consequences.

## 4   Historical examples

In the following we present brief sketches of historical episodes to illustrate the point that in the development of mathematics from the 19th to the 20th century one can identify a decrease of emphasis of bottom-up characterizations and an increased reliance on top-down characterizations. Many of the historical developments that led to the emergence of modern mathematics have been presented and discussed elsewhere, and this is not the place to

---

[21]See the discussion of Dedekind in (Sieg and Schlimm, 2005).

[22]See (Moore, 2008).

add another such study. Instead, we adduce examples from a wide range of developments to support our claim.

The framework introduced in the previous section allows us to view seemingly very different approaches as instances of the bottom-up view. A great example is Gauss' view on the justification of new mathematical objects. Fraenkel describes it as follows:

> Gauss adopts a decidedly realist standpoint [...] according to which an extension of a given domain of numbers is only justified, if it is possible to intuitively associate with the new entities that are to be accepted other things or concepts, which have already gained general acceptance—for example, on the basis of spatial experience or spatial intuition. (Fraenkel 1920; quoted from Volkert 1986, p. 40.)

Hamilton's work on number pairs also fits right into this characterization. And the so-called 'formalism' of late 19th century mathematicians like Heine can also be understood as an instance of this general approach: In order to justify the existence of the natural numbers they are themselves reduced to something 'more concrete,' namely, written symbols. This way of proceeding was famously criticized by Frege, but he was himself working with a reductionist goal in mind (i.e., aiming at a bottom-up account), only to different kinds of objects, namely logical ones.

These views stand in stark contrast to the 'top-down' approach of Dedekind and Hilbert, who are proponents of the modern view of mathematics, according to which mathematical objects are regarded as being determined purely by their descriptions. Such an exclusive reliance on the relations expressed in axioms was a demand also formulated by Pasch—in his famous *Vorlesungen über neuere Geometrie* (1882):

> Indeed, if geometry is to be really deductive, the deduction must be independent of the [sc. bottom-up] *meaning* of geometrical concepts, just as it must be independent of the diagrams; only the *relations* specified in the propositions and definitions employed may legitimately be taken into account. (Pasch, 1882, p. 98)

While Pasch himself made this demand for the sake of gap-free deductions and did not regard mathematical objects to be defined in this way, it was soon employed as a methodological desideratum also for definitions. An early expression of this way of proceeding is given in the opening paragraph of Dedekind's *Was sind und was sollen die Zahlen?* (1888):

> In what follows, I understand by *thing* every object of our thought. In order to be able easily to speak of things, we designate them by symbols, e.g., by letters [...]. A thing is completely determined by all that can be affirmed or thought about it. (Dedekind, 1888, p. 44)

Reck refers to this as Dedekind's 'principle of determinateness' and he considers it to be a crucial component of Dedekind's 'logical structuralism' (Reck, 2003, pp. 394 & 400). A similar formulation of this principle is expressed by Hilbert:

> [...] by the set of real numbers we do not have to imagine, say, the totality of all possible laws according to which the elements of a fundamental sequence can proceed, but rather—as just described— a system of things whose mutual relations are given by the *finite and closed* system of axioms I–IV, and about which new statements are valid only if one can derive them from the axioms by means of a finite number of logical inferences. (Hilbert 1900, p. 184; quoted from Ewald 1996, p. 1095.)

A mathematical notion whose characterization has changed dramatically in the course of the 19th century is that of a function. Originally conceived as a particular, rule-based relation between numbers, it gained more and more generality in the hands of Dirichlet and Dedekind, until it was defined purely in set-theoretic terms. For a wide range of different characterizations of 'function,' see (Volkert, 1986, 55–57).

A similar development can be identified in abstract algebra, nowadays considered a prime example of axiomatically characterized structures. However, it also began as the study of concrete sets of given objects. A 'group,' for example, was defined by Galois as a set of substitutions that is closed under composition; and group-theoretic constructions were always made in terms of substitutions. Only gradually the relational structure was emphasized and taken as the essential aspect of the theory. See (Wussing, 1984) for the general development of group theory, and (Schlimm, 2008) for a particular episode that nicely illustrates our main point.

In traditional geometry, its elements were construed as abstract, usually obtained by some sort of Aristotelian abstraction. With the development of projective geometry, 'points at infinity' or 'ideal points' were introduced, but at first they were treated with the same skepticism that had been directed at the negative and imaginary numbers before. The reduction of these new geometric objects (i.e., the definition of them in terms of 'real' points and lines) was considered to be a great achievement. This sentiment is expressed, for example, in Torretti's remark on Pasch's treatment of projective geometry (Pasch, 1882):

> From a philosophical point of view, Pasch's most remarkable feat is the introduction of the ideal elements of projective geometry using only the ostensive concepts of point, segment, and flat surface and the empirically justifiable axioms $S$ and $E$. (Torretti, 1978, p. 213)

Only with Hilbert's groundbreaking *Foundations of Geometry* (1899) the idea of implicit definitions of mathematical structures slowly gained general acceptance.

## 5  Visualizing mathematical objects

The historical transition discussed above has also had effects on the use of visualizations in mathematics, to which we turn our attention next.[23] We maintain, with Plato and Aristotle, that mathematical objects, as abstract entities, cannot be directly visualized. Euclid's definitions of a point as 'that which has no part' and of a line as a 'breadthless length' (Heath, 1909, p. 153) clearly hint at the ontological and epistemological difficulties that mathematical objects pose, but also at the problem of their accessibility to the senses. Both Plato and Aristotle agreed that these objects are not to be found in our physical world. They disagreed on the accounts of where mathematical object live and how they are related to the things we see with our eyes. Recall Aristotle's account (discussed in more detail above): the mathematical objects are idealizations of physical objects, obtained through a process of abstraction. Thus, even if there is no mathematical sphere— sometimes called 'perfect' to flag its ideal character—in the physical world, there are objects in our world that resemble such spheres to some degree. Such physical objects, imperfect instantiations as they are, can nonetheless be regarded as visualizations of their abstract counterparts. We can easily see and touch spherical objects and also imagine them; if we stretch our imagination just a little bit, we can imagine these objects to be perfectly smooth and spherical, and thus we arrive at a representation of a mathematical sphere. This representation is not identical to the sphere, but closer to it in the relevant respects than any physical object could be.[24]

In sum, some mathematical objects can be construed as idealizations of physical objects, and, accordingly, some objects are easier to visualize than others. We may refer to these as *elementary* objects, and among them we find the geometric notions of point, line, circle, square, cube, sphere, etc., the notions of natural and real numbers, and the modern notion of sets. Many philosophers have limited their discussions of the nature of mathematics to these objects and thus may have been influenced by the particular character of these kinds of objects into thinking that all of mathematics can be built up in this way.[25] However, with the increased reliance on top-down

---

[23] In this programatic sketch, we are unable to do justice to all the ramifications of the topic of visualization. For a more complete picture, the reader might want to look at other studies, like the contributions by Marcus Giaquinto or Ken Manders in (Mancosu, 2008).

[24] See Klein's discussion of the limits of our imagination in (Klein, 1893).

[25] The bottom-up generation of mathematical concepts via conceptual metaphors is presented in (Lakoff and Núñez, 2000).

characterizations of mathematical structures, this conception of visualization soon reaches its limits and becomes untenable as being applicable for all mathematical objects.

The relation between visual representations and mathematical objects has been a topic of debate among mathematicians themselves. In particular with the growing emphasis on rigor in the 19th century the use of diagrams was more and more scrutinized.[26] However, such representations play very different roles in mathematical practice that can be distinguished: a) Visualizations as means to mathematical *understanding* and *education*; b) visualizations as heuristics for mathematical *inferences*; c) visualizations as *justifications* of mathematical inferences; d) visualizations as vehicles for mathematical *creativity*.

Let us briefly illustrate these different roles of visualizations. Consider a formulation of Hilbert's first axiom of Euclidean geometry in the language of first-order logic, '$\forall x \forall y \exists z \; P(x) \land P(y) \land L(z) \land \text{on}(x,z) \land \text{on}(y,z)$.' To understand such a symbolic expression as a geometric statement the primitive terms have to be interpreted and given meaning. Reformulated into English, the statement then becomes 'between any two points there is a line.' Since the words 'points,' 'line,' and 'between' are familiar to us, we immediately understand the statement (or, at least, we think we do). Thus, the familiar terms with their associated visual representations allow us to grasp the content of a proposition much more easily. Accordingly, complex geometric propositions are often visualized using diagrams.[27] Once a proposition has been represented by a diagram, the graphical information can also be exploited for making inferences. For example, if you draw a triangle with an additional straight line going through one of its sides not at a vertex, that line, if drawn sufficiently long, will also go through one of the other two sides of the triangle. This can be easily verified using a diagram. However, diagrams can also be misleading and thus lead to incorrect proofs. Because of this, and the renewed interest in mathematical rigor in the 19th century, the tendency of rejecting the use of diagrams for licensing inferences became stronger (Mancosu, 2005).[28]

Nevertheless, it was commonly agreed that visual representations are very helpful tools in the process of forming new conjectures, since representations can suggest previously unseen connections and completely new directions of research. That multiplying and juxtaposing modes of representation leads to a 'productive ambiguity' that is crucial in the development of science and mathematics has been argued by Grosholz (2007).

---

[26] See (Mancosu, 2005).

[27] For a colorful example, see (Byrne, 1847).

[28] We wish to note, however, that there is danger of oversimplifying these quite complex movements within the mathematical community. Category theory, e.g., which freely and deliberately embraces diagrams and their properties (and therefore sometimes dubbed 'archery'), can serve as an antidote to such oversimplications.

In our analysis of some developments of mathematics we found a general tendency towards introducing mathematical objects in a top-down fashion, and away from the more traditional bottom-up fashion. This move correlates with a change of the role that visualizations play in mathematics. On the one hand, their justificatory power was called into question by the more urgent demands for increased rigor. On the other hand, however, visualizations became more important in their role as vehicles for promoting mathematical understanding. Since the structures defined in a top-down fashion are initially more abstract, a need was felt to provide some substance to flesh them out, and this substance was often furnished by visualizations. Examples are Klein's collections of mathematical models, or, more recently, computer visualizations of fractals. In other words, the (purely linguistic) combination of mathematical properties can lead to conceptions for which no visualization immediately springs to mind: For example, a space in which more than one parallel to a line through a given point exist, and Weierstrass' continuous, but nowhere differentiable, curves. However, mathematicians were not satisfied with this situation and put much effort into finding ways of relating these new notions to others, which were previously available, for example Beltrami's, Klein's, and Poincaré's models for non-Euclidean geometry. Thus, in general we think it is incorrect to say that the amount of and the need for visualizations has decreased in modern mathematics, but rather that the roles they play in mathematical practice have been clarified and have changed.

While bottom-up constructions of new mathematical objects were favored in the 19th century, it is characteristic for modern, 20th century mathematics to rely heavily on top-down characterizations. This focus on linguistic descriptions (axioms) went alongside the demands for more rigor in mathematical argumentations and it also provided the means for extending the limits of what is possible. For example, the simple construal of a space as $\mathbb{R}^3$ quickly led to the question of the nature of $\mathbb{R}^4$, and was generalized to $\mathbb{R}^n$, which allowed for the possibility of $\mathbb{R}^\infty$. Thus, hitherto unthinkable generalizations became possible and were being pursued. As consequences of these developments, visualizations lost their role as warrants of mathematical deductions and more abstract structures became the objects of mathematical investigations for which Aristotelian abstraction no longer works.

# Bibliography

Awodey, S. (1996). Structure in mathematics and logic: A categorical perspective. *Philosophia Mathematica*, 4(3):209–237.

Awodey, S. (2004). An answer to G. Hellman's question 'Does category theory provide a framework for mathematical structuralism?'. *Philosophia Mathematica*, 12(1):54–64.

Barnes, J., editor (1984). *The Complete Works of Aristotle: The Revised Oxford Translation.* Princeton University Press, Princeton NJ. 2 vols.

Beman, W. W., editor (1901). *Richard Dedekind. Essays on the Theory of Numbers.* Open Court Publishing, Chicago IL.

Bourbaki, N. (1950). The architecture of mathematics. *American Mathematical Monthly*, 57(4):221–232.

Byrne, O. (1847). *The first six books of the elements of Euclid, in which coloured diagrams and symbols are used instead of letters for the greater ease of learners.* William Pickering, London.

Cassirer, E. (1910). *Substanzbegriff und Funktionsbegriff. Untersuchungen über die Grundfragen der Erkenntniskritik.* Bruno Cassirer, Berlin.

Cooper, J. M., editor (1997). *The Complete Works of Plato.* Hackett Publishing, Indianapolis IN.

Cranefield, P. F. (1970). On the origin of the phrase *nisi est in intellectu quod non prius fuerit in sensu. Journal of the History of Medicine and Allied Sciences*, 25(1):77–80.

Dedekind, R. (1888). *Was sind und was sollen die Zahlen?* Vieweg, Braunschweig. Page numbers refer to the English translation in (Beman, 1901).

Diels, H. and Kranz, W., editors (1952). *Die Fragmente der Vorsokratiker.* Weidmann, Berlin. 3 vols; first 1903, 6th ed. 1952, later editions are just reprints.

Drobisch, M. W. (1875). *Neue Darstellung der Logik nach ihren einfachsten Verhältnissen: mit Rücksicht auf Mathematik und Naturwissenschaft.* Leopold Voss, Leipzig, 4 edition.

Ewald, W. (1996). *From Kant to Hilbert: A Source Book in the Foundations of Mathematics.* Clarendon Press, Oxford.

Ferreirós, J. and Gray, J., editors (2006). *The Architecture of Modern Mathematics.* Oxford University Press, Oxford.

Fraenkel, A. A. (1920). Zahlbegriff und Algebra bei Gauß. In Klein, F., Brendel, M., and Schlesinger, L., editors, *Materialien fur eine wissenschaftliche Biographie von Gauss*, Nachrichten von der Gesellschaft der Wissenschaften zu Göttingen. Mathematisch-Physikalische Klasse. Beiheft.

Fricke, R. and Vermeil, H., editors (1922). *Felix Klein. Gesammelte mathematische Abhandlungen. Zweiter Band. Anschauliche Geometrie, Substitutionsgruppen und Gleichungstheorie, zur mathematischen Physik.* Springer, Berlin.

Gray, J. (2004). Anxiety and abstraction in nineteenth-century mathematics. *Science in Context*, 17(1/2):23–47.

Gray, J. (2008). *Plato's Ghost: the Modernist Transformation of Mathematics.* Princeton University Press, Princeton NJ.

Grosholz, E. R. (2007). *Representation and Productive Ambiguity in Mathematics and the Sciences.* Oxford University Press, Oxford.

Heath, T. L., editor (1909). *The Thirteen Books of Euclid's Elements. Volume I: Introduction and Books i & ii.* Cambridge University Press, Cambridge.

Hellman, G. (2003). Does category theory provide a framework for mathematical structuralism? *Philosophia Mathematica*, 11(2):129–157.

Hilbert, D. (1900). Über den Zahlbegriff. *Jahresbericht der Deutschen Mathematiker Vereinigung*, 8:180–194. English translation in (Ewald, 1996, pp. 1092–1095).

Kant, I. (1781). *Kritik der reinen Vernunft.* Johann Friedrich Hartknoch, Riga. 2nd edition 1787; page numbers refer to the English translation in (Smith, 1929).

Klein, F. (1893). On the mathematical character of space-intuition and the relation of pure mathematics to the applied sciences. *Evanston Colloquium.* Lecture held in Evanston, Ill., on September 2, 1893; reprinted in (Fricke and Vermeil, 1922, Chapter 46, pp. 225–231).

Lakoff, G. and Núñez, R. E. (2000). *Where Mathematics Comes From. How the Embodied Mind Brings Mathematics into Being.* Basic Books, New York NY.

Larvor, B. (2010). Albert Lautman: Dialectics in mathematics. *Philosophiques*, 37(1).

Lear, J. (1982). Aristotle's philosophy of mathematics. *Philosophical Review*, 91(2):161–192.

Lewin, K. (1931). The conflict between Aristotelian and Galilean modes of thought in contemporary psychology. *Journal of General Psychology*, 5:141–177.

Mancosu, P. (2005). Visualization in logic and mathematics. In Mancosu, P., Jørgensen, K. F., and Pedersen, S. A., editors, *Visualization, Explanation and Reasoning Styles in Mathematics*, volume 327 of *Synthese Library*, pages 13–30. Springer, Dordrecht.

Mancosu, P., editor (2008). *The Philosophy of Mathematical Practice*. Oxford University Press, Oxford.

Marcovich, M., editor (2001). *Heraclitus. Greek text with a short commentary*, volume 2 of *International Pre-Platonic studies*. Academia, Sankt Augustin, 2nd edition.

Moore, G. H. (2008). The emergence of open sets, closed sets, and limit points in analysis and topology. *Historia Mathematica*, 35:220–241.

Mueller, I. (1970). Aristotle on geometrical objects. *Archiv für Geschichte der Philosophie*, 52(2):156–171.

Pasch, M. (1882). *Vorlesungen über Neuere Geometrie*. B.G. Teubner, Leipzig.

Reck, E. H. (2003). Dedekind's structrualism: An interpretation and partial defense. *Synthese*, 137:369–419.

Reck, E. H. and Price, M. P. (2000). Structures and structuralism in contemporary philosophy of mathematics. *Synthese*, 125:341–383.

Rehn, R. (1982). *Der logos der Seele. Wesen, Aufgabe und Bedeutung der Sprache in der platonischen Philosophie*. Meiner, Hamburg.

Schlimm, D. (2008). On abstraction and the importance of asking the right research questions: Could Jordan have proved the Jordan-Hölder Theorem? *Erkenntnis*, 68(3):409–420.

Shapiro, S. (2000). *Thinking about Mathematics: The Philosophy of Mathematics*. Oxford University Press, Oxford.

Sieg, W. and Schlimm, D. (2005). Dedekind's analysis of number: Systems and axioms. *Synthese*, 147(1):121–170.

Smith, N. K., editor (1929). *Immanuel Kant. Critique of Pure Reason.* Macmillan, London.

Spade, P. V. (1985). A survey of mediaeval philosophy. Lectures Notes, University of Indiana, Bloomington.

Torretti, R. (1978). *Philosophy of Geometry from Riemann to Poincaré.* D. Reidel, Dodrecht.

Volkert, K. T. (1986). *Die Krise der Anschauung.* Vandenhoeck & Ruprecht, Göttingen.

Wussing, H. (1969). *Die Genesis des abstrakten Gruppenbegriffes.* VEB Deutscher Verlag der Wissenschaften, Berlin.

Wussing, H. (1984). *The Genesis of the Abstract Group Concept.* MIT Press, Cambridge MA. Translation of Wussing (1969).

# The cognitive basis of arithmetic

Helen De Cruz[1], Hansjörg Neth[2], Dirk Schlimm[3]*

[1] Centre for Logic and Analytic Philosophy, Katholieke Universiteit Leuven, Kardinaal Mercierplein 2, 3000 Leuven, Belgium

[2] Max-Planck-Institut für Bildungsforschung, Lentzeallee 94, 14195 Berlin, Germany

[3] Department of Philosophy, McGill University, 855 Sherbrooke Street West, Montreal QC, H3A 2T7, Canada

E-mail: `helen.decruz@hiw.kuleuven.be`; `neth@mpib-berlin.mpg.de`; `dirk.schlimm@mcgill.ca`

## 1  Introduction

Arithmetic is the theory of the natural numbers and one of the oldest areas of mathematics. Since almost all other mathematical theories make use of numbers in some way or other, arithmetic is also one of the most fundamental theories of mathematics. But numbers are not just abstract entities that are subject to mathematical ruminations—they are represented, used, embodied, and manipulated in order to achieve many different goals, e.g., to count or denote the size of a collection of objects, to trade goods, to balance bank accounts, or to play the lottery. Consequently, numbers are both abstract and intimately connected to language and to our interactions with the world.

In the present paper we provide an overview of research that has addressed the question of how animals and humans learn, represent, and process numbers. The interrelations among mathematics, the world, and the cognitive capacities that are frequently discussed in terms of mind and brain have been the subject of many theories and much speculation. Figure 1a shows that the four basic concepts that anchor this discussion (mathematics, world, mind, brain) enable six possible binary relationships (four edges and two diagonals), each of which raises fundamental philosophical questions. Traditionally, philosophy of mathematics focuses on the triangle between mind, mathematics, and the world (Figure 1b, $\Phi$), asks how mushy minds can grasp abstract numerical concepts, wonders about the nature of mathematical truth, and is puzzled by "the uncanny usefulness of mathematical concepts" (Wigner, 1960). In contrast, psychologists and their colleagues from cognitive science and neuroscience investigate the relationship between mind and brain and its relation to the world, that is further sub-divided into

*The authors would like to thank Bernd Buldt, Benedikt Löwe, and Rachel Rudolph for many valuable comments on an earlier draft of this paper. The authors would like to thank the *Wissenschaftliches Netzwerk* PhiMSAMP funded by the *Deutsche Forschungsgemeinschaft* (MU 1816/5-1) for travel support.

Benedikt Löwe, Thomas Müller (*eds.*). *PhiMSAMP. Philosophy of Mathematics: Sociological Aspects and Mathematical Practice.* College Publications, London, 2010. Texts in Philosophy 11; pp. 59–106.

*Received by the editors:* 13 November 2009; 22 February 2010.
*Accepted for publication:* 23 February 2010.

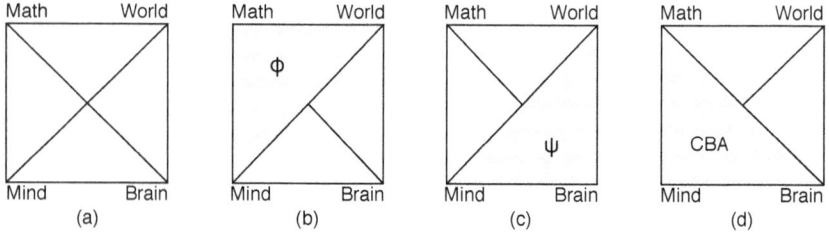

FIGURE 1. Schematic views of possible perspectives on the cognitive basis of arithmetic.

physical and social environments (Figure 1c, $\Psi$). From this perspective, the ability to understand and solve mathematical problems is just one accidental topic among many others, and is typically overshadowed by more prominent issues of perception, categorization, language, memory, judgment, and decision making.

As a departure from both of these traditional perspectives, this paper on the cognitive basis of arithmetic focuses on the manifold relations between mathematics, mind, and brain (Figure 1d, CBA). To illuminate this triangle, we shall cross many disciplinary boundaries and collect past and present insights from philosophy, animal learning, developmental psychology, cultural anthropology, cognitive science, and neuroscience. Although mathematics consists of far more than arithmetic and certainly involves cognitive faculties that extend beyond the ones discussed here (like reasoning with diagrams and infinite objects), we restrict ourselves to the cognitive basis of arithmetic. Such a foundation will provide the groundwork for a more comprehensive understanding of mathematics from a cognitive perspective.

**Some basic terminological distinctions.** By *numbers* we mean the abstract entities that are denoted by number words like 'seventeen' or numerals like '42.' The properties of these objects are studied by mathematicians. In contrast, what we encounter with our senses are *collections* of things, which are also called *numerosities*.[1] These are discrete, concrete numerical quantities of objects, like a pile of peas or the musicians in a band called 'The Beatles.' The magnitude or size of such collections are cardinal numbers,

---

[1] In ordinary parlance one also speaks of *sets* of objects, but we shall try to avoid this term, because it should not be confused with the mathematical notion of set. Some philosophers play down this difference, in order to account for our epistemic access to sets (Maddy, 1992, pp. 59–61).

|                        | one | two | three | four | five | six |
|------------------------|-----|-----|-------|------|------|-----|
| (a) Tally system:      | I   | II  | III   | IIII | ⅢⅢ   | ⅢⅢ I |
| (b) Roman numerals:    | I   | II  | III   | IIII | V    | VI  |
| (c) Greek alphabetic:  | α   | β   | γ     | δ    | ε    | ϝ   |
| (d) Arabic decimal:    | 1   | 2   | 3     | 4    | 5    | 6   |
| (e) Binary digits:     | 1   | 10  | 11    | 100  | 101  | 110 |

|                        | seven | eight | nine | ten | eleven | twelve |
|------------------------|-------|-------|------|-----|--------|--------|
| (a) Tally system:      | ⅢⅢ II | ⅢⅢ III | ⅢⅢ IIII | ⅢⅢ ⅢⅢ | ⅢⅢ ⅢⅢ I | ⅢⅢ ⅢⅢ II |
| (b) Roman numerals:    | VII   | VIII  | VIIII | X   | XI     | XII    |
| (c) Greek alphabetic:  | ζ     | η     | ϑ    | ι   | ια     | ιβ     |
| (d) Arabic decimal:    | 7     | 8     | 9    | 10  | 11     | 12     |
| (e) Binary digits:     | 111   | 1000  | 1001 | 1010 | 1011  | 1100   |

TABLE 1. Different representational systems to represent numbers. Note that the letter for the Greek alphabetic numeral for 6 is the now obsolete Greek letter *digamma*, and that we represent the Roman numerals without the 'subtractive' notation (i.e., representing 4 as IV, instead of IIII) that was introduced in the Middle Ages.

while ordinal numbers indicate positions in an ordered sequence (e.g., first, second, third, etc.).[2]

Numbers can be represented in multiple ways, and it is important to distinguish between *external* and *internal* representations. The two main external representations are *numerical* and *lexical notation systems* (Chrisomalis, 2004). The latter are sequences of numeral words in a language, either written or spoken, with a distinctive phonetic component, while the former are sequences of *numeral phrases* or simply *numerals*, themselves consisting of a group of elementary *numeral signs* or *symbols*. In our familiar decimal place-value numeral system, these are also called *digits*. Clearly, different numerals can denote the same number: Each column in Table 1 denotes the same numerical quantity. In ordinary parlance a numerical notation system is also referred to as a 'number system' and authors frequently use terms like 'Roman numbers' to refer to Roman numerals. We might also slip into this habit, if the context provides sufficient information to prevent ambiguity. Internal representations of numbers are how numbers are represented 'in the head', which can refer to the level of neurons in the brain, but also to a higher, more abstract conceptual or representational level. Note that in the psychological literature a *concept* is typically understood to be a mental entity, which is not necessarily so in the philosophical literature. Indeed, in the tradition of analytic philosophy concepts are expressly

---

[2]The related, but somewhat different, technical notions of cardinal and ordinal numbers in set theory are beyond the scope of this paper.

*not* considered to be mental entities.[3] Thus, cognitive scientists often refer to mental *representations* as ways in which a particular system of numerals is mapped to internal concepts and also mention mental *processes* as operations that translate between external and internal formats and that manipulate internal representations.

During the following discussion it is useful to keep in mind the distinction between different levels of *numerical competence.* The most basic level consists in the ability to recognize and distinguish small numerosities. The second level, that of *counting*, involves mastering at least an initial segment of a numerical or lexical notation system and the ability to systematically map numerosities to that system in a one-to-one fashion. The realization that there is no greatest number and that the numeral systems are potentially infinite, i.e., that the process of counting can be continued indefinitely, can be considered a further step in the acquisition of knowledge about numbers. *Arithmetic competence* begins with the ability to perform basic computations, i.e., to correctly apply the operations of addition, subtraction, multiplication, and division.[4] In analogy to the distinction between internal and external notation systems, we can also distinguish between internal and external arithmetic. As we shall see in Section 4.3, however, there is evidence that these systems are closely related. Finally, *logical reasoning* about numbers and the ability to *prove* and understand arithmetic theorems, e.g., that there are infinitely many prime numbers, constitutes the most advanced level of numerical competence.

**Mathematical, philosophical, and psychological perspectives on arithmetic.** Our above characterization of arithmetic as the theory of the natural numbers is one that a mathematician would provide. From this point view, practicing arithmetic mainly involves establishing properties of numbers by means of proofs, resulting in discoveries like Euclid's theorem about the infinity of prime numbers and conjectures like Fermat's Last Theorem. In the late nineteenth century arithmetic was even considered to be the most fundamental mathematical theory to which all others should be reduced (see Klein, 1895). In modern mathematics, however, there are no restrictions on the methods used for studying numbers; and while mathematicians operate with numerical representations, they usually do not worry too much about them, since they are interested in establishing relationships between numbers at an abstract level.

---

[3]The analytic tradition follows Frege's distinction between concepts as logical entities and ideas as psychological entities (Frege, 1884, pp. xxi–xxii). We return to this issue in Section 5.

[4]The difficulties involved in learning the basic arithmetical operations and how these are related to particular systems of numerals are discussed in (Lengnink and Schlimm, 2010, this volume).

The metaphysical nature of numbers has been a topic of philosophical discussion since the ancient Greeks, most famously by Pythagoras, Plato, and Aristotle. Surprising and counterintuitive results in the mathematics of the nineteenth century kindled an interest in foundational questions.[5] In order to secure these foundations, Frege (1884) attempted to reduce the concept of number to purely logical notions while Dedekind (1888) and Peano (1889) provided axiomatizations of the natural number structure. These developments greatly influenced the practice and understanding of mathematics. Indeed, many contemporary philosophers characterize mathematics as the science of *structures* (Shapiro, 2000). According to this view it makes no sense to regard individual numbers in isolation; instead, they must be regarded as positions within a natural number structure. This raises the important question of how we can have access to and knowledge of such an abstract and infinite structure. Traditionally, philosophers also have paid very little attention to the representations of numbers, except to motivate their accounts of our epistemic access to them. In general, they seem to be content to mention *ad hoc* accounts based on anecdotal evidence and to outsource these investigations to cognitive scientists.[6]

Since mathematical reasoning is often considered to be a fundamental human ability, the relatively young discipline of cognitive science has shown great interest in exploring it, with its main concern being how the brain or the mind processes numerical reasoning (see, e.g., Dehaene, 1997; Butterworth, 1999; Lakoff and Núñez, 2000).

One of the basic issues concerns the relation between basic numerical processing and the use of language. In particular, cognitive scientists investigate how the levels of arithmetic competence are related to various internal and external representations. Theory formation in cognitive science often goes hand in hand with the gathering of empirical data on arithmetical abilities of animals and infants, as well as on the use of number words in different cultures (see also François and Van Kerkhove, 2010, this volume). Thus, cognitive scientists are primarily concerned with the lower levels of numerical competence and with internal representations and rarely discuss how their empirical findings relate to higher-level mathematical abilities.

**Overview.** In the following sections we discuss a wide range of empirical findings on the cognitive foundations of arithmetic from a variety of scientific disciplines and perspectives. Of particular interest with regard to the phylogenetic and ontogenetic developments of numerical abstractions are the mathematical abilities of animals as well as those of infants (Sections 2.2

---

[5]See (Buldt and Schlimm, 2010, this volume), for a general discussion of these developments.

[6]See, e.g., (Maddy, 1992, pp. 50–74), (Resnik, 1997, Ch. 11), and (Shapiro, 2000, pp. 279–280).

and 2.3). The relation between these abilities and the use of number words is studied through investigations of the numerical abilities of cultures with only a limited repertoire of number words (Section 2.4). After discussing research on the localization of arithmetical processes in the brain (Section 3), we shall turn to questions about arithmetic notation (Section 4). In particular, we shall ask questions like: How do notations facilitate or constrain simple and complex arithmetic computations? What is the relationship between external notations and mental calculations? What is the impact of resources provided by the computational environment? To address these more theoretical issues we need to consider fundamental aspects of arithmetic notations. We shall provide some terminological distinctions and historical context for the comparative study of number systems. We then sketch a computational method that allows us to illustrate and quantify the trade-offs between specific numeration systems and the internal and external processes they require for performing calculations. Ultimately, we argue for a more nuanced view of the merits and faults of particular numeration systems and for a more careful analysis of the connections between internal and external representations in arithmetic reasoning. A comprehensive analysis of mathematical practice will have to study the complex interplay among representational systems, their biological and psychological bases, and their linguistic and cultural manifestations.

## 2  Developing arithmetic

### 2.1  Intuitive arithmetic

Up to the eighteenth century, philosophers of mathematics were primarily intrigued by the relationship between human cognition and the abstract objects that mathematical entities seem to be. Our apparent epistemic access to such objects needed an explanation. In the dialogue *Meno*, Plato proposed that our knowledge of geometry actually stems from recollecting (ἀνάμνησις) forms that we knew from before we were born. Descartes and Kant also thought that geometry derives from 'innate' knowledge—Kant's argument from geometry was an ambitious attempt to demonstrate that our cognitive capacities are reflected in Euclidean geometry.[7] Similar claims were made about numbers and arithmetic. Leibniz, for example, argued that mathematical knowledge must be innate, because it pertains to necessary truths rather than contingent facts. Nevertheless, he believed that this

---

[7] *Innateness* is a notion associated with a complex, shifting range of meanings. Today, under the influence of cognitive ethology, it has a distinctly biological meaning (as in genetically determined, or developmentally invariant) that it did not originally possess. As it is beyond the scope of this paper to present a detailed discussion of nativism in philosophy and psychology, suffice it to say that 'innate' for these authors was something akin to the notion of *a priori*.

innate knowledge needed to elicited through education: "The truths about numbers are in us; but still we learn them" (Leibniz, 1765, p. 85). Elsewhere he likens innate knowledge to veins in marble that outline a shape to be un-covered by a sculptor: our innate knowledge is uncovered through learning (Leibniz, 1765, p. 52). By contrast, Locke (1690) argued that numerical cognition can be traced back to perceptual knowledge. The number one, for him, is an idea, i.e., a mental representation due to perceptual input: "Amongst all the ideas we have, as there is none suggested to the mind by more ways, so there is none more simple, than that of unity, or one" (Locke, 1690, Book II, Ch. XVI). This "simplest and most universal idea" (Locke, 1690, Book II, Ch. XVI) can then be taken as a starting point to make other numbers; for example, by repeating the number one, we end up with larger natural numbers.

Since the late nineteenth century, philosophers of mathematics have turned away from examining the relationship between cognition and mathe-matics, focusing instead on formal properties and foundational ideas, such as how the natural numbers can be derived from set theory. Recently, however, philosophers of mathematics have taken a renewed interest in epistemic is-sues, primarily driven by the increased focus on mathematical practice, i.e., on mathematics as a human activity.

The emphasis on formal aspects of mathematics, such as proofs, is a recent phenomenon of Western culture that seems absent in other cultures with rich mathematical traditions, such as China, India, and the medieval Arabic world. Even Western mathematics up to the eighteenth century was result-driven, with proofs subservient to methods for solving specific mathematical problems. Today, intuitions have not disappeared in math-ematical practice, as Thurston (1994) observes: mathematicians are born and encultured in a rich fabric of pre-existing mathematical procedures and concepts. Some of these ideas are akin to living oral traditions in that they have never been published but yet are tacitly accepted by the math-ematical community. Mathematicians have accorded a privileged role to intuition as a source of creativity. In their influential account of how math-ematicians work, Davis and Hersh (1981, p. 399) go as far as to say: "[T]he study of mental objects with reproducible properties is called mathematics. Intuition is the faculty by which we can consider or examine these (internal, mental) objects."

Where does mathematical intuition come from? As we shall see in Sec-tion 2.3, some developmental psychologists argue for an innate basis of mathematical knowledge. A growing body of experimental literature indi-cates that infants can predict the outcomes of simple numerical operations. The study of numerical cognition in animals (see Section 2.2) predates this literature, again providing evidence of animals' successes in estimating cardinalities, comparing numbers of different magnitudes, and predicting the outcomes of arithmetic operations. Complementary to this is neuroscientific evidence (see Section 3), which shows that some areas of the human brain are consistently involved in arithmetical tasks, strengthening the case for evolved, numerical competence. Finally, in Section 4, we shall argue that humans also draw on their external environment to make mathematical problems more tractable. Thus, mathematical cognitive processes can be situated both internally (inside the head) and externally (in the world).

## 2.2   Animals' arithmetic

Examining the numeric competence of non-linguistic creatures presents a methodological challenge: In the absence of language the evidence for arithmetic abilities or their underlying representations has to be inferred from overt behavior. As failed attempts at meeting this challenge have lead to famous misattributions, popular accounts of animal arithmetic (e.g., Dehaene, 1997; Shettleworth, 1998) often begin with the cautionary tale of Clever Hans. Clever Hans was a horse that lived in the early 1900s and appeared to have astonishing arithmetic abilities. Among various verbal and calendar-related feats, Hans could add, subtract, multiply, divide, and even work out fractions, indicating the results by tapping his hoof. The skeptical inquiry of Oskar Pfungst (1907) revealed that Hans indeed was clever, but his abilities consisted in detecting the subtle cues that his questioners or audience inadvertently would provide. Even after debunking Hans' alleged abilities, Pfungst was unable to refrain from providing signals that the horse could use. Thus, the story of Clever Hans teaches an important lesson to comparative psychology: To prevent observer-expectancy effects, the number senses of animals and pre-verbal infants ought to be probed either without experimenter intervention or in double-blind designs in which neither the examined creature nor the experimenter is aware of the correct answer. More generally, we have to be cautious not to over-interpret the abilities of animals by anthropomorphizing them. Whenever animals—including humans—show surprising arithmetic abilities we need to distinguish between ingenious trickery, natural competence, and the results of extensive training.

After several decades of deep skepticism there has been a resurgence of research efforts to probe the arithmetic abilities of animals by behavioral means. In 1993 a prominent researcher concluded enthusiastically that "the common laboratory animals order, add, subtract, multiply, and divide representatives of numerosity [...]. Their ability to do so is not surprising if number is taken as a mental primitive [...] rather than something abstracted by the brain from sense data only with difficulty and long experience" (Gallistel, 1993, p. 222). We shall organize our discussion of animals' abilities according to different levels of numerical competence.

**Numerosity discrimination.** Using an operant conditioning paradigm that required rats to press $n$ times on a lever $A$ before obtaining a treat by pressing another lever $B$ once, Mechner (1958) demonstrated that rats can associate rewards with a specific number of repeated actions. Interestingly, the rodents never performed fully error-free and the variance of their actual runs increased as $n$ increased (from a minimum of 4 to a maximum of 16). As premature switches were punished (by not obtaining a reward at all) the rats' number of responses were skewed toward over-estimates, rather than under-estimates.

Because the number $n$ of lever presses was confounded with the time $t$ it took to perform these actions a rival explanation of the rats' alleged ability to discriminate between different numbers was that they could have used duration as a cue to estimate number. However, Mechner and Guevrekian (1962) ruled out this alternative account by depriving rats of water for different periods of times. Whereas thirsty animals pressed the lever much faster, their degree of deprivation had little effect on the number of responses. Meck and Church (1983) later showed that rats spontaneously attend to both the number and duration of a series of discrete events.

**Counting.** Capaldi and Miller (1988) provided evidence that rats count the number of rewarded trials. By randomly exposing them to sequences of trials RRRN and NRRRN (where R stands for a rewarded and N for a non-rewarded trial) rats learned that they could expect to be rewarded on three trials. A much slower speed on the last (N) trial of both sequences shows that rats no longer counted on being rewarded after having accumulated three rewards on earlier trials. Importantly, rats readily transfer their counts to other types of food and even integrate their counts across different types of food, suggesting that their internal counts are abstract rather than tied to concrete events. As counting the types and amount of food items obtained from a particular patch is a fundamental part of animal foraging such abilities may not come as a complete surprise (Shettleworth, 1998). But it is easily overlooked that systematically exploiting many food resources requires some basic—and possibly implicit—method for keeping track of both time and number.

Merely identifying and counting numerical quantities does not necessarily require an abstract concept of counting or number. As a possible mechanism Dehaene (1997) suggests the metaphor of an analog accumulator that gathers the amount of some continuous variable (like water) rather than discrete quantities (like pebbles). By incrementing and decrementing such an accumulator animals would possess an approximate representation of numerical quantities that would allow for basic comparisons, as well as elementary additions and subtractions. A fuzzy or noisy boundary of the elementary counting unit implies that larger quantities get increasingly imprecise. The resulting consequences that two numbers are more easily distinguished when they are further apart and that two numbers of a fixed distance are harder to discriminate as they get larger are known as the *distance* and *magnitude* effects in both animal and human experiments on number comparisons.

To model the identification of a number of objects from visual or auditory perception Dehaene and Changeux (1993) developed a neuronal network model that relies on number-detecting neurons. Despite its simplicity, this model can account for the detection and discrimination of numerosities in animals and pre-verbal infants without assuming any ability to count explicitly.

**Abstract and symbolic representations.** We just saw that comparing and counting numerosities does not yet imply the mastery of an abstract concept of number (see Shettleworth, 1998, p. 369). However, there is also evidence that rats can abstract from sensory modalities and add discrete events. Church and Meck (1984) trained rats to discriminate between two vs. four tones and two vs. four light flashes by teaching them the regularities $ll \rightarrow L$, $llll \rightarrow R$, $tt \rightarrow L$, and $tttt \rightarrow R$, where lowercase $l$ and $t$ stand for flashes of light or tones, and uppercase $L$ vs. $R$ correspond to pressing either a left or right lever, respectively. What happens if rats that have learned those contingencies are confronted with a stimulus configuration of $lltt$? Despite the double dose of stimuli that individually required a $L$ response to "twoness" throughout the training phase, the rodents now pressed the right lever $R$, indicating that they instinctively added $2 + 2 = 4$. To emphasize the significance of this finding, Dehaene (1997) compares it to a fictitious experiment that trained rats to discriminate both between red and green objects and between square and circular shapes. Surely it would seem surprising if presenting a red square evoked the response for green and circular objects. Both our own intuition and the rats of (Church and Meck, 1984) suggest that discrete events are to be integrated in an additive fashion, rather than by a merging process that combines perceptual properties like color and shape.

To address the question of whether animals can associate and manipulate numeric symbols we have to turn to parrots and chimpanzees. Pepperberg (1994) trained the African grey parrot Alex to vocally label collections of 2 to 6 simultaneously presented homogeneous objects and showed that he could then identify quantities of subsets in heterogeneous collections. For instance, Alex would be shown a collection of blue and red keys and cups and then identified the number of blue cups with an overall accuracy of over 80%. Boysen and Berntson (1989) demonstrated that their chimpanzee Sheba could assign Indo-Arabic[8] numerals to collections of objects and *vice versa*. In addition, Sheba could add up small numbers of oranges or numerals (up to a total of 4) when they were hidden in different locations.[9]

This section has shown that an assessment of the numeric competence of animals needs to strike a balance between two extremes: On the one hand, abundant credulity or naïve enthusiasm about animals' numeric feats would overlook that the numeric competence of non-human animals is fundamentally different from that demonstrated by humans. Almost always any abstract mastery of numeric symbols is the result of extensive training, is specific to a few numbers (i.e., difficult to generalize), and remains notoriously error-prone, particularly with numbers beyond 6 or 8. On the other hand, a refusal to acknowledge that lower animals can distinguish, count, and represent numerical quantities in some way would border on species chauvinism. There is no reason in principle why the perception of numerosity ought to be more complex than that of color, shape, or spatial orientation. And as detecting the amounts of prey, predators, or potential mates conveys a clear advantage for survival we should not be surprised that evolution has endowed non-human animals with at least some rudimentary number sense.

## 2.3   Infants' arithmetical skills

Prior to the late 1970s, developmental psychologists interested in the domain of numerical competence almost exclusively examined the development of explicit counting and exact positive integer representation during the preschool years. The early focus on explicit skills was partly due to methodological limitations (how to study cognition in infants) and partly due to firm conceptions about the cognitive foundations of arithmetical skills. Piaget's 1952 seminal work places the development of arithmetical skills late in cognitive development, between 5 and 12 years of age. Piaget thought that children must first master abstract reasoning skills, such as transitive reasoning or one-to-one correspondence. A problem with this framework is that it assumes that abstract reasoning skills are psychologically primitive for understanding number. The attraction of this view is that features such as one-to-one correspondence do play an important role in foundational work on mathematics, such as attempts to reduce arithmetic to set theory. However, it may be a category mistake to take that which is primitive in the development of formal arithmetic as psychologically primitive. As we shall see, infants and young children have some understanding of number which develops independently of other abstract reasoning skills. In 1978,

---

[8]See Section 4.2 for this terminology.

[9]See the review chapters by Boysen (1993) and Rumbaugh and Washburn (1993) for more details on the numeric competence of monkeys and chimpanzees.

Gelman and Gallistel published an influential monograph on arithmetical skills in preschoolers. From then on, the road was open for developmental psychologists to examine numerical capacity in infants. The early focus on explicit number representations has given way to the study of a broad domain of mathematical skills that are related to quantities, including the exact and immediate counting of small numerosities (subitizing), relative numerical judgments, and approximate systems of counting of larger sets (estimation). In this short review, we shall focus on arithmetic skills.

In a pioneering series of experiments, Karen Wynn (1992) tested the ability of five-month-olds to perform addition and subtraction on small quantities. To probe her subjects' capacities, she relied on the *looking time procedure* and the *violation of expectation paradigm*. The looking time procedure aims to probe cognitive abilities with a minimum of task demands. Clearly, infants cannot speak, so any test to probe infant knowledge is necessarily non-verbal (as was also the case with animals in Section 2.2). Moreover, human infants are motorically helpless (e.g., they are unable to release objects intentionally until 9 months of age), so one cannot rely on tasks that involve manual dexterity—this is importantly different from animal studies, which frequently require the subject to perform some particular action (e.g., pecking, pressing a lever). The violation of expectation paradigm exploits the propensity of humans and other animals to look longer at unexpected than at expected events. Our knowledge of the world enables us to make predictions of how objects will behave. For example, we expect coffee to remain in a stationary cup, but to flow out of a cup in which holes were drilled. When something happens that violates these predictions, we are surprised. Prior to the test trials, infants are exposed to habituation or familiarization trials to acquaint them with various aspects of the test events. With appropriate controls, evidence that infants look reliably longer at the unexpected than at the expected event is taken to indicate that they (1) possess the expectation under investigation, (2) detect the violation in the unexpected event, and (3) are surprised by this violation. The term 'surprise' is used here simply as a short-hand descriptor to denote a state of heightened attention or interest caused by an expectation violation.

In one of Wynn's experiments, a group of infants watched a $1 + 1$ operation: a Mickey Mouse doll was placed on a display stage, a screen rotated upwards to temporarily hide it from view, a hand entered the display stage with another identical looking doll, and placed it behind the screen. Then the screen was lowered to reveal either the possible outcome $1 + 1 = 2$, or the impossible outcome $1 + 1 = 1$. The infants looked significantly longer at the impossible outcome than at the possible one, suggesting to Wynn that they expected the outcome of $1 + 1$ to be 2. Similarly, they gazed longer at $2 - 1 = 2$ than at $2 - 1 = 1$. A methodological problem with the looking time procedure is that one cannot be sure what causes the longer looking times. Wynn (1992) favored an account in terms of fairly advanced conceptual cognition, namely that infants possess the ability to reason about number and perform arithmetical operations.

Wynn's interpretation is not the only possible way to account for these data. It is equally possible that the results are caused by lower-level cognitive capacities, such as a preference for visual stimuli that are familiar. For instance, Cohen and Marks (2002) proposed that the infants' longer looking time could be explained by a familiarity preference: they looked longer at one doll in the case of $1 + 1 = 1$ or 2, because during habituation, when the infants were familiarized with the setup, they saw one doll. Similarly, for the case of $2 - 1 = 2$ or 1, they looked longer at two dolls since that is what they saw during the familiarization trials. Importantly, developmental psychologists who probe innate knowledge do not exclude this possibility—indeed, they attempt to minimize familiarization effects by designing controls. Several subsequent experiments in independent labs (e.g., Kobayashi et al., 2004) have attempted to control for these alternative explanations, such as placing the puppets on rotating platforms, or familiarizing the infants equally with one, two, and three puppets. The results of these studies have supported Wynn's original experiment, and by controlling for lower-level cognitive accounts, have made the case for early developed numerical skills stronger.

Still, it is important to note that translating the experimental setup into mathematical notation can be misleading; it is not evident that Wynn's experiments show that infants are capable of operations that are equivalent to the mathematical notions of addition and subtraction. For instance, Uller et al. (1999) have argued that the experiments show that infants represent the objects that are being added and subtracted not as integers, but as object-files. According to this view, an object-file of two entities is represented as follows: there is an entity, and there is another entity numerical distinct from it, and each entity is an object, and there is no other object, i.e.,

$$(\exists x)(\exists y)\{(\text{object}[x] \ \& \ \text{object}[y])$$
$$\& \ x \neq y \ \& \ \forall z(\text{object}[z] \rightarrow [z = x] \vee [z = y])\}.$$

This conception of numerosity is different from formal mathematical notions, but it is compatible with the empirical data on infants.

Later studies have probed whether infants can reason with larger quantities, and predict the outcomes of arithmetical operations that yield an absence of objects. McCrink and Wynn (2004), for example used a similar setup to Wynn's original experiment to investigate whether or not infants of 10 months of age can predict that $5 + 5 = 10$ and not 5, and that $10 - 5 = 5$

and not 10. This time, they used a computer-animated setup so that they could adequately control for total surface area (e.g., in the case of $5 + 5$, either ten objects of the same size as the original objects were shown, or five very large objects). The fact that infants could reliably predict these results strengthens the view that ten-month-olds can predict the outcomes of additions and subtractions of items in a visual display over larger numbers than previously studied, even when other factors such as the size of objects are controlled for.

The case of operations that yield an absence of objects reveals some limitations of this intuitive arithmetic. Wynn and Chiang (1998) used a looking-time experiment to show eight-month-olds subtraction events which had outcomes of no items (e.g., $1 - 1 = 0$). In contrast to the earlier experiments, the infants' looking time did not differ between the expected, correct result of $1 - 1 = 0$ and the incorrect, surprising outcome of $1 - 1 = 1$. This might suggest that infants have difficulties representing zero as a cardinal number. Unrelated experiments with chimpanzees yield similar results: although these animals can learn to distinguish between numbers up to 9 with good accuracy, they keep on confusing zero with very small natural numbers (1 and 2; Biro and Matsuzawa, 2001). These results are in tune with observations of mathematical practice in history and across cultures: most indigenous mathematical systems do not have a zero, neither as a placeholder symbol nor as a number.

Other limitations on infant arithmetic are related to working memory. In one study, Feigenson et al. (2002) presented infants of 10 and 12 months with the choice between two opaque buckets whose contents they were unable to see. In each of them, a number of crackers were dropped, one by one, so that at the end of each trial the buckets contained different numbers of crackers. After presentation, the subjects were allowed to crawl to the bucket of their choice to retrieve the crackers. Although the infants could successfully choose between 3 vs. 2 (i.e., they realized that $1+1 < 1+1+1$), they performed at chance level in the two versus four and three versus six conditions, despite the highly discriminable ratio between the quantities. Control tests ensured that this experiment cannot be explained by movement complexity. Possibly, working memory demands are a limiting factor: it is perhaps difficult to keep two collections with more than three objects each hidden from view in working memory. Indeed, as will be discussed in Section 3, humans rely on a host of complementary resources when doing arithmetic, including spatial representations, verbal labeling, finger representations, and imagined motion.

The results from studies with nonhuman animals and infants suggest that humans are furnished with an unlearned, early-developing capacity to perform simple arithmetical operations. However, one may wonder whether these studies can tell us anything relevant about arithmetic as a culturally elaborated skill. Some authors (e.g., Rips et al., 2008) remain deeply skeptical about the role of such evolved competencies in formal mathematical reasoning. It is indeed possible that similarities between infants' performance on some tasks and adult mathematical knowledge are superficial, and that there is no overlap between intuitive and formal mathematical concepts. Although we can *describe* Wynn's (1992) experiment in mathematical terms as $1 + 1 = 1$ or 2, does this mean that babies actually know that $1 + 1 \neq 1$? It may be problematic to use symbolic notation to describe such events, given that symbolic notations themselves influence mathematical cognition, a point that will be developed in more detail in Section 4. Non-numerical factors like language and verbal memory play an important role in elementary mathematics education, as is demonstrated by the memorizing of exact addition facts like $5 + 7 = 12$ or multiplication facts like $7 \times 9 = 63$. Young children also rely extensively on fingers and hands when they add and subtract.

Notwithstanding their sometimes problematic interpretation of the results, cognitive scientists offer the best hope of explaining our epistemic access to mathematical objects. Several lines of evidence indicate a causal connection between the early development of numerical skills and formal numerical competence. Halberda et al. (2008) found that children who are better at estimating numerical magnitudes (e.g., guessing the number of dots on a screen) also achieve better results in mathematics at school. Thus, approximate numerical skills are important for the development of more formalized ways of manipulating numbers such as symbolic arithmetic. Indeed, a study by Barth et al. (2006) found that both adults and preschoolers can perform additions and subtractions approximately, without the use of symbolic aids. In one of their experiments, inspired by Wynn's procedure, the preschoolers were shown a large number of blue dots. Then the blue dots were covered by a screen, and some more blue dots were shown to go hiding behind the screen. The children were then asked whether there were less or more blue dots compared to a set of visible red dots. The subjects answered well above chance level, indicating that approximate addition over large numbers develops prior to extensive training on arithmetical principles. Moreover, developmental dyscalculia, a disruption in the normal development of mathematical skills in some children, is correlated with an inability to grasp the concept of numerosity (Butterworth, 2005). Molko et al. (2003) studied the brain structure of subjects with developmental dyscalculia and found that their intraparietal sulci (which, as we shall see in Section 3, is implicated in numerical cognition) showed abnormal structural properties.

In sum, the evidence reviewed here strongly indicates that human infants possess elementary numerical skills. Combined with the evidence of numerical skills in animals, one could make a case for an evolutionary basis of numerical cognition. The importance of numerical knowledge in everyday decision making, such as foraging or forming groups, makes this evolutionary origin quite plausible.

## 2.4   Arithmetic in few-number cultures

As early as 1690, the philosopher John Locke (1690) mentioned possible effects of a limited numerical vocabulary on numerical cognition: "Some Americans I have spoken with (who were otherwise of quick and rational parts enough) could not, as we do, by any means count to 1000; nor had any distinct idea of that number." These Americans were the Tououpinambos, a culture from the Amazon forest in Brazil, who "had no names for numbers above 5." Although Locke thought that the absence of count words limited their ability to reason about large cardinalities, he mentioned that they could reckon well to twenty, by "showing their fingers, and the fingers of others who were present." He thus argues that count words are "conducive to our well-reckoning," but not strictly necessary for it (Locke, 1690, all citations from Book II, ch. XVI). By contrast, Alfred Russell Wallace, co-discoverer with Darwin of the principle of natural selection, believed that count words were essential for numerical cognition, in particular arithmetic: "if, now, we descend to those savage tribes who only count to three or five, and who find it impossible to comprehend the addition of two and three without having the objects actually before them, we feel that the chasm between them and the good mathematician is so vast, that a thousand to one will probably not fully express it" (Wallace, 1871, p. 339). The question of the role of language in arithmetic became the focus of recent experimental psychological studies in cultures with few number words, in particular the Pirahã and the Mundurukú, two cultures from the Amazon forest with an extremely limited number vocabulary.[10]

The Pirahã (Gordon, 2004) have only three words that consistently denote cardinality, 'hói', 'hoí' and 'baágiso'. These terms are not used as count words, but rather as approximations of perceived magnitude (not just cardinality). For example, the word 'hói' is used to denote single objects, but also as a synonym for small (as in a small child). 'Hoí' is used to denote a few items or a medium quantity, and 'baágiso' is used for large items or large quantities of items. One can ask 'I want only *hói* fish' to denote one fish, but one cannot use this phrase to ask for one very large fish, except as a joke (Everett, 2005). The imprecision of the Pirahã count words was recently demonstrated in a series of experiments (Frank et al., 2008a) in which

---

[10]The elaboration of mathematical ideas differs considerably between cultures. For an extensive discussion of ethnomathematics, see (François and Van Kerkhove, 2010), this volume.

Pirahã subjects were simply asked to say how many objects they saw. If the objects were presented in an *increasing* order, from 1 to 10 items, the subjects consistently said 'hói' for one item and 'hoí' for two items. For more than two items, some subjects said 'hoí' or 'baágiso'. By contrast, if the objects were presented in a *decreasing* order, the subjects said both 'hoí' or 'baágiso' for objects up to 7, and some claimed to see 'hói' starting at 6 items. Some years earlier, Gordon (2004) confronted Pirahã with a battery of experiments to test numeracy, such as probing the capacity to place objects into a one-to-one correspondence and memory for specific numbers of items. Their capacity to reason about exact magnitudes was severely compromised, especially for numerosities that are above the subitizing range ($n > 4$). An example of a matching task required that the subject draw as many lines as were presented to him or her by the experimenter. The accuracy dropped linearly as the target number of lines increased. After 7 items, none of the participants drew the correct number of lines. In one of the experiments that probed memory for numerosity, the participants witnessed a quantity of nuts being placed in a can, and then being withdrawn one by one. After each withdrawal, the subjects responded as to whether the can still contained nuts or was empty. This task proved extremely difficult, as the responses dropped to chance level between 4 and 5 items.

Authors who have studied Pirahã do not agree on the implications of these experiments on the role of external symbolic systems for numerical cognition. Gordon (2004, p. 498) claimed that his study "represents a rare and perhaps unique case for strong linguistic determinism." In contrast, Frank et al. (2008a) showed that Pirahã performed relatively well on tasks that did not involve memory, such as matching tasks (e.g., matching a number of objects to those that an experimenter showed them), by employing strategies that involve making one-to-one correspondences. These results suggest that count words do not create number concepts, but rather concur with Locke's view that they are "conducive to our well-reckoning." Similar results have been obtained with people from other non-numerate cultures, such as Australian aboriginal children who speak languages with few count words. In these studies, the children could even solve division problems if they could use one-to-one matching, such as dividing six or nine play-doh discs between three puppets. They simply dealt discs to each puppet one by one, until all discs were divided (Butterworth et al., 2008).

To better tease apart the role of language and other cultural factors, Frank et al. (2008b) conducted experiments with American college students that were very similar to those presented to the Pirahã. These tasks involved both one-to-one matching tasks and memory tasks. In the meantime, the participants performed a task that made it impossible for them to rely on subvocal counting. Apparently, the ability to perform one-to-one

matchings was relatively unimpaired by the inability to count, but memory
for numbers, as in the nuts-in-a-can task, was severely compromised (subjects answered correctly only 47 % of the time). In addition to language,
other external tools may explain the limited numerical skills in the Pirahã.
For example, Everett (2005) noted that they do not have individual names
for fingers (e.g., ring finger, index), but collectively refer to their fingers
as 'hand sticks'. In many cultures, finger counting plays a crucial role in
the development of number concepts; the fact that words for 'one', 'four',
and 'five' in many Indo-European languages are related to words for fingers
(or digits) is indicative of this. If fingers are not differentiated, this might
impair the formation of exact magnitude concepts, or vice versa.

The Mundurukú is another Amazonian culture with few number words
(up to five), which are likewise used in an approximate fashion: pũg ('one'),
xep-xep ('two'), e-ba-pũg (literally: 'your arms and one'), e-ba-dip-dip (literally, 'your arms and two', pũg-pog-bi (literally 'a handful' or 'a hand').
The approximate nature of these quantities is illustrated by the fact that the
use of these terms is inconsistent when Mundurukú subjects have to denote
three or more items. For example, when five dots are presented, the subjects
respond pũg-pog-bi in only 28 % of the trials, and e-ba-dip-dip in 15 % of
the trials. Above five, the Mundurukú do have words to denote numerosities, but these terms have very little consistency. Subjects refer to 10 items
using the expressions ade ma ('really many'), adesũ ('not so many') and
xep xep pog-bi ('two hands') (Pica et al., 2004). Pica et al. (2004) studied
the effects of this limited vocabulary on arithmetic, revealing an interesting
discrepancy between exact and approximate arithmetic. Mundurukú exact
arithmetic proved to be highly compromised. For example, in one study, the
subjects predicted how many objects would be left in a can after several had
been removed. Although the results were small enough to be named with
their number vocabulary (e.g., $6 - 4 = 2$), they were unable to predict them.
In contrast, Mundurukú subjects did very well on approximate arithmetical
tasks, where they were asked whether the addition of two large collections
of dots (e.g., 16 and 16) in a can was smaller or bigger compared to given
number of dots (e.g., 40). In this task, which involved quantities far above
their count range, they did as well as French numerate adults.

Another aspect of numerical cognition that is clearly affected by external
representations are questions regarding the distance between internal representations of different numbers, i.e., the shape of the 'mental number line'.
Several studies have shown that young children (Siegler and Booth, 2004)
and non-human animals (Nieder and Miller, 2003) represent numbers on a
logarithmic, rather than a linear mental number line. In brief, a logarithmic
mental number line is one where estimations of numerosities conform to the
natural logarithms (ln) of these numbers. This typically leads to an overes-

timation of the distance between small numbers, such as 1 and 2, where the psychological distance is typically judged to be much larger than between larger numbers like 11 and 12. Young children make characteristic errors when plotting numbers on a scale. Siegler and Booth, for example, gave five- to seven-year-olds a number line with 0 at the left side and 100 at the right. Younger children typically place small numbers too far to the right. For example, they tend to place the number 10 in the middle of the scale, which is roughly in accordance with a logarithmic representation. As children become older, their number lines look more linear. From these results, Siegler and Booth (2004) conclude that our intuitive number representation is logarithmic, and that it becomes more linear when children learn to manipulate exact quantities. Dehaene et al. (2008) adapted this experiment in an elegant fashion to a study with Mundurukú participants, presenting them with a line with one dot to the left, and ten dots to the right. Then, the Mundurukú were given a specific numerical stimulus, either as a number of tones, or as a number word in Portuguese or Mundurukú. In all cases, the best fit of the responses was logarithmic, not linear. As the authors of this study acknowledge, language cannot be the sole factor responsible for linear numerical representations in Western people, as the Mundurukú responded logarithmically, regardless of the language or format in which the numbers were presented. Perhaps other external representations, such as rulers or the practice of measurement, can explain this change.

Taken together, these results suggest that approximate arithmetic relies less on external tools such as language than exact arithmetic. The animal, neuroimaging, and infant studies demonstrate that our intuitive numerical competence allows for approximate arithmetical tasks. External representational systems, such as fingers, count words, and numerical notation systems, serve to enhance exact numerical cognition that ventures beyond the range of our intuitive capabilities.

# 3   Arithmetic and the brain

## 3.1   Lesion studies

Neuropsychological studies offer the opportunity to study the neural correlates that underlie our capacity to perform arithmetical operations. What neural structures enable us to comprehend and compute with numbers? Are there differences between approximate arithmetic and exact arithmetic? How are external media, such as symbolic notation systems, reflected in the brain? The oldest method to study the neural basis of arithmetic relies on an examination of the effects of brain lesions on various cognitive tasks. This methodology was developed in the later decades of the 19th century when physicians like Broca and Wernicke noticed that specific lesions, i.e., patterns of brain damage led to an inability to speak. Such lesions can be the result of an external injury or a stroke (a blood-clot which momentarily deprives part of the brain of oxygen and nutrients), leading to specific patterns of cognitive impairment. Indirectly, one can infer from the correlation between damage to a given brain area $X$ and loss of a certain cognitive function $a$, that $X$ and $a$ are functionally correlated.

Early studies by Gerstmann (1940) showed that patients with damage to the left inferior parietal lobule (a subsection of the parietal lobe) often had marked impairments in mathematical cognition. Lesions in this area often leave a patient unable to perform very simple arithmetical operations such as $3 - 1$ or $8 \times 9$. However, these lesions usually also affect other domains of cognition. This is exemplified in Gerstmann's syndrome (Gerstmann, 1940), a neurological condition that is associated with damage to the parietal lobe, and that is characterized by an inability to perform arithmetic, count, and do other numerical tasks, as well as by difficulties in writing (agraphia), the inability to recognize one's own fingers (finger agnosia), and left-right confusion (Chochon et al., 1999). The fact that loss of mathematical function is often accompanied by finger agnosia, agraphia, and left-right confusion might be due to the fact that lesions usually damage several adjacent functionally specialized brain areas. In that case, the cognitive functions are not really related, but their damage coincides because the areas correlated with them are in close anatomical proximity. Alternatively, one could take these findings as support for the view that finger counting, writing, and spatial skills play an important role in numerical processing. Evidence for this latter interpretation comes from several modern studies that impair finger cognition in an experimentally controlled and reversible way. In these repetitive transcranial magnetic stimulation (rTMS) experiments, brain activity was briefly disrupted in areas important for finger cognition, including the left intraparietal lobule (Sandrini et al., 2004) and the right angular gyrus (Rusconi et al., 2005). In both studies, disrupting finger cognition led to a marked increase in reaction time when subjects solved arithmetical operations. This suggests that finger recognition remains an important part of adult numerical cognition, even when we no longer count on our fingers.

Lesion studies have also examined whether or not language is essential for mathematical tasks. This has given rise to a nuanced picture. First, it seems that language, especially verbal memory, is more important for multiplication and addition than for division and subtraction. Lemer et al. (2003) assessed the differential contributions of brain areas specialized in language and number for diverse arithmetical operations. In their study, they examined a patient with a verbal deficit (caused by lesions in the left temporal lobe), and another patient with a numerical deficit (with a focal lesion in the left parietal lobe), but intact verbal skills. The authors hypothesized that language would play an especially important role in arithmetical tasks in which verbal memory is important, such as multiplication (due to memorization of multiplication facts like $5 \times 7 = 35$) and addition. By contrast, since we do not store subtraction facts in verbal memory, this capacity should be less affected by the loss of language. As predicted, Lemer et al. (2003) found that the patient with the language impairment performed worse on multiplications than on subtraction, whereas the patient with numerical impairments exhibited the reverse pattern. Thus it seems that verbal memory can play an important role in the performance of arithmetical tasks in the adult human brain.

Another study (Varley et al., 2005) probed whether language may be important for numerical cognition on a more deep, structural level. Parallels between recursive structures in mathematics and grammar have suggested to some authors that the generative power of grammar may provide a general cognitive template and a specific constituting mechanism for 'syntactic' mathematical operations involving recursiveness and structure dependency, such as the computation of arithmetical operations involving brackets, e.g., $50 - ((4 + 7) \times 4)$. Indeed, Hauser et al. (2002) argue that a domain-general and uniquely human capacity for recursion underlies our capacity for mathematics.[11] More specifically, they state that "Humans may be unique [...] in the ability to show open-ended, precise quantificational skills with large numbers, including the integer count list. In parallel with the faculty of language, our capacity for number relies on a recursive computation" (p. 1576). To test this relationship between language and numerical cognition, Varley et al. (2005) examined three severely agrammatic patients (i.e., people with an inability to comprehend and make grammatical sentences) on several numerical tasks, including multiplication tables and bracket operations. Despite their lack of grammar, all three men performed excellently on these tasks, solving problems like $80 - ((6 + 14) \times 2)$ accurately. One of the problems specifically examined the preservation of recursive capacities in the absence of grammar: it required the patients to come up with numbers smaller than 2, but larger than 1. Although none of the patients was capable of generating recursive linguistic expressions, they could solve these problems, coming up with numbers like $1, 1.9, 1.99, 1.999, \ldots$. From this, the authors conclude that, at least in the mature adult brain, the non-linguistic neural circuits that deal with recursive structure in mathematics are functionally independent of language. However, it does not follow that language is unimportant for the *development* of mathematical competence. For instance, Donlan et al. (2007) showed that eight-year-old children with specific language impairments (i.e., children with language impairments but

---

[11] A recursion consists of a few simple cases or objects, and rules to break down complex cases into simpler ones, e.g., my (full) brother is my blood relation (base case), anyone who is a blood relation of this brother is also a blood relation (recursion).

overall normal intelligence in other domains) are developmentally delayed for several numerical tasks compared to children without language impairments: as many as 40 % failed to count to twenty, and they showed problems in understanding the place-value system. By contrast, these children had no problems understanding high-level principles of arithmetic, such as commutativity.

## 3.2   Neuroimaging studies and EEG experiments

A more direct way to study which regions of the brain are involved in performing specific tasks is provided by functional neuroimaging techniques. All neuroimaging techniques exploit the fact that although the whole brain is always active, not every part is equally active. Regions that are more active require more energy (glucose) and oxygen. Neuroimaging techniques measure differential brain-activation after presentation of a relevant stimulus, and compare these activations to a carefully chosen control stimulus. If this effect is constant across subjects and if it is reproducible, the cerebral parts that are more active after presentation of the test stimulus compared to a control stimulus are taken as neural correlates for the task that the stimulus probes. The most frequently used neuroimaging technique for probing numerical competence is functional Magnetic Resonance Imaging (fMRI), which relies on strong magnetic fields to measure differences in oxygen-levels in cerebral blood flow. A problem with most neuroimaging techniques is that while they have a relatively good spatial resolution (i.e., they give a relatively accurate map of differential brain activity), they have a relatively poor temporal resolution (i.e., they are slow and may not pick up transient patterns of brain activity). By contrast, electroencephalography (EEG) scans, which measure electric activity in the brain through electrodes on the scalp, can pick up subtle and quick changes in brain activity, but have poor spatial resolution, as only areas at the surface of the brain can be accurately measured. EEG scans can be used to measure the specific response of the brain for a given task; these task-related patterns of electric brain activity are termed Event Related Potentials (ERPs).

Dehaene et al. (1999) investigated the relative importance of language and non-linguistic approximate representations of number in two brain-imaging studies: one with high temporal resolution (ERPs) and one with high spatial resolution (fMRI). First, they conducted a behavioral experiment with Russian-English bilinguals. The subjects were taught a series of exact or approximate sums of two-digit numbers in one of their languages, either Russian or English. The test condition consisted of a set of new additions. This was either an exact condition, in which they had to choose the correct sum from two numerically close numbers, or an approximate condition, in which they had to estimate the result and select

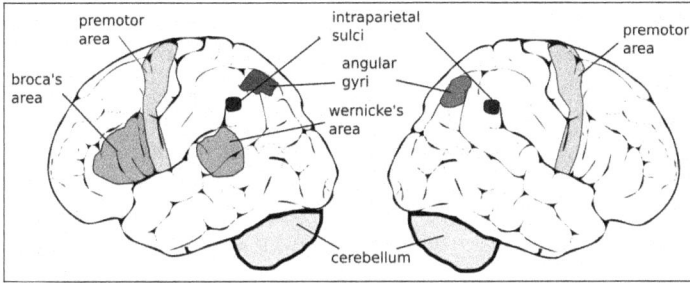

FIGURE 2. Regions of interest mentioned in the text, left hemisphere shown on the left.

the closest number. After training, response time and accuracy improved in both types of tasks. However, when tested in the exact condition, subjects performed much faster in the teaching language than in the untaught language. In contrast, for the approximate condition, there was no cost in response time when switching between languages. To the authors, this suggested that exact arithmetical facts are stored in a language-specific format; each new addition is separately stored from neighboring magnitudes, e.g., $9 + 1$ would be stored differently from $9 + 2$. Because there was no cost in the approximate condition when switching between languages, the authors assumed that number is also stored in a language-independent format (Dehaene et al., 1999, p. 971). The authors examined whether this apparent behavioral dissociation is the result of distinct cerebral circuits. In fMRI, the bilateral parietal lobes showed greater activation for the approximate task than for exact calculations. In the approximate task, the most active areas were the bilateral horizontal banks of the intraparietal sulci (IPS) (see Figure 2). Additional activation was found in the left dorsolateral prefrontal cortex and in the left superior prefrontal gyrus, as well as in the left cerebellum, the left and right thalami, and the left and right precentral sulci. Most of these areas fall outside of the areas associated with language. Exact calculations elicited a distinctly different pattern of brain activation, which was strictly left-lateralized in the inferior frontal lobe. Smaller activations were also noted in the left and right angular gyri. Previous studies have shown that the left inferior frontal lobe plays a critical role in verbal association tasks. Together with the left angular gyrus, this region may constitute a network involved in the language-dependent coding of exact addition facts (Dehaene et al., 1999).

Several studies since then have confirmed that the intraparietal sulci of both hemispheres, but predominantly of the left, are active during arith-

metic and other numerical tasks. This is the case even in the absence of
explicit tasks, for example, when subjects are only required to look passively
at Indo-Arabic digits, or to listen to number words spoken out loud (Eger
et al., 2003). The IPS seems to be an important neural correlate for numerical cognition, regardless of the format in which it is presented. This finding
is confirmed by several studies that measure the firing rate (i.e., electric
activity) of single neurons in monkeys. Tudusciuc and Nieder (2007) found
that neurons in the intraparietal sulci of monkeys were sensitive to differences in numerosity, line length or both. The neurons were optimally tuned
to a specific quantity (e.g., two items) and gradually showed less activity as
the presented numerosity deviated from this preferred quantity. By focusing on which neural correlates are constant across numerical tasks, we have
left open the question of whether the use of symbolic notations and other
external tools affect numerical cognition at the neural level.

An intriguing fMRI study by Tang et al. (2006) provides indirect support for the role of symbolic representation in numerical cognition.[12] In
this study, both native English speakers and native Chinese speakers solved
arithmetical operations. Although the IPS were active in both groups, they
exhibited marked differences in other brain areas. Whereas the English
speakers had a stronger activation in perisylvian, language-related areas
(such as Broca and Wernicke's areas), the Chinese speakers showed an
enhanced response in premotor areas, involved in the planning of motor
actions. The authors offered a possible reason for this: whereas English
speakers learn arithmetical facts in verbal memory (e.g., when they learn
multiplication tables), Chinese speakers rely on the abacus in their schooling. These differences in schooling might still be reflected in arithmetical
practice, with English speakers mentally relying on language-based strategies, and Chinese speakers on motor-based strategies.

Taken together, neuropsychological studies indicate that numerical cognition relies on an interplay of cognitive skills that are specific to number
(primarily located in the IPS) and cognitive skills from other domains, including language, finger cognition, and motor skills. Such findings indicate
that numerical cognition is a complex skill, which involves a variety of capacities that are coordinated in very specific ways.

## 4    The role of notation in arithmetic

In Sections 2.4 and 3 we have described some connections between intuitive
arithmetic notions and language, and have seen how the use of a lexical
numeral system greatly affects people's basic arithmetic abilities. We now
turn to the role of notation and its relation to computations. We shall ar-

---

[12]For a more detailed discussion on the relationship between extended mind and mathematics, we refer to (Johansen, 2010), this volume.

| 4 | 9 | 2 |
|---|---|---|
| 3 | 5 | 7 |
| 8 | 1 | 6 |

FIGURE 3. Arranging the digits 1 to 9 into cells of a 3-by-3 magic square reveals that the game of number scrabble (see text for details) is isomorphic to tic-tac-toe.

gue for the thesis that mental arithmetic is an interplay between internal and external representations. The properties of external media and characteristics of the representational format profoundly affect the process of manipulating and computing the solutions to arithmetic problems.

## 4.1 Representational effects

It is hard to overestimate the relevance of representations for problem solving. As all deductive inferences (e.g., in mathematics) are essentially changes of representation, an extreme argument for the crucial role of representations is that solving a problem is nothing but a change in representation, or "solving a problem simply means representing it so as to make the solution transparent" (Simon, 1996, p. 132). Simon illustrates his claim by the game of number scrabble: Two players alternate in choosing a unique number from 1 to 9. The player who first manages to select a triple of numbers that sum to 15 wins the game.

Most people find this game rather abstract and have difficulties in choosing numbers strategically. However, when the game is represented as in Figure 3, it becomes apparent that number scrabble is structurally identical to the game of tic-tac-toe, in which players alternatively pick a cell of a 3-by-3 grid and win by first occupying three cells on a straight line. The spatial re-representation makes it easier to 'see' that some numbers (e.g., 5) are more valuable for winning than others, as they are part of more potential solutions and allow for the obstruction of more opponent moves.

The phenomenon that different representations of a problem can greatly change their level of difficulty is referred to as a *representational effect* and problems that are identical except for the their surface representation are called *isomorphs* (e.g., Kotovsky et al., 1985; Kotovsky and Simon, 1990). Our introductory distinction between numbers and numerical notation systems (see Table 1 on page 61) illustrated that many alternative representational systems can represent the same entities.

From a cognitive standpoint the distinction between computational and informational equivalence is important for characterizing alternative representations (Simon, 1978; Larkin and Simon, 1987). Two representations are *informationally equivalent* if they allow the same information to be represented, but they are *computationally equivalent* if in addition any information that can be inferred 'easily and quickly' from one representation can also be inferred 'easily and quickly' from the other. Larkin and Simon (1987) admit that these definitions are inherently vague, but they are still useful in describing the effects of different numeric representations on the ease or difficulty of arithmetic calculations. Two systems are informationally equivalent with regard to a set of tasks if they both allow the same tasks to be performed. They are additionally computationally equivalent if the relative difficulty of tasks is the same no matter which representation is used (Bibby and Payne, 1996; Payne, 2003).

The distinction between the two different types of equivalence can readily be applied to numeration systems. The representational systems illustrated in Table 1 are all informationally equivalent, as every natural number can be unambiguously expressed in each system. Nonetheless, the different systems differ in their computational properties. Whereas the tally system makes the summation of two numbers a simple matter of combining their respective number of elementary strokes, the Indo-Arabic decimal system requires the recognition of different symbol shapes and the retrieval of arithmetic facts from memory.

Particular representational systems can be exploited in different ways to reveal properties of the represented entities. For instance, judging whether a particular number is even or odd requires a cumbersome counting process when it is represented in tallies (e.g., ||||||||||), but only requires looking at the last digit when the same number is represented in binary or decimal notation (e.g., 1001 or 9, respectively). As another example, a natural number represented in decimal notation is divisible by nine when the sum of its digits is divisible by nine. Importantly, these computational shortcuts are not only dependent upon particular representations, but they are only available when both the meaning of the elementary symbols and their arithmetic properties are *known* to and actively *used* by the problem solver.

The terminology introduced above allows us to state that different number systems are isomorphs for the mathematical theory of arithmetic. Differences between their usefulness for solving particular problems are representational effects, i.e., phenomena in which representational systems that are informationally equivalent lack computational equivalence. Importantly, any judgment about task difficulty needs to consider the triad of the specific arithmetic task at hand, the representational system used, and the problem solver's mental (representational and computational) capacities.

The crucial question to be addressed in the following sections is: What causes or explains representational effects? As the definition of computational equivalence included a reference to the ease of operations that can be internal or external we need to consider the internal and external mechanisms involved in solving a problem to explain the genesis of representational effects. For instance, the ease of tic-tac-toe when compared to number scrabble can possibly be explained by the claim that visuo-spatial representations are more compatible with the cognitive mechanisms of ordinary human beings than the arithmetic properties of small numbers. But the relative ease or difficulty of operations also depend on the task to be performed. Note that a game of number scrabble could be communicated much more concisely (e.g., during a telephone call) than the identical game of tic-tac-toe. For an analogous argument about the usefulness of number systems we need to investigate their properties and computational demands on cognition for particular tasks.

## 4.2   Computations with different notation systems

Traditional accounts of number systems (e.g., Dantzig, 1954; Ifrah, 1985) draw a basic distinction between additive and place-value systems. Moreover, it is usually claimed that the positional system is superior, because it allows for more complex calculations, like the application of multiplication algorithms.[13] However, assessing the strengths and weaknesses of different notational systems is not as simple as this suggests and demands a much more nuanced analysis.

**Cipherization.** By contrasting several numerical notation systems, in particular the Greek alphabetic system of numerals (see Table 1c), which is *not* positional, and the Babylonian sexagesimal place-value system, Boyer (1944) argued that it is not so much the use of the principle of place-value that guarantees the ease of computation, but 'independent representation' or 'cipherization.' Characteristic for the latter is that individual symbols, that are brief and easy to write and read, are introduced to represent numbers in a concise way, avoiding the frequent repetition of basic symbols. What makes the Babylonian system cumbersome to use, despite the fact that it is a place-value system, is that is uses an additive system to represent numbers less than 60. Thus, for example, 57 is essentially represented by ⟨⟨⟨⟨⟨ ||||||||. By contrast, the Greek alphabetic system (see Table 2) represents numbers up to 999 by combining 27 different elementary symbols and uses at most three different symbols per number. Since the Greek system is not positional, it does not need a symbol to mark an empty position (zero), but—in contrast to the system of Roman numerals—its basic symbols never occur repeatedly in a numeral. As a consequence, numbers are represented by even fewer symbols on average than in our familiar decimal place-value notation. For example, the number 208 is written as 'ση', and 400 simply as 'υ'. As in the Roman system to be discussed below, larger numbers require new symbols or a systematic scheme for modifications (see Boyer, 1944).

---

[13]See, e.g., (Menninger, 1969, p. 294), (Ifrah, 1985, p. 431), and (Dehaene, 1997, p. 98).

|      | 1 | 2 | 3 | 4 | 5 | 6 | 7 | 8 | 9 |
|------|---|---|---|---|---|---|---|---|---|
| ×1   | α | β | γ | δ | ε | ϝ | ζ | η | ϑ |
| ×10  | ι | ϰ | λ | μ | ν | ξ | ο | π | ϟ |
| ×100 | ρ | σ | τ | υ | φ | χ | ψ | ω | ϡ |

TABLE 2. The Greek alphabetic system of numerals. The numerals for 6, 90, and 900 are the now obsolete symbols digamma (ϝ), koppa (ϟ), and san (ϡ) (see Ifrah, 1985).

**Typology for numerical notation.** Boyer's observations have been taken up by Chrisomalis (2004), who developed a two-dimensional typology for numerical notation, with intraexponential (e.g., cumulative or ciphered) and interexponential structuring (e.g., additive or positional) as the two independent dimensions. In this terminology the Greek alphabetic and the Roman system differ with regard to their intraexponential representations: The former is ciphered, while the latter is cumulative. Using this classification he was able to discern some patterns in the evolutionary change of number systems that go beyond the traditional view that takes the development of numeral systems to be linear, always from additive systems to positional ones.

**Trade-offs.** We already mentioned that the succinctness of the Greek alphabetic system is achieved through the use of 27 elementary symbols to represent the numbers from 1 to 999. Thus, the Greek system requires more symbols to be memorized, and more single addition and multiplication facts to be learned than our familiar decimal system. We see here clearly a trade-off between the complexity of an external representation system and the cognitive capacities (in this case recognition memory) that are required to master it. Arguably the best way to evaluate such differences would be to familiarize oneself equally thoroughly with multiple systems and then conduct the same computations in different systems. Unfortunately, such comparisons have rarely been done—presumably due to considerable practical limitations. However, the historian of mathematics Paul Tannery reported in 1882 that he practiced the Greek alphabetic system and "found that this notation has practical advantages which he had hardly suspected before, and that the operations took little longer with Greek than with modern numerals" (Boyer, 1944, pp. 160 f.). Unfortunately, Tannery did not provide more details about his experiences.

   In the following, two different numerical notation systems are presented in detail, and some trade-offs with regard to the ease of computations and the cognitive abilities needed to master each system are discussed.

**The decimal number system.** It has been claimed that the numerical competence of the average teenager in today's Western societies exceeds that of an educated adult in antiquity by far (Nickerson, 1988, p. 198). Moreover, due to the nearly universal adoption of the decimal place-value system and our high degree of familiarity with it from early childhood it is hard for us to recognize it as a cultural convention. Despite its universality today, it originated in India and was only introduced to Europe in the 10th century C. E. by way of the Arabic world. We shall refer to it as the Indo-Arabic system.[14]

An important characteristic of the Indo-Arabic system is its economy of expression: Ten elementary symbols suffice to represent every conceivable natural number. This is possible because the basic values of a symbol and its position in the numeral sequence jointly determine its value. Hence, the *order* of digits in a numeral carries magnitude information and cannot be changed inconsequentially. Note how the value represented by the symbol '4' in the numeral 420 is 400, whereas the same symbol signifies 40 in the numeral 42. The fact that only numbers smaller than ten can be expressed within the numerical 'alphabet' of single symbols makes the Indo-Arabic system a *base-10* or *decimal* number system. This base, in turn, is responsible for the operational decade effects, which we discuss in Section 4.3. In general, in a place-value system with base $p$, the value of a numeral phrase containing $n+1$ symbols $a_n a_{n-1} \ldots a_2 a_1 a_0$, where each $a_i$ is an elementary symbol, is equal to $\sum_{i=0}^{n}(a_i \times p^i)$, i.e.,

$$(a_n \times p^n) + (a_{n-1} \times p^{n-1}) + \cdots + (a_2 \times p^2) + (a_1 \times p) + a_0.$$

It becomes clear from this notation that the *position* of a symbol, which is denoted by the subscript, must also be taken into account when performing basic arithmetic operations. For example, *addition* of two numerals $a_n \ldots a_0$ and $b_m \ldots b_0$ in a place-value system, amounts to adding two polynomials, which results in

$$((a_n + b_n) \times p^n) + ((a_{n-1} + b_{n-1}) \times p^{n-1}) + \cdots + ((a_1 + b_1) \times p) + (a_0 + b_0).$$

Note that only the $a$s and $b$s that are at the same position, i.e., that have the same index, are added. Moreover, one has to take special care of the fact that every single digit of the result must be smaller than the base $p$. This is done by 'carrying' into the next position if $(a_i + b_i) \geq p$. Analogously, *multiplication* of two numerals in a place-value system amounts to

---

[14]Most commonly this system is known as the Hindu-Arabic system, but has also been referred to as 'Western,' since the digits have nothing to do with the Hindu religion and are different from Arabic scripts (Chrisomalis, 2004). A still better name might be the 'indic' system of numerals, as was suggested to us by Brendan Gillon. See (Menninger, 1969) and (Ifrah, 1985) for historic accounts of number system development.

multiplication of two polynomials, and the value of the result of multiplying
the two numerals $a_n \ldots a_0$ and $b_m \ldots b_0$ is

$$\sum_{i=0}^{n} \sum_{j=0}^{m} ( a_i \times b_j \times p^{i+j} ).$$

From this form of representation two things become evident: that $m \times n$
operations (i.e., multiplications of two single digits) are necessary, and that
one has to keep track of the *position* $(i + j)$ of the intermediate terms
$a_i \times b_j$ in the resulting numeral. In the various multiplication algorithms
for computations on paper this is achieved by the careful positioning of the
intermediate terms (e.g., in columns).

   The purpose of our presentation of the familiar Indo-Arabic system in
these rather abstract terms is to bring to the fore its inner complexities,
which are usually hidden from us due to our familiarity with it. Most educated adults will be able to instantly 'read off' the number denoted by the
numeral 4711, without any deliberations, but the situation is very different
for children or if we replaced the familiar symbols with unfamiliar ones.
This shows that we have internalized the recognition and transformation of
Indo-Arabic numerals in a very effective way and our familiarity with this
system hides the underlying complexity of such processes. In turn, the fact
that a numeral system looks unfamiliar to us should not play a role in its
assessment.[15]

   Place-value systems enable efficiency of representation in two important
ways: Their use of only a finite number of basic symbols (in general, a base-
$n$ system requires $n$ different symbols) and the relatively short length of
their numerical phrases. The small number of symbols reduces the mental
efforts needed to interpret and write the numerals as well as the number
of basic facts that have been stored in long-term memory. (See Nickerson, 1988; Zhang and Norman, 1993, 1995, for a more detailed treatment of
these issues.) On the other hand, the fact that the position of symbols has
to be kept track of in computations increases the complexity of the algorithms that are needed. Dealing with columns and carries, which are both
devices to keep track of positions, are the major stumbling blocks children
face in learning to compute with Indo-Arabic numerals (see  Lengnink and
Schlimm, 2010).

   One final issue about place-value numeral system concerns the choice of
the base. This choice is essentially arbitrary, though it has been speculated
that it may be due to the "anatomical accident" (Ifrah, 2000, p. 12) of
the human body having ten fingers (see also Section 2.4, page 76). Ifrah
also suggests that from a purely computational perspective a base of 11

---

[15]This was noted also by Anderson (1956), but see (Menninger, 1969, p. 294).

could in some circumstances be better (by virtue of being a prime number), and for trade purposes a base-12 system would yield benefits, as it would allow for more even divisors. A system with base 60 would fare even better in this last respect, and the ancient Babylonians did in fact use such a system. As another extreme, a base-2 (or binary) system would lead to the smallest number of basic symbols for a positional system, but also greatly increases the length of the numerals. Binary and hexadecimal (base-16) systems are in common usage in engineering and computer science, and their properties illustrate additional trade-offs between representational efficiency and implementation requirements. Thus, even the choice of the value of the base involves substantial trade-offs.

**The system of Roman numerals.** The system of Roman numerals is a purely additive system, in which each elementary symbol in a numeral phrase has a fixed value.[16] For example, the symbols I, V, X, L, C, D, M, stand for 1, 5, 10, 50, 100, 500, and 1000, respectively. The value of the numeral is then obtained by simply adding the values of its constituents. Thus, for a numeral $a_0 a_1 \ldots a_n$, the value obtained is:

$$a_0 + a_1 + \cdots + a_n. \tag{1}$$

Comparing this with the process required for reading off the value of an Indo-Arabic numeral (in Equation 4.2 above) emphasizes the simplicity of additive systems. This internal simplicity comes at the cost of requiring a potentially infinite amount of elementary symbols if all natural numbers are to be represented. However, this theoretical limitation may be extenuated by the fact that only numbers up to a certain limit are used in practice.

A quantitative route for the evaluation of different notational systems in terms of their cognitive demands has been taken by Schlimm and Neth (2008). Using a computational cognitive modeling approach (along the lines of Payne et al., 1993), they analyzed algorithms for addition and multiplication with Roman and Indo-Arabic numerals in order to quantify the trade-offs between basic perceptual-motor operations and (short-term and long-term) memory requirements. For their comparisons, they modeled the common paper-and-pencil algorithms for addition and multiplication with the Indo-Arabic numerals, collecting information regarding the number of symbols used, single perceptual activities (e.g., reading a symbol), attentional shifts (to the next symbol in an array or to some absolute position on paper), memory usage (retrieval of addition and multiplication facts from long term memory, remembering of intermediate results, internal computations, etc.), and output activities (writing or deleting symbols). They also

---

[16]The 'subtractive notation,' according to which 4 is represented as IV and not as IIII in Roman numerals, was introduced only in the Middle Ages and is not discussed here.

devised paper-and-pencil algorithms for addition and multiplication with Roman numerals, and despite common prejudices, these algorithms were found not to be more difficult for humans to execute than those for the Indo-Arabic numerals (see Schlimm and Neth, 2008, for a description of the algorithms).

The analysis by Schlimm and Neth (2008) of the elementary information processes employed in the computations revealed that addition with Indo-Arabic numerals requires knowledge of many basic addition facts (like '2 + 3 = 5'), but that only few simplification rules (like 'IIIII → V') need to be mastered for additions with Roman numerals. The fact that a Roman numeral is on average longer than the Indo-Arabic numeral of the same value, has the effect that many more individual steps (perceptions, attention shifts, write operations) have to be carried out, but that these put little strain on working memory. The authors also noticed that the Indo-Arabic algorithms are highly optimized in order to reduce external computations and thus employ more internal ones, whereas the Roman algorithm they devised made heavy use of external representations. Thus, not only the different numeral systems themselves, but also the different computational strategies that they require, have considerable effects on the ease and speed with which both systems are used.

As mentioned above, one of the main disadvantages of the Roman numeral system in comparison with the Indo-Arabic one is the on average longer length of its numerals, which results in lengthier computations. However, it is important to keep in mind that this only holds on average (assuming that uniform ranges of natural numbers are used), since, for particular numbers, the Roman numerals can be shorter than the corresponding ones in the Indo-Arabic system. As an example, compare M with 1000. One could make a case that it is exactly these kinds of (round) numbers that are used most frequently in practice.

**Computing with artifacts.** We have argued above that notions of problem difficulty or computational equivalence between two problems require a reference to the machinery or mechanism involved in solving the problems (see Section 4.1, page 83). Our emphasis on the actual computational process also revealed the crucial role of external resources. For instance, the paper-and-pencil algorithms for arithmetic computations with Roman numerals discussed in Schlimm and Neth (2008) could also be carried out on an abacus, whereby the computations would be simplified considerably. This observation naturally leads to the consideration of the use of artifacts for computations. Examples of such artifacts are the digits on one's hands or toes, sand tablets, paper and pencil, the abacus, but also cash registers, pocket calculators, and modern computers. The interactions between such devices and arithmetical practice are manifold: On the one hand, each device operates with a particular representation of numbers, so that this representation affects the mechanical computations. For instance, one of the biggest challenges when constructing the first 'analytical engine' was to devise an automatic mechanism to carry the tens, which is needed even for simple additions such as $9 + 2 = 11$:

> "The most important part of the Analytical Engine was undoubtedly the mechanical method of carrying the tens. On this I laboured incessantly, each succeeding improvement advancing me a step or two. [...] At last I came to the conclusion that [...] nothing but teaching the Engine to foresee and then to act upon that foresight could ever lead me to the object I desired..." (Babbage, 1864, p. 114, Ch. VIII)

Babbage's conundrum has nothing to do with the mathematical features of addition, and everything to do with the arbitrary properties of the base-ten place-value notation system. Not only do the properties of notations constrain the design of calculators, but the availability of calculators can also influence our arithmetic abilities. For instance, the negative effects of the widespread availability and use of pocket calculators on students' mathematical abilities have been widely discussed.[17] While we cannot go into further detail about these interactions, they illustrate the subtle interplay between cultural (technological) and cognitive (mental) operations.

All the analyses we have described so far concern the ease with which external numeric representations are processed during computation. But arithmetic is often done 'in the head' suggesting that constraints imposed by external (perceptual or motor) processes should not apply. However, we shall argue that notation, in particular the Indo-Arabic decimal system, exerts some subtle effects on arithmetic performance, even when the tasks are performed primarily mentally.

## 4.3   Notation and mental arithmetic

If a particular notation constrains the design of mechanical devices, what are the effects of adopting the Indo-Arabic decimal system on the 'machinery' of mental arithmetic? The very notion of 'mental' arithmetic might suggest that any effects are limited to the translation between input and output formats, as in the simple three-stage view of problem solving attributed to Craik (1943). He claims that, after some initial translation process, operations are carried out in some medium of thought (mentalese) before the final answer is returned. A more embodied and embedded view of cognition (as promoted in different ways by Clark, 1997; Wilson, 2002; Neth et al., 2007) would argue that this simple model needs to be elaborated with a

---

[17]See (Dehaene, 1997, pp. 134–136) and (Butterworth, 1999, p. 350), but more systematic studies are needed.

more interactive view of arithmetic that might allow the environment, including external representations, a more potent role in shaping mental representations and processes. However, most psychological studies of mental arithmetic have ignored notational effects and concentrate only on the representation of number and of number processing independently of specific numeral systems.

A notable exception is the existing research on *representational effects* in the psychological literature on mental arithmetic. One body of research on that topic analyzes the influence of particular representational systems on the ease with which mental operations can be carried out. For example, in comparing the Indo-Arabic number system with Roman numerals Nickerson (1988) and Zhang and Norman (1993, 1995) point out that each system selectively facilitates different subprocesses. Zhang and Norman analyze such differences in terms of which computational constraints are enforced by the notation, and which need to be maintained mentally. The thrust of this work is more theoretical than empirical. Some empirical work on the effects of different number systems has shown that notational effects can be measured on internal operations as well as on interactive read-write processes.

Gonzalez and Kolers (1982) showed that the reaction times of very simple addition tasks (with sums below 10) were influenced by notation. They used a verification/rejection task in which participants were presented with equations like IV+2 =VI, displayed in a variety of mixtures of Roman and Indo-Arabic numerals, and showed that the notation used affected the slope of problem-size effects. On this basis, they suggested that different mental operations were applied to Roman and Indo-Arabic numerals, so that the notation was exerting an effect even on internal transformation processes (Gonzalez and Kolers, 1987). Other examples of research directed at representational effects in arithmetic include investigations of linguistic number name effects (see, e.g., Miller, 1992), based on discrepancies in the regularities of number names in different languages.

More recently, Campbell and Fugelsang (2001) presented simple addition problems in a verification task in either Indo-Arabic digit or English number word format and monitored participants' adding strategies. As participants were less likely to retrieve results, but rather resorted to calculation when facing number word problems, and this difference increased with problem size, they concluded that presentation format does have an impact on central aspects of cognitive arithmetic. In a similar vein Nuerk et al. (2001) suggested on the basis of a number comparison paradigm that tens and units might not be represented on a single continuous mental number line.

The work of LeFevre et al. (1996) links two pervasive issues in experimental studies of arithmetic skills. First, even relatively simple arithmetic tasks involve a choice between alternative strategies. For instance, the problem $3 + 6$ can be solved by retrieving the answer 9 from memory or by using an incremental counting procedure to generate the answer. If some stepwise algorithm is used one might start with the number 3 and count up 6 units from it, or, more efficiently, one might reverse the left-right order of addends to start with the larger number 6 and count up 3 from that (see Siegler and Shrager, 1984; Siegler, 1987; Siegler and Lemaire, 1997; Shrager and Siegler, 1998; Siegler and Stern, 1997, for developmental studies of strategy selection).

The phenomenon that problems involving larger numbers (e.g., $4 + 5$) are generally solved more slowly than those with smaller numbers ($4 + 3$) is known as the *problem size effect* (Parkman and Groen, 1971; Groen and Parkman, 1972; Campbell, 1995; Zbrodoff, 1995; Geary, 1996). While explanations for the problem size effect remain controversial, LeFevre et al. (1996) have suggested that it may be related to the issue of strategy choice. If different participants resorted, at different points, to counting strategies (which show a linear relationship between addend size and count duration) then a problem size effect could be explained as a methodological consequence of averaging over an entire group of participants. (See also Siegler, 1987, for a similar point.)

Interestingly, a related analysis of addition tasks is sensitive to the properties of the Indo-Arabic decimal number system. LeFevre et al. (1996) report that when the sum of a pair of digits exceeds ten, an adder is more likely to use a counting rather than a fact retrieval strategy. According to this account, sums greater than ten (e.g., $6 + 7$) are sometimes *decomposed* into two stages: up to the decade ($6 + 4$), and beyond ($10 + 3$). Moreover, Geary (1996) reports evidence that even adults frequently use decomposition as a back-up strategy, particularly on larger-valued addition problems (with sums exceeding ten). If this strategy is indeed widespread, then it makes an interesting prediction, namely, that those additions that sum to ten are likely to be the most practiced of all additions. In this case, one would predict that sums to ten, or more generally, sums reaching decade boundaries will be even easier than smaller sums. Neth (2004) investigated this hypothesis by letting participants add up sequences of random single-digit numbers and measuring the time for each individual addition (see also Neth and Payne, 2001). The results show clear *decade effects* in mental addition. So-called complements (two addends adding up to a round sum, e.g., $16 + 4$) are added faster than sub-complements (e.g., $16 + 3$), which, in turn, are faster than super-complements (e.g., $16 + 5$). As the first of these results in particular cannot be explained by problem size effects, the duration of mental operations is influenced by arbitrary properties of a numeric notation. In line with this argument is the observation that post-complements (e.g., $20 + 6$) are computed faster than any other type of addition, presumably because such operations do not require any actual addition, but can be achieved by mere replacement of the unit digit 0 by the current addend 6. Thus, the neural gears of our minds seem to be affected by properties of our decimal notation just like Babbage's analytical engine required some special ingenuity to carry the tens.

In this section we highlighted various trade-offs between the richness of a representational system (in terms of its basic set of symbols), its demands on human cognition (e.g., the need to memorize symbol meanings and rules for symbol manipulation), and the resulting potential for algorithmic computations. This potential not only depends on the mental machinery of the human mind but is modulated by the availability of external tools (like paper and pencil, an abacus, or an electronic calculator). More generally, the difficulty of any arithmetic problem crucially depends on the relations among the specific task to be solved, the representational system used to grasp and frame it, and the internal and external capacities and resources that are available and required to solve it.

## 5   Conclusion

As was mentioned briefly in Sections 1 and 2.1, philosophers of mathematics in general, and analytic philosophers in particular, have shown great reservations toward taking seriously the work of psychologists on mathematical reasoning. This is possibly due to the influence of Frege's arguments against psychological accounts of mathematical objects, which he deemed either unsatisfactory or subjective. Since then anti-psychologistic tendencies have been popular in philosophy of mathematics, so that philosophers have shunned the idea that any psychological insights might be relevant to their enterprise. The distinction made between the contexts of discovery and justification (Reichenbach, 1938), together with the view that philosophy has nothing to say about the former, has further ingrained this attitude (see Schlimm, 2006, for a general discussion of these developments). Objections to these sentiments were raised mainly by mathematicians, who tried to get a better understanding of their practices, and who were fully aware that their activities had strong psychological components.[18]

Frege's emphasis on mathematical objects as logical entities led to a static view of mathematics and a focus on the ontological nature of mathematical objects, which dominated philosophy of mathematics for a long time. With the turn toward history and mathematical practice that was pushed by the work of Lakatos (1976), philosophy of mathematics began to show more interest in the cognitive foundations of mathematical reasoning. Nevertheless, the relationship between mathematics and cognition is still

---

[18]See, e.g., (Klein, 1926, p. 152) and (Hadamard, 1945).

tenuous, as can be seen by the reserved interest of philosophers in the first attempt to relate cognitive psychology to mathematical practice (Lakoff and Núñez, 2000).

One point that we have stressed in this presentation is that notation is relevant for understanding how we deal with numbers and that even mental arithmetic is best understood as an interplay between internal and external representations. This suggests that mathematical notation also plays an important role in our understanding of higher mathematics and mathematical practice in general. Indeed, if notation is an irreducible part of our mathematical cognition, this might have important consequences for philosophy of mathematics, in particular for our epistemic access to mathematical objects.

While philosophers can learn from psychologists about the cognitive underpinnings of mathematics, psychologists can also learn from philosophers, in particular when it comes to conceptual clarification. The basic terminological distinctions presented in Sections 1 and 4 are not always adhered to in psychology, where we frequently find the terms 'number,' 'number concept,' and sometimes even 'numeral' used interchangeably. We have also seen that what cognitive scientists mean by 'arithmetic' (simple computations with mostly natural numbers) is not necessarily what philosophers or mathematicians take it to be (e.g., the study of properties of prime numbers). And indeed, mathematics extends far beyond differentiating between heaps of discrete objects, and counting up to small numbers. Therefore, the question arises whether empirical results about the ways in which humans or lower animals deal with small numerosities tell us anything at all about high-level mathematics. Related to this are issues regarding the nature of infinity. Many mathematical objects—like the set of natural numbers, the continuum, the functions in analysis, and the lines in geometry—are essentially infinite. How human beings are able to reason about such structures is still largely a mystery from a cognitive perspective. This opens new opportunities for cognitive science, which hitherto has mainly dealt with elementary numerical cognition, to investigate the cognitive underpinnings of more complex mathematical thought.

At the beginning of this paper, we presented our aim as illuminating the relationship between mathematics, mind, and brain from various perspectives (recall Figure 1d on page 60). We now realize that even our interdisciplinary journey could not do full justice to all the nuances of our topic. Any comprehensive treatment of the cognitive basis of arithmetic will have to include specifications of the world in which mathematical problem solving is situated. Such references to the context must not only include descriptions of the physical environment (e.g., a characterization of the precise task to be solved, the shape of our bodies, and the availability of artifacts),

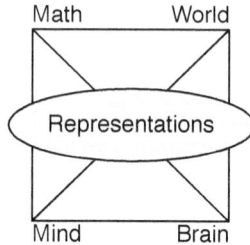

FIGURE 4. Representations shape and are shaped by the relations between mathematics, world, mind, and brain.

but also aspects of the social, linguistic, and historical context in which particular problems occurred. As we have seen, representations play a key role in the explanation of the various relationships between our four main reference points of mathematics, mind, brain, and world (Figure 4). Due to their mediating role representations not only shape our interactions with mathematical problems and constructs (e.g., by determining the difficulty of mathematical tasks), but they are themselves adapted and shaped by the need to solve specific mathematical problems in particular environments. The theory of arithmetic and the development of numerical notations as tools to solve mathematical tasks cannot meaningfully be studied in isolation from the physical, social, historical, psychological, and biological context in which theory and tools were conceived and applied. We hope that future work in this direction will further illuminate the cognitive basis of arithmetic and of mathematics as a whole.

## Bibliography

Anderson, W. F. (1956). Arithmetical computations in Roman numerals. *Classical Philology*, 51(3):145–150.

Babbage, C. (1864). *Passages from the Life of a Philosopher*. Longman, Roberts, & Green, London.

Barth, H., La Mont, K., Lipton, J., Dehaene, S., Kanwisher, N., and Spelke, E. (2006). Non-symbolic arithmetic in adults and young children. *Cognition*, 98(3):199–222.

Bibby, P. A. and Payne, S. J. (1996). Instruction and practice in learning to use a device. *Cognitive Science*, 20:539–578.

Biro, D. and Matsuzawa, T. (2001). Use of numerical symbols by the chimpanzee (*Pan troglodytes*): Cardinals, ordinals, and the introduction of zero. *Animal Cognition*, 4:193–199.

Boyer, C. B. (1944). Fundamental steps in the development of numeration. *Isis*, 35(2):153–168.

Boysen, S. T. (1993). Counting in chimpanzees: Nonhuman principles and emergent properties of number. In Boysen, S. T. and Capaldi, E. J., editors, *The development of numerical competence: Animal and human models*, pages 39–59. Lawrence Erlbaum Associates, Hillsdale NJ.

Boysen, S. T. and Berntson, G. G. (1989). Numerical competence in a chimpanzee (*Pan troglodytes*). *Journal of Comparative Psychology*, 103(1):23–31.

Buldt, B. and Schlimm, D. (2010). The loss of vision—how mathematics turned blind while it learned to see more clearly. In Löwe, B. and Müller, T., editors, *PhiMSAMP. Philosophy of Mathematics: Sociological Aspects and Mathematical Practice*, volume 11 of *Texts in Philosophy*, pages 39–58, London. College Publications.

Butterworth, B. (1999). *The Mathematical Brain*. Macmillan, London.

Butterworth, B. (2005). The development of arithmetical abilities. *Journal of Child Psychology and Psychiatry*, 46(1):3–18.

Butterworth, B., Reeve, R., Reynolds, F., and Lloyd, D. (2008). Numerical thought with and without words: Evidence from indigenous Australian children. *Proceedings of the National Academy of Sciences of the United States of America*, 105:13179–13184.

Campbell, J. I. D. (1995). Mechanisms of simple addition and multiplication: A modified network-interference theory and simulation. *Mathematical Cognition*, 1:121–164.

Campbell, J. I. D. and Fugelsang, J. (2001). Strategy choice for arithmetic verification: effects of numerical surface form. *Cognition*, 80(3):21–30.

Capaldi, E. J. and Miller, D. J. (1988). Counting in rats: Its functional significance and the independent cognitive processes that constitute it. *Journal of Experimental Psychology: Animal Behavior Processes*, 14(1):3–17.

Chochon, F., Cohen, L., van de Moortele, P. F., and Dehaene, S. (1999). Differential contributions of the left and right inferior parietal lobules to number processing. *Journal of Cognitive Neuroscience*, 11:617–630.

Chrisomalis, S. (2004). A cognitive typology for numerical notation. *Cambridge Archaeological Journal*, 14(1):37–52.

Church, R. M. and Meck, W. H. (1984). The numerical attribute of stimuli. In Roitblat, H. L., Bever, T. G., and Terrace, H. S., editors, *Animal cognition*, pages 445–464. Lawrence Erlbaum Associates, Hillsdale NJ.

Clark, A. (1997). *Being There: Putting brain, body, and world together again*. MIT Press, Cambridge MA.

Cohen, L. B. and Marks, K. S. (2002). How infants process addition and subtraction events. *Developmental Science*, 5:186–212.

Craik, K. J. W. (1943). *The nature of explanation*. Cambridge University Press, Cambridge.

Dantzig, T. (1954). *Number, the Language of Science*. Macmillan, New York NY, 4th, revised and augmented edition.

Davis, P. J. and Hersh, R. (1981). *The Mathematical Experience*. Birkhauser, Boston MA.

Dedekind, R. (1888). *Was sind und was sollen die Zahlen*. Vieweg, Braunschweig. Page numbers refer to the edition in (Fricke et al., 1932); English translation in (Ewald, 1996, pp. 787–833).

Dehaene, S. (1997). *The number sense: How the mind creates mathematics*. Oxford University Press, New York NY.

Dehaene, S. and Changeux, J. P. (1993). Development of elementary numerical abilities: A neuronal model. *Journal of Cognitive Neuroscience*, 5(4):390–407.

Dehaene, S., Izard, V., Spelke, E., and Pica, P. (2008). Log or linear? Distinct intuitions of the number scale in Western and Amazonian indigene cultures. *Science*, 320(5880):1217–1220.

Dehaene, S., Spelke, E. S., Pinel, P., Stanescu, R., and Tsivkin, S. (1999). Sources of mathematical thinking: Behavioral and brain-imaging evidence. *Science*, 284:970–974.

Donlan, C., Cowan, R., Newton, E., and Lloyd, D. (2007). The role of language in mathematical development: Evidence from children with specific language impairments. *Cognition*, 103(1):23–33.

Eger, E., Sterzer, P., Russ, M. O., Giraud, A.-L., and Kleinschmidt, A. (2003). A supramodal number representation in human intraparietal cortex. *Neuron*, 37:1–20.

Everett, D. (2005). Cultural constraints on grammar and cognition in Pirahã. *Current Anthropology*, 46:621–634.

Ewald, W. (1996). *From Kant to Hilbert: A Source Book in the Foundations of Mathematics*. Clarendon Press, Oxford.

Feigenson, L., Carey, S., and Hauser, M. D. (2002). The representations underlying infants' choice of more: Object files versus analog magnitudes. *Psychological Science*, 13:150–156.

François, K. and Van Kerkhove, B. (2010). Ethnomathematics and the philosophy of mathematics (education). In Löwe, B. and Müller, T., editors, *PhiMSAMP. Philosophy of Mathematics: Sociological Aspects and Mathematical Practice*, volume 11 of *Texts in Philosophy*, pages 121–154, London. College Publications.

Frank, M., Everett, D., Fedorenko, E., and Gibson, E. (2008a). Number as a cognitive technology: Evidence from Pirahã language and cognition. *Cognition*, 108:819–824.

Frank, M., Fedorenko, E., and Gibson, E. (2008b). Language as a cognitive technology: English-speakers match like Pirahã when you don't let them count. In Love, B., McRae, K., and Sloutsky, V., editors, *Proceedings of the 30th Annual Conference of the Cognitive Science Society*, pages 439–444. Cognitive Science Society, Austin, TX.

Frege, G. (1884). *Grundlagen der Arithmetik*. Wilhelm Koeber, Breslau.

Fricke, R., Noether, E., and Ore, O., editors (1932). *Richard Dedekind: Gesammelte mathematische Werke*, volume 3. Vieweg & Sohn, Braunschweig.

Fricke, R. and Vermeil, H., editors (1922). *Felix Klein. Gesammelte mathematische Abhandlungen. Zweiter Band. Anschauliche Geometrie, Substitutionsgruppen und Gleichungstheorie, zur mathematischen Physik*. Springer, Berlin.

Gallistel, C. R. (1993). A conceptual framework for the study of numerical estimation and arithmetic reasoning in animals. In Boysen, S. T. and Capaldi, E. J., editors, *The development of numerical competence: Animal and human models*, pages 211–223. Lawrence Erlbaum Associates, Hillsdale NJ.

Geary, D. C. (1996). The problem-size effect in mental addition: Developmental and cross-national trends. *Mathematical Cognition*, 2(1):63–93.

Gelman, R. and Gallistel, C. R. (1978). *The Child's Understanding of Number*. Harvard University Press, Cambridge MA.

Gerstmann, J. (1940). Syndrome of finger agnosia, disorientation for right and left, agraphia and acalculia. *Archives of Neurology and Psychiatry*, 44:398–408.

Gonzalez, E. G. and Kolers, P. A. (1982). Mental manipulation of arithmetic symbols. *Journal of Experimental Psychology: Learning, Memory, and Cognition*, 8(4):308–319.

Gonzalez, E. G. and Kolers, P. A. (1987). Notational constraints on mental operations. In Deloche, G. and Seron, X., editors, *Mathematical disabilities: A cognitive neuropsychological perspective*, pages 27–42. Lawrence Erlbaum Associates, Hillsdale NJ.

Gordon, P. (2004). Numerical cognition without words: Evidence from Amazonia. *Science*, 306:496–499.

Groen, G. J. and Parkman, J. M. (1972). A chronometric analysis of simple addition. *Psychological Review*, 79(4):329–343.

Hadamard, J. (1945). *The Psychology of Invention in the Mathematical Field*. Princeton University Press, Princeton NJ.

Halberda, J., Mazzocco, M. M., and Feigenson, L. (2008). Individual differences in non-verbal number acuity correlate with maths achievement. *Nature*, 455(7213):665–668.

Hauser, M. D., Chomsky, N., and Fitch, W. T. (2002). The faculty of language: What is it, who has it, and how did it evolve. *Science*, 298:1569–1579.

Ifrah, G. (1981). *Histoire universelle des chiffres*. Seghers, Paris.

Ifrah, G. (1985). *From One to Zero: A Universal History of Numbers*. Viking Penguin Inc., New York NY. Translation of (Ifrah, 1981).

Ifrah, G. (2000). *The Universal History of Numbers. The Computer and the Information Revolution*. The Harvill Press, London. Translation of (Ifrah, 1981).

Johansen, M. (2010). Embodied strategies in mathematical cognition. In Löwe, B. and Müller, T., editors, *PhiMSAMP. Philosophy of Mathematics: Sociological Aspects and Mathematical Practice*, volume 11 of *Texts in Philosophy*, pages 179–196, London. College Publications.

Klein, F. (1895). Über Arithmetisierung der Mathematik. *Nachrichten der Königlichen Gesellschaft der Wissenschaften zu Göttingen.* Heft 2; reprinted in (Fricke and Vermeil, 1922, pp. 232–240).

Klein, F. (1926). *Vorlesungen über die Entwicklung der Mathematik im 19. Jahrhundert. Teil 1*, volume 24 of *Die Grundlehren der mathematischen Wissenschaften.* Springer, Berlin.

Kobayashi, T., Hiraki, K., Mugitani, R., and Hasegawa, T. (2004). Baby arithmetic: One object plus one tone. *Cognition*, 91:B23–B34.

Kotovsky, K., Hayes, J. R., and Simon, H. A. (1985). Why are some problems hard? Evidence from Tower of Hanoi. *Cognitive Psychology*, 17(2):248–294.

Kotovsky, K. and Simon, H. A. (1990). What makes some problems really hard: Explorations in the problem space of difficulty. *Cognitive Psychology*, 22(2):143–183.

Lakatos, I. (1976). *Proofs and Refutations: The Logic of Mathematical Discovery.* Cambridge University Press, Cambridge. Edited by John Worrall and Elie Zahar.

Lakoff, G. and Núñez, R. E. (2000). *Where Mathematics Comes From. How the Embodied Mind Brings Mathematics into Being.* Basic Books, New York NY.

Larkin, J. H. and Simon, H. A. (1987). Why a diagram is (sometimes) worth ten thousand words. *Cognitive Science*, 11(1):65–100.

LeFevre, J. A., Sadesky, G. S., and Bisanz, J. (1996). Selection of procedures in mental addition: Reassessing the problem size effect in adults. *Journal of Experimental Psychology: Learning, Memory, and Cognition*, 22(1):216–230.

Leibniz, G. W. (1765). Nouveaux essais sur l'entendement humain. In Raspe, R. E., editor, *Oeuvres philosophiques latines et françaises de feu Mr de Leibnitz, tirées des ses Manuscrits qui se conservant dans la Bibliothèque royale à Hanovre*, Amsterdam/Leipzig. Jean Schreuder. Page numbers refer to the edition in (Remnant and Bennet, 2001).

Lemer, C., Dehaene, S., Spelke, E., and Cohen, L. (2003). Approximate quantities and exact number words: Dissociable systems. *Neuropsychologia*, 41:1942–1958.

Lengnink, K. and Schlimm, D. (2010). Learning and understanding numeral systems: Semantic aspects of number representations from an educational perspective. In Löwe, B. and Müller, T., editors, *PhiMSAMP. Philosophy of Mathematics: Sociological Aspects and Mathematical Practice*, volume 11 of *Texts in Philosophy*, pages 235–264, London. College Publications.

Locke, J. (1690). *An essay concerning humane understanding.* The Basset / Edw. Mory, London. Page numbers refer to (Nidditch, 1975).

Maddy, P. (1992). *Realism in Mathematics.* Clarendon Press, Oxford.

McCrink, K. and Wynn, K. (2004). Large-number addition and subtraction by 9-month-old infants. *Psychological Science*, 15:776–781.

Mechner, F. (1958). Probability relations within response sequences under ratio reinforcement. *Journal of the Experimental Analysis of Behavior*, 1(2):109–121.

Mechner, F. and Guevrekian, L. (1962). Effects of deprivation upon counting and timing in rats. *Journal of the Experimental Analysis of Behavior*, 5(4):463–466.

Meck, W. H. and Church, R. M. (1983). A mode control model of counting and timing processes. *Journal of Experimental Psychology: Animal Behavior Processes*, 9(3):320–334.

Menninger, K. (1969). *Number words and number symbols: A cultural history of numbers.* MIT Press, Cambridge MA.

Miller, K. F. (1992). The nature and origins of mathematical skills. In Campbell, J. I. D., editor, *What a Number is: Mathematical Foundations and Developing Number Concepts*, pages 3–38. Elsevier Science Publishers, Amsterdam.

Molko, N., Cachia, A., Rivière, D., Mangin, J.-F., Bruandet, M., Le Bihan, D., Cohen, L., and Dehaene, S. (2003). Functional and structural alterations of the intraparietal sulcus in a developmental dyscalculia of genetic origin. *Neuron*, 40:847–858.

Neth, H. (2004). *Thinking by Doing. Interactive Problem Solving with Internal and External Representations.* PhD thesis, School of Psychology, Cardiff University.

Neth, H., Carlson, R. A., Gray, W. D., Kirlik, A., Kirsh, D., and Payne, S. J. (2007). Immediate interactive behavior: How embodied and embedded cognition uses and changes the world to achieve its goals. In McNamara, D. S. and Trafton, J. G., editors, *Proceedings of the 29th Annual Conference of the Cognitive Science Society*, pages 33–34, Austin TX. Cognitive Science Society.

Neth, H. and Payne, S. J. (2001). Addition as interactive problem solving. In Moore, J. D. and Stenning, K., editors, *Proceedings of the 23rd Annual Conference of the Cognitive Science Society*, pages 698–703, Mahwah NJ. Lawrence Erlbaum Associates.

Nickerson, R. S. (1988). Counting, computing, and the representation of numbers. *Human Factors*, 30(2):181–199.

Nidditch, P. H., editor (1975). *John Locke. An Essay Concerning Human Understanding*. Oxford University Press, Oxford.

Nieder, A. and Miller, E. K. (2003). Coding of cognitive magnitude: Compressed scaling of numerical information in the primate prefrontal cortex. *Neuron*, 37:149–157.

Nuerk, H. C., Weger, U., and Willmes, K. (2001). Decade breaks in the mental number line? Putting the tens and units back in different bins. *Cognition*, 82(1):25–33.

Parkman, J. M. and Groen, G. J. (1971). Temporal aspects of simple addition and comparison. *Journal of Experimental Psychology*, 89(2):335–342.

Payne, J. W., Bettman, J. R., and Johnson, E. J. (1993). *The Adaptive Decision Maker*. Cambridge University Press, Cambridge.

Payne, S. J. (2003). User's mental models: The very ideas. In Carroll, J. M., editor, *HCI Models, Theories, and Frameworks: Towards a Multidisciplinary Science*. Morgan Kaufmann, San Francisco CA.

Peano, G. (1889). *Arithmetices principia, nova methodo exposita*. Bocca, Torino. Quoted after the English translation in (van Heijenoort, 1967, pp. 83–97).

Pepperberg, I. M. (1994). Numerical competence in an African gray parrot (*Psittacus erithacus*). *Journal of Comparative Psychology*, 108:36–36.

Pfungst, O. (1907). *Das Pferd des Herrn von Osten (Der Kluge Hans). Ein Beitrag zur experimentellen Tier- und Menschen-Psychologie*. Johann Ambrosius Barth, Leipzig. English translation in (Pfungst, 1911).

Pfungst, O. (1911). *Clever Hans (The horse of Mr. von Osten): A contribution to experimental animal and human psychology.* Henry Holt, New York NY. Translated by C. L. Rahn.

Piaget, J. (1952). *The Child's Conception of Number.* Norton, New York NY.

Pica, P., Lemer, C., Izard, V., and Dehaene, S. (2004). Exact and approximate arithmetic in an Amazonian indigene group. *Science,* 306(15):499–503.

Reichenbach, H. (1938). *Experience and Prediction. An analysis of the foundations and the structure of knowledge.* University of Chicago Press, Chicago IL.

Remnant, P. and Bennet, J. F., editors (2001). *Leibniz. New essays on human understanding.* Cambridge Texts in the History of Philosophy. Cambridge University Press, Cambridge.

Resnik, M. D. (1997). *Mathematics as a Science of Patterns.* Clarendon Press, Oxford.

Rips, L., Bloomfield, A., and Asmuth, J. (2008). From numerical concepts to concepts of number. *Behavioral and Brain Sciences,* 31:623–642.

Rumbaugh, D. M. and Washburn, D. A. (1993). Counting by chimpanzees and ordinality judgments by macaques in video-formatted tasks. In Boysen, S. T. and Capaldi, E. J., editors, *The development of numerical competence: Animal and human models,* pages 87–106. Lawrence Erlbaum Associates, Hillsdale NJ.

Rusconi, E., Walsh, V., and Butterworth, B. (2005). Dexterity with numbers: rTMS over left angular gyrus disrupts finger gnosis and number processing. *Neuropsychologia,* 43:1609–1624.

Sandrini, M., Rossini, P. M., and Miniussi, C. (2004). The differential involvement of inferior parietal lobule in number comparison: A rTMS study. *Neuropsychologia,* 42:1902–1909.

Schlimm, D. (2006). Axiomatics and progress in the light of 20th century philosophy of science and mathematics. In Löwe, B., Peckhaus, V., and Räsch, T., editors, *Foundations of the Formal Sciences IV,* volume 3 of *Studies in Logic,* pages 233–253, London. College Publications.

Schlimm, D. and Neth, H. (2008). Modeling ancient and modern arithmetic practices: Addition and multiplication with Arabic and Roman numerals.

In Love, B., McRae, K., and Sloutsky, V., editors, *Proceedings of the 30th Annual Meeting of the Cognitive Science Society*, pages 2097–2102, Austin TX. Cognitive Science Society.

Shapiro, S. (2000). *Thinking about Mathematics: The Philosophy of Mathematics*. Oxford University Press, Oxford.

Shettleworth, S. J. (1998). *Cognition, Evolution, and Behavior*. Oxford University Press, New York NY.

Shrager, J. and Siegler, R. S. (1998). SCADS: A model of children's strategy choices and strategy discoveries. *Psychological Science*, 9(5):405–410.

Siegler, R. S. (1987). The perils of averaging data over strategies: An example from children's addition. *Journal of Experimental Psychology: General*, 116(3):250–264.

Siegler, R. S. and Booth, J. L. (2004). Development of numerical estimation in young children. *Child Development*, 75:428–444.

Siegler, R. S. and Lemaire, P. (1997). Older and younger adults' strategy choices in multiplication: Testing predictions of ASCM using the choice/no-choice method. *Journal of Experimental Psychology: General*, 126(1):71–92.

Siegler, R. S. and Shrager, J. (1984). Strategy choice in addition and subtraction: How do children know what to do? In Sophian, C., editor, *Origins of cognitive skills*, pages 229–293. Lawrence Erlbaum Associates, Hillsdale NJ.

Siegler, R. S. and Stern, E. (1997). Conscious and unconscious strategy discoveries: A microgenetic analysis. *Journal of Experimental Psychology: General*, 127(4):377–397.

Simon, H. A. (1978). On the forms of mental representation. In Savage, C. W., editor, *Perception and Cognition. Issues in the Foundations of Psychology*, volume IX of *Minnesota Studies in the Philosophy of Science*, pages 3–18. University of Minnesota Press, Minneapolis MN.

Simon, H. A. (1996). *The Sciences of the Artificial*. MIT Press, Cambridge MA, 3rd edition.

Tang, Y., Zhang, W., Chen, K., Feng, S., Ji, Y., Shen, J., Reiman, E., and Liu, Y. (2006). Arithmetic processing in the brain shaped by cultures. *Proceedings of the National Academy of Sciences of the United States of America*, 103:10775–10780.

Thurston, W. P. (1994). On proof and progress in mathematics. *Bulletin of the American Mathematical Society*, 30(2):161–177.

Tudusciuc, O. and Nieder, A. (2007). Neuronal population coding of continuous and discrete quantity in the primate posterior parietal cortex. *Proceedings of the National Academy of Sciences of the United States of America*, 104:14513–14518.

Uller, C., Carey, S., Huntley-Fenner, G., and Klatt, L. (1999). What representations might underlie infant numerical knowledge? *Cognitive Development*, 14:1–36.

van Heijenoort, J., editor (1967). *From Frege to Gödel. A Source Book in Mathematical Logic, 1879–1931*. Harvard University Press, Cambridge MA.

Varley, R. A., Klessinger, N. J. C., Romanowski, C. A. J., and Siegal, M. (2005). Agrammatic but numerate. *Proceedings of the National Academy of Sciences of the United States of America*, 102:3519–3524.

Wallace, A. R. (1871). *Contributions to the Theory of Natural Selection*. Macmillan, London.

Wigner, E. (1960). The unreasonable effectivenss of mathematics in the natural sciences. *Communications in Pure and Applied Mathematics*, 13(1):1–14.

Wilson, M. (2002). Six views of embodied cognition. *Psychological Bulletin and Review*, 9(4):625–636.

Wynn, K. (1992). Addition and subtraction by human infants. *Nature*, 358:749–750.

Wynn, K. and Chiang, W. (1998). Limits to infants' knowledge of objects: The case of magical appearance. *Psychological Science*, 9:448–455.

Zbrodoff, N. J. (1995). Why is $9 + 7$ harder than $2 + 3$? Strength and interference as explanations of the problem-size effect. *Memory and Cognition*, 23(6):689–700.

Zhang, J. and Norman, D. A. (1993). A cognitive taxonomy of numeration systems. In Kintsch, W., editor, *Proceedings of the 15th Annual Conference of the Cognitive Science Society*, pages 1098–1103, Hillsdale NJ. Lawrence Erlbaum Associates.

Zhang, J. and Norman, D. A. (1995). A representational analysis of numeration systems. *Cognition*, 57:271–295.

# Revolutions in mathematics. More than thirty years after Crowe's "Ten Laws". A new interpretation

Karen François[1,*] and Jean Paul Van Bendegem[1,2]

[1] Centre for Logic and Philosophy of Science, Vrije Universiteit Brussel, Pleinlaan 2, 1050 Brussel, Belgium

[2] Centre for Logic and Philosophy of Science, Universiteit Gent, Blandijnberg 2, 9000 Gent, Belgium

E-mail: karen.francois@vub.ac.be; jpvbende@vub.ac.be

## 1 Introduction

The subject of this paper is whether or not revolutions occur in mathematics. We do not attempt to resolve this problem by arguing for or against the existence of revolutions in mathematics. We want to look at the discussions of the past thirty years from a meta-level. Furthermore, instead of settling the discussion, we want to show what the useful aspects are that these discussions have produced to enrich a study of mathematical practices. We are convinced that research on mathematical revolutions does indeed contribute to such a study.

The paper consists of two major parts. First, we shall give an overview of the discussion about mathematical revolutions and its results. Second, we shall present the way in which Crowe can be fruitfully used by a rather bold interpretation as a basis and framework for a study of the evolving practice of mathematics in all its variety.

## 2 Overview

Following Kuhn's seminal work on scientific revolutions, attempts have been made to apply the Kuhnian framework to mathematics. The debate within the community of mathematicians is even more passionate and tumultuous than within the sciences in general. One could say that Kuhn himself has unchained some kind of revolution within the philosophy of science by putting an end to the classical image of the continuous, linear and steady growth of sciences. After Kuhn, the image of growth of sciences has obtained a more humane face. And of course this didn't come smoothly. The community of philosophers and historians of science appeared to be at least touched by it. The influence of *The Structure of Scientific Revolutions* lingered until

*The first author should like to thank the *Wissenschaftliches Netzwerk* PhiMSAMP funded by the *Deutsche Forschungsgemeinschaft* (MU 1816/5-1) for travel support.

Benedikt Löwe, Thomas Müller (*eds.*). *PhiMSAMP. Philosophy of Mathematics: Sociological Aspects and Mathematical Practice*. College Publications, London, 2010. Texts in Philosophy 11; pp. 107–120.

*Received by the editors:* 12 October 2008; 13 March 2009; 15 September 2009.
*Accepted for publication:* 15 September 2009.

many years later after its publication, in the subjects of symposia and on postscripts (Kuhn, 1970).

Within the community of mathematicians, philosophers and historians of mathematics, the debate is at least as tumultuous and is characterised by explicit oppositions. Most diverging opinions exist about the so-called revolutions within mathematics, ranging from philosophers who construct their reasoning and their definition of a revolution in such a way that mathematics can escape from a revolutionary structure, to philosophers who cite numerous facts from the history of mathematics (following Thomas Kuhn) to demonstrate that mathematics does not differ from the revolutionary nature that characterises the sciences. Kuhn himself did not really address the history of mathematics and those that do may argue that the history of mathematics indeed has experienced revolutions. The extreme positions within the debate are represented by respectively Michael Crowe's "*Ten 'laws' concerning patterns of change in the history of mathematics*" (1992) and by Joseph Dauben's "*Conceptual revolutions and the history of mathematics. Two studies in the growth of knowledge*" (1992b).[1] Dauben is one of the philosophers who discussed the history of mathematics. He holds that there are transitions in mathematical development that are 'critical' or 'discontinuous' enough in order to become worthy of the label of 'revolution'. Dauben holds that revolutions within mathematics cannot be 'Kuhnian' in nature. The revolutionary quality of Cantorian set theory, non-Euclidean geometries or Einstein's Relativity Theory, for example, must be found in another explanation that differs from the normal science-anomaly-revolution scheme that typifies a Kuhnian revolution. Cantor's creation of transfinite numbers transformed mathematics by enlarging its domain from finite to infinite numbers and the conceptual step from transfinite sets to transfinite numbers represents a shift or a transformation that revolutionized mathematics (Dauben, 1992b, p. 57). Non-Euclidean geometries did indeed not as much as replace the Euclidean geometry and Einstein's Relativity Theory did not replace Newton's mechanics, but they did affect, surpass and relegate the domain of mathematics considerably (Van Kerkhove, 2005, p. 211).

The discussion on the revolutionary nature of mathematics is studied by Caroline Dunmore—as part of her Ph.D. thesis—which resulted in the argumentation that revolutions may occur in meta-mathematics, but not in mathematics proper. Mathematics is conservative on the object-level and revolutionary on the meta-level (Dunmore, 1992, p. 211).[2] One could argue that this distinction between meta-mathematics and mathematics proper is

---

[1] Cf. (Dauben, 1992a, p. 206) and (Van Bendegem, 2004, p. 230).

[2] Caroline Dunmore finished her Ph.D. thesis at King's College London by June 1987. Her supervisor was Donald Gillies, the editor of the collection on revolutions in mathematics (Gillies, 1992).

not really satisfying for the question of whether revolutions occur in mathematics. How could a meaningful revolution occur in meta-mathematics that does not in some way refer to or affect mathematics? Dauben (1992a,b) argues that the introduction of transfinite set theory transformed much more than the meta-mathematical level. The same can be said about the introduction of complex numbers, Non-Euclidean geometry, the discovery of incommensurable magnitudes by the Greeks, and so on. In all of these cases, there were certainly meta-level changes that were revolutionary, but this affected both the practice of mathematicians and the content of mathematics.

If today we would be surprised by a denial of scientific revolutions within the growth of sciences, within mathematics this denial still appears to be obvious. Mathematics has traditionally been considered as the science (if a science at all) that can attribute to itself the status of generating absolute certainties, resulting in the idea that what once has been proven, will remain so until eternity (Pourciau, 2000, pp. 297–298; Van Kerkhove, 2005, p. 210).

During the nineteenth and a substantial part of the twentieth century, mathematics was considered as the science where revolutions never occur; in so far it was considered as a science at all. Mathematics was taken to be the science that only accumulated positive knowledge without revolutionary transitions and without rejection of the knowledge that was based on old structures. The house of cards that mathematicians are building and forever extending can, apparently, never be rebuilt. These ideas were expressed by among others the German mathematician and historian of mathematics Hermann Hankel in 1871, by Claude Bernard in 1927, by George David Birkhoff in 1934 and by Clifford Truesdell in 1968 (Dauben, 1992a, pp. 205–206). As a consequence, Crowe must protect the house of cards by developing a system in which he himself describes the notion 'revolution' in such a restrictive way that it is not applicable to mathematics. In his definition, an essential property of a revolution (in general) is the fact that existing entities (be it a king, a constitution, a theory, a concept or a mathematical structure) are knocked over, pulled down, and incontrovertibly renounced. The tenth law indeed states that '*Revolutions never occur in mathematics*' (Crowe, 1992, p. 19).

When Cohen started writing his famous book *Revolution in Science* (Cohen, 1985) he could not neglect the case of mathematics since he was not convinced by the statement that revolutions never occur in mathematics. Some examples from the history of mathematics (e.g. Descartes and Cantor) were fruitful examples to explore the discussions concerning whether or not revolutions in mathematics occur. For Cohen, the denial of the occurrence of revolutions in mathematics is one of the reasons why he considers the topic of revolutions in mathematics in his *Revolution in Science* (Cohen, 1985).

The topic of revolutions in mathematics must, however, be considered
here because a number of mathematicians and historians of mathematics have denied that revolutions can ever occur in mathematics
(Cohen, 1985, p. 489).

It is not the purpose in his article to try to settle the question of whether
or not revolutions have actually occurred in mathematics. However it is
of great interest to explore the ways in which some major innovations in
mathematics have been considered as revolutions. In his *Revolution in
Science* revolutions in mathematics are not generally discussed, although
some attention is paid to the Cartesian revolution in mathematics (Cohen,
1985, pp. 505–507) and mention is made of the revolutionary aspects of
the nineteenth century probability theory and statistics and the radically
new set theory of Georg Cantor (Cohen, 1985, pp. 319–324). Cohen mentioned the unpublished lecture notes on history of mathematics from 1982
by Hawkins who states that revolutions in mathematics occurring when the
method of solving mathematical problems are radically changed on a large
scale. In this sense, a revolution occurred in the seventeenth century with
Descartes as the central figure in initiating this evolution in mathematics.
Hawkins describes—with respect to content and influence—Descartes's *Geometry* (1637) as the major work in the transition from ancient to modern
mathematics. Indeed, the work of Descartes did not involve a 'rejection' of
ancient mathematics (in the sense that for example Euclid's *Elements* were
declared false) but his work did involve a rejection of the methods by which
ancients solved problems. Descartes—and his contemporaries—introduced
new methods. Mathematical problems should be reduced to the symbolic
form of equations and the equations should be used to effect the resolution.
Hence, the introduction of these new methods altered the nature of the
problems posed and ultimately radically altered the scope and content of
mathematics. This sounds indeed revolutionary.

Cohen mentioned other philosophers, mathematicians and historians of
mathematics who discerned revolutions in mathematics, e.g. Fontenelle (the
first author who applied the word revolution to the history of mathematics),
Kant, Cantor, Bell, Kline, Dyson, Mandelbrot, Dauben, among others. As
there is a difference among mathematicians concerning the occurrence of
revolutions in the domain of mathematics, Cohen mentioned also the work
of Fourier, Hankel, Claude Bernard, Truesdell, Boyer, Raymond Wilder and
Crowe who stated that in mathematics there has never yet been a revolution
in mathematics.

For Donald Gillies (1992) who situates himself rather in the direction
of Dauben, the whole debate turns around a semantic discussion wherein
the meaning of the notion 'revolution' is obviously crucial. 'Revolution' is
a concept that belongs to the framework of political theory and therefore it

is used in sciences and mathematics as a metaphor. As a consequence, the
debate on the existence of revolutions is a semantic discussion that slides
away from its original theme (Corry, 1993, p. 95). Not only the concept
'revolution' is undetermined, even the concept 'mathematics' is within the
discussion subject to fluctuations (McCleary, 1993, p. 995). Even more, next
to the fluctuations in the interpretation of what is the object of mathematics and what belongs to the meta-level, there is the discussion about which
mathematical paradigm applies. According to Pourciau (2000), the whole
debate takes place within a specific particular paradigm, which he calls the
"classical paradigm". The classical paradigm determines which mathematical assumptions are true or false and by doing so logical inconsistencies are
excluded (Pourciau, 2000, p. 300). Within the classical paradigm, the discussion then takes place at the semantic level. A good example within this
discussion is of course the definition of revolution itself by Crowe, as cited
above. However, Crowe makes the following remark on the phenomenon
of non-Euclidean geometry: this would certainly have resulted in a revolutionary change in views as to the nature of mathematics, which meant
a revolution within the 'philosophy of mathematics'. Later on, Dummore
will use precisely this argument to situate revolutions at the level of values
related to the nature of the mathematical objects. Crowe will nevertheless stick to the fact that this is no revolution 'within' mathematics itself
(Dauben, 1992a, p. 207). Joseph Dauben, on the other hand, explores the
history of mathematics to demonstrate where revolutions actually do occur
within mathematics. He has studied in depth a number of historical developments and states clearly that he is a defender of the idea that revolutions
do exist within mathematics.[3]

> To the question of whether or not revolutions occur in mathematics,
> my answer is an emphatic "yes". (Dauben, 1992a, p. 229).

Finally, there is Pourciau (2000) who believes that the discussion between Crowe and Dauben on the notion of revolution, only existed at the
semantic level. Crowe and Dauben represent two distinguished positions
within the history of mathematics. Crowe proposed as a law that revolutions never occur in mathematics, while Dauben maintained that such
revolutions do occur and gave examples. However, to Pourciau, no 'real'
controversy exists. The so-called controversy exists only at the semantic
level, the level of the meaning of the term revolution. When they interpret
the word revolution in the same way, as a Kuhnian revolution, where results from the old paradigm are rendered meaningless or untrue in the new

---

[3]Dauben studied among others the independent discovery of the infinitesimal calculus
by Newton and Leibniz in the 17th century, the calculus by Cauchy in the 19th century
and the creation of the non-standard analysis (NSA) by Abraham Robinson in the 20th
century (Dauben, 1992a).

paradigm, then they agree that Kuhnian revolutions are inherently impossible in mathematics (Pourciau, 2000, p. 299). Pourciau himself wants to go beyond this 'semantic' discussion how to redefine revolution so that it might become a useful notion in mathematics. Pourciau defends that 'nothing in the nature of mathematics logically prevents a Kuhnian revolution' (Pourciau, 2000, p. 301). Therefore he has to explain what a Kuhnian revolution means. Referring to Kuhn's *The Structure of Scientific Revolutions* (Kuhn, 1970, p. 92) a revolution is characterized by 'noncumulative developmental episodes' in which a scientific community, having fallen into crisis, rejects its established paradigm and chooses a new paradigm to guide its specific practice. In such a noncumulative paradigm shift, some truths of the old paradigm become unintelligible, unresolved or false under the new paradigm. Those old truths do not translate into new truths and are lost in the new paradigm. There are no translations of the old conception which are true statements in the new conception. Pourciau now defends the idea that a noncumulative shift is not in contradiction with the nature of mathematics. He states that revolutions in mathematics are logically possible (if one wants to question also the classical paradigm), actually possible and historically possible. The case put forward by Pourciau is the development of the activity called intuitionist mathematics as proposed by the Dutch mathematician L. E. J. Brouwer in 1907. This mathematical system is in Pourciau's view incommensurable with classical mathematics—or would be as it is depicted ad a failed revolution in the course of history. According to Brouwer, the seat of mathematical activity resides in the individual mind which has a primordial intuition of time—in terms of its continuity—but also its discreteness, that is, its falling apart into a 'twoity'—the now and the past—in the passing of an instant. From this basic intuition, the mind of a human being, step by step, builds mathematics. These steps of human beings are finite. Hence there are no completed infinite constructions. Because every intuitionist assertion points to the completion of a finite construction, no infinite object can be the result of such a human construction. As a consequence certain classical theorems are no longer theorems when translated to intuitionist terms. Pourciau elaborated on the law of the excluded middle. Other theorems cease to be meaningful statements, e.g. when explicitly appealing to infinite expansions or sets.

What Pourciau brings to the discussion of whether or not revolutions occur in mathematics is that he goes beyond the semantic discussion on the notion of revolution. Referring to Kuhn's notion of revolutions, Pourciau upholds the logical possibility of a revolutionary mathematical transition into intuitionism. He also advances the thesis that Brouwer's intuitionism failed to convert the classical community due to contingent factors such as injudicious technical moves by and strategic choices against Brouwer

(Van Kerkhove, 2005, p. 214). With the example of Brouwer's intuitionism, Pourciau demonstrates how a possible Kuhnian revolution has even been avoided in the course of history (Pourciau, 2000, p. 303).

Actually we see a development where a deepening of the history of mathematics can reveal new perceptions. The American historian of mathematics Crowe seems to be with his *Ten 'laws'* one of the last ones to try to preserve the traditional ideas about mathematical knowledge. In his footsteps we also see the work of Caroline Dunmore who considers revolutions on the object-level of mathematics as impossible, but who does accept them at the meta-level of mathematics. The meta-level then consists of the meta-mathematical values that exist within a community, concerning the purpose and the method of mathematical objects, and of the general values concerning their nature (cf. the debate on the ontological status of mathematical objects).

Whatever one's position in the debate, one must note that any judgment on Crowe's tenth 'law' *Revolutions never occur in mathematics* depends on the meaning of the concepts of revolution and of mathematics. This observation has already been mentioned by Crowe himself who stressed the fact that the preposition *in* is crucial, for, as a number of examples make clear, revolutions may occur in mathematical nomenclature, symbolism, meta-mathematics, metaphysics of mathematics, methodology, standards of rigor, historiography of mathematics, values and mathematics and philosophy of mathematics, but that does not imply that they occur *in* mathematics itself. Here the question can be put forward that if revolutions can occur in mathematical nomenclature, symbolism, meta-mathematics, metaphysics of mathematics, methodology, standards of rigor, historiography of mathematics, etc. . . . , what is left *in* mathematics when we strip all those aspects of mathematics away.

So, where does that leave us? With nothing but an inconclusive debate that should preferably be abandoned? Or is a different perspective required? As it happens, there is an important source, that has not been mentioned so far, and that could provide a creative way out: the works of Imre Lakatos.[4] It is indeed rather striking that in the revolutions-in-mathematics discussion his name is rarely mentioned (notwithstanding the fact that Donald Gillies was his pupil). Part of the key of a possible answer to this puzzle is that Lakatos clearly made a distinction between development and growth in the sciences on the one hand and in mathematics on the other hand. It is sufficient to look at the titles of some of his papers on (the philosophy of) mathematics to see that his views are not a variation on Kuhnian themes,

---

[4]We are referring here to the "mathematical" Lakatos and not to the "scientific" one. The latter Lakatos is first and foremost a pupil of Karl Popper, but the former one is heir to the works of George Pólya, including its Hegelian flavour. This however does not exclude a possible application of Popper's approach to the growth of mathematics, as Teun Koetsier (1991) has shown.

but rather a new vocabulary on its own: "A renaissance of empiricism in the recent philosophy of mathematics?", "What does a mathematical proof prove?" (not published during his lifetime), and "The method of analysis-synthesis" (although the title is not his own, it fits the content quite well).[5] And, of course, not to be forgotten, the seminal "Proofs and Refutations".

No talk about (grand) revolutions is to be found here but rather a detailed analysis of microfeatures of the history and development of mathematical proofs. A proof is proposed, it is refined, hidden lemmas are revealed, made explicit, incorporated, a new proof proposed and the process can be repeated. It is quasi-empiricist because a process of proofs *and refutations* is at work or, as the subtitle of "Proofs and Refutations" states: The Logic of Mathematical Discovery.

So, in what sense, could Lakatos's work present a way out? It is undeniably true that the distance between the macrolevel where revolutions are to be situated and the microlevel reduced to the search and improvement of mathematical proofs, is indeed immense and seems hardly bridgeable. But what if the microlevel is enriched, what if more elements of what mathematicians do in their daily business, are added so that a richer mathematical picture emerges, for we believe that such additional elements are an integral part of mathematics? Might such an "intermediate" description not make it easier to help to decide the matter whether yes or no revolutions occur in mathematics? The bold conjecture we put forward in the next section is that Crowe's Laws can be read as such bridge principles.

## 3    Crowe as a basis and outline

In this section we propose to have a second look at Crowe's laws and to investigate whether they could serve as a basis and outline to study mathematical practices. Our proposal is meant to stimulate discussions about and research of the full mathematical practices by interpreting these *laws* as statements about different types of (local) changes in mathematics. In other words, they are worthy of study not with a view to their justification or refutation within some philosophical framework but solely with a view to obtain a more complete descriptive study of mathematical practices. Each 'law' can be considered as a topic to set up a comparison between mathematics and the sciences, thus illuminating the relevant features that such a full-fledged theory of mathematical practices should study. Let us first of all recapitulate Crowe's ten laws (Crowe, 1992, pp. 16–19):

1. New mathematical concepts frequently come forth not at the bidding, but against the efforts, at times strenuous efforts, of the mathematicians who create them.

---

[5] All these papers can be found in Worrall and Currie (1978).

2. Many new mathematical concepts, even though logically acceptable, meet forceful resistance after their appearance and achieve acceptance only after an extended period of time.

3. Although the demands of logic, consistency, and rigour have at times urged the rejection of some concepts now accepted, the usefulness of these concepts has repeatedly forced mathematicians to accept and to tolerate them, even in the face of strong feelings of discomfort.

4. The rigour that permeates the textbook presentations of many areas of mathematics was frequently a late acquisition in the historical development of those areas, and was frequently forced upon, rather than actively sought by, the pioneers in those fields.

5. The 'knowledge' possessed by mathematicians concerning mathematics at any point in time is multilayered. A 'metaphysics' of mathematics, frequently invisible to the mathematician yet expressed in his writings and teaching in ways more subtle than simple declarative sentences, has existed and can be uncovered in historical research or becomes apparent in mathematical controversy.

6. The fame of the creator of a new mathematical concept has a powerful, almost a controlling, role in the acceptance of that mathematical concept, at least if the new concept breaks with tradition.

7. New mathematical creations frequently arise within, and depend in the mind of their creators upon, contexts far larger than the preserved content of these creations; yet these contexts, for all their original importance, may impede or even prohibit the acceptance of the creations until they are removed by the mathematical community.

8. Multiple independent discoveries of mathematical concepts are the rule, not the exception.

9. Mathematicians have always possessed a vast repertoire of techniques for dissolving or avoiding the problems produced by apparent logical contradictions, and thereby preventing crises in mathematics.

10. Revolutions never occur in mathematics.

As said, the first thing that is in our view absolutely striking about these ten statements, exception made perhaps and somewhat obviously for the tenth law, is that they highlight essential features of mathematical practices seen as a complete process. Not merely the end results are referred to, as, e.g., in the 7th law—no mention shall be made about the creator or the creation process in the end result—but nearly every aspect of what a mathematician practices when doing mathematics. The origin of and resistance to new concepts, the fact that multiple discoveries occur over and over again, the elimination of anomalies, etc., all refer to moments in the mathematical process, different from the final moment or end result.

The second thing that is absolutely striking is that, if in the first nine laws "mathematical" and "mathematics" is replaced respectively by "scientific" and "science", few would disagree with the resulting laws: the importance of the growth of scientific concepts, the resistance to new scientific concepts, the acceptance of new scientific concepts, the curious phenomenon of multiple discoveries, the objectivity of scientific laws that do not refer to specific scientists, the different techniques that allow to deal with inconsistencies and contradictions in scientific theories and so on. Of course, the question then forces itself upon us: if that is indeed the case, why then "deny" that mathematical revolutions occur as well? The first nine laws make no distinction between science and mathematics, and revolutions in science are generally accepted by philosophers of science. As nearly all philosophers see mathematics as different from science, this implies that some basic, fundamental or essential property needs to be found that can make such a distinction, to start with on the level of mathematical and scientific practices.

The Lakatosian key, as it were, to the solution is, we believe, to be found in the 4th law. The development of rigour makes reference to the most important (though not unique) element in mathematical practice, viz. proofs, as Lakatos rightly emphasized. Of course, one might cautiously claim that this law talks about mathematical proofs as there should be a law on such entities. After all, missing out proofs would be a serious shortcoming. As we already made one bold conjecture, let us add another one to it: the ten laws Crowe formulated, exception made for the tenth law itself, should all be interpreted *with reference to* the role they play in the process of finding, constructing, rejecting, and/or refining proofs. In different words, we claim the following: mathematical practice is composed of a set of different types of activities, but they are all related to the one core activity, which is the "proof business" with its possibly revolutionary dynamics. What follows is a closer examination.

The first three laws all refer to (mathematical) concepts. One of the most important elements in the process of constructing a proof for a given statement is the search for the "right" set of concepts to be used in the proof. A perhaps somewhat trivial example is this: give an easy proof that the equation $12445454545\, x^2 + 789878823823\, x - 989086789237921 = 0$ has no solutions in the integers. One could, of course, follow the standard procedure for calculating the solutions of a quadratic equation, but, once one realizes that the core concept to be used here is "odd-even", then it is immediately obvious. The sum of the two terms involving $x$ is always even as the coefficients are both odd, whereas the last term is odd, hence the sum can never be zero, QED. This simple example also shows that the questions what concepts are relevant in what context, is itself a hard problem. And, of course, it will happen that some concept is truly useful in the framework of a proof, but meets resistance if the concept is considered as a concept on its own. One thinks immediately of the introduction of the square roots of negative numbers tolerated within the context of polynomial equations of third degree, but considered "silly" outside of it (an example, by the way, that Crowe himself refers to, cf. p. 16), as Cardano himself testifies. When discussing the famous square root of $-15$, he refers to "mental tortures that one has to put aside" (Cardano, 1545, p. 219) and the famous quote that "so progresses arithmetic subtlety the end of which, as is said, is as refined as it is useless" (Cardano, 1545, p. 221).

The fourth law and the ninth law form a nearly obvious pair from the viewpoint of proofs. Start from the ideal situation from a formal logical position. Proofs in that context, it is generally understood, are represented by an ordered list of formulas, involving premises, axioms, outcomes of the application of formal rules, and such that the last formula of the list is the theorem to be proved. We all know, though perhaps some of us are not all too keen to emphasize it, that "real" mathematical proofs hardly ever reach this high quality level. (The usual claim is that the first chapters of any introductory book in whatever area of mathematics satisfy this standard, but from the third chapter onwards the standard is left behind). Seen from this angle, the fourth law sets a standard, so difficult to meet, that is far more interesting to develop subtle techniques that allow the mathematician to deal with problems such as the occurrence of contradictions, where the logical ideal has nothing to contribute, except the rule that inconsistencies should quite simply not occur.

The sixth, seventh, and eight laws refer, in this interpretation, to the social organization of, what we propose to call, a proof community (thus, by the way, making a distinction with a scientific community). A trivial observation: proof implies statement to be proved. The next question is equally obvious: what statement are we talking about? In less banal terms: what mathematical problems are interesting enough to merit our attention and to invite our attempts to (dis)prove it? Why, e.g., so much attention for the Riemann hypothesis (RH)? Among the manifold reasons that can be produced, let us just mention one: there is at this moment already a wealth of statements of the form "If the RH holds, then $\Phi$". Imagine the day that RH is proved: all of a sudden, all these theorems transform into quite simply "$\Phi$". It needs expert mathematicians—"of fame", to use

Crowe's expression—to identify these interesting statements and, hence, the importance of the 1900 speech of David Hilbert, outlining the research agenda for a century to come. Given this rough picture, it follows that, if such figures as Klein, Hilbert or Poincaré "declare" that this or that problem should be solved, then a host of mathematicians will actually take up the invitation. Little wonder that multiple discoveries are made. Of course, much more needs to be said here, but we do believe that these laws too can be seen as relevant to proof practices.

This leaves us finally with the fifth law, perhaps the most intriguing one. From the proof viewpoint, it seems obvious that different layers are required: the search for a proof and related concepts involves heuristics, proof search methods. As in many cases such techniques become interiorized, it is quite understandable why there is so much talk about mathematical intuition (whether given by birth as God's gift or as a genetic code, or the result of a long and hard training process or, more likely, a combination of both). To identify interesting problems, be they internal to mathematics or with a view to possible applications, requires having a broad picture of the mathematical domain, how it relates to other areas of scientific interest, and how mathematics connects to society at large. Just one example: almost every mathematician knows the "feeling" (really no other term for it) that a specific proof of some statement is not the "right" proof. It is, e.g., too clumsy, too long-winded, or somehow arbitrary. Perhaps this idea is no more than a particular state of mind of the mathematician, but, at the same time, it does feed the idea that the proof is "out there" somewhere to be "discovered". Thus, this proof practice "feeds" a particular metaphysical view that supports and stimulates the mathematician's search and that definitely should be taken into account.

We repeat that, at this stage of our argument, no effective theory of mathematical practice has been presented but rather a list of ingredients, necessary to make the undertaking succeed. We hope to elaborate further on these matters; more specifically, we believe that the approach of Reviel Netz (1999) and the background furnished by Jerry A. Fodor (1983), roughly identifiable as "cognitive history of mathematics" will prove to be quite fruitful.

## 4   Conclusion

Let us summarize our findings. In the first place, we demonstrated the curious nature of the discussion about the occurrence of revolutions in mathematics. We noted that the different positions in the debate are highly dependent on the meaning of both terms, viz. mathematics and revolution. Although the impression might be that the discussion seems to be rather sterile, we defend the thesis that the fruitfulness of the research concerning revolutions and mathematics consists in the fact that the discussion need not settle the basic question "Are there or are there not revolutions in mathematics?", but that it stimulates the study of mathematical practices.

In the second place, as a positive piece of evidence for the above claim, the work of Crowe is used as such a frame of reference to study mathematical practices. Each of his famous ten 'laws' can be considered as a topic to set up a comparison between mathematics and the sciences as we showed by emphasizing the importance of mathematical proofs and all the elements in mathematical practice it ties up with. Although the laws in themselves are equally applicable to mathematics and the sciences, the characteristic element is that they can be interpreted in mathematics as all related to proof, whether in the context of the search and/or creation of proofs or of formulation, presentation, and importance.

We reiterate the observation at the end of the previous section. We have now brought together a set of ingredients to prepare the dish that will be called "A full-fledged theory of mathematical practice". We are quite sure that without any of the ingredients, the result will be disappointing. It remains to be shown that they are sufficiently good "chefs" out there, who will guarantee a splendid meal.

# Bibliography

Bridoux, A., editor (1953). *Descartes. Œuvres et lettres*. Editions Gallimard, Paris.

Cardano, G. (1545). *The Great Art or the Rules of Algebra*. Johannes Petreius, Nuremberg. Quoted from the edition in (Witmer, 1968).

Cohen, B. (1985). *Revolution in Science*. Belknap Press of Harvard University Press, Cambridge MA.

Corry, L. (1993). Kuhnian issues, scientific revolutions and the history of mathematics. *Studies in History and Philosophy of Science*, 24(1):95–117.

Crowe, M. (1992). Ten 'laws' concerning patterns of change in the history of mathematics. In Gillies, D., editor, *Revolutions in Mathematics*, pages 15–20. Clarendon Press, Oxford.

Dauben, J. (1992a). Are there revolutions in mathematics? In Echeverria, J., Ibarra, A., and Mormann, T., editors, *The Space of Mathematics, Philosophical, Epistemological, and Historical Explorations*, pages 205–229. Walter de Gruyter, Berlin.

Dauben, J. (1992b). Conceptual revolutions and the history of mathematics: Two studies in the growth of knowledge. In Gillies, D., editor, *Revolutions in Mathematics*, pages 49–71. Clarendon Press, Oxford.

Descartes, R. (1637). *La Géometrie*. Jan Maire, Leiden. Quoted from the edition in (Bridoux, 1953).

Dunmore, C. (1992). Meta-level revolutions in mathematics. In Gillies, D., editor, *Revolutions in Mathematics*, pages 209–225. Clarendon Press, Oxford.

Fodor, J. (1983). *The Modularity of Mind. An Essay on Faculty Psychology*. Bradford Books, London.

Gillies, D., editor (1992). *Revolutions in Mathematics*. Clarendon Press, Oxford.

Koetsier, T. (1991). *Lakatos' Philosophy of Mathematics. A Historical Approach*, volume 3 of *Studies in the History and Philosophy of Mathematics*. North-Holland, Amsterdam.

Kuhn, T. (1970). *The Structure of Scientific Revolutions*. The University of Chicago Press, Chicago IL, second edition.

McCleary, J. (1993). Review of (Gillies, 1992). *Science*, 259(5097):995–996.

Netz, R. (1999). *The Shaping of Deduction in Greek Mathematics*, volume 51 of *Ideas in Context*. Cambridge University Press, Cambridge.

Pourciau, B. (2000). Intuitionism as a (failed) Kuhnian revolution in mathematics. *Studies in History and Philosophy of Science*, 31(2):297–329.

Van Bendegem, J. P. (2004). The creative growth of mathematics. In Gabbay, D., Rahman, S., Symons, J., and Van Bendegem, J. P., editors, *Logic, Epistemology, and the Unity of Science*, volume 1 of *Logic, Epistemology, and the Unity of Science*, pages 229–255. Kluwer Academic, Dordrecht.

Van Kerkhove, B. (2005). *Naturalism and the Foundations of Mathematical Practice. A Metaphilosophical Essay*. PhD thesis, Faculty of Letters and Philosophy, Vrije Universiteit Brussel.

Witmer, T., editor (1968). *Girolamo Cardano: The Great Art or the Rules of Algebra*. The MIT Press, Cambridge MA.

Worrall, J. and Currie, G., editors (1978). *Imre Lakatos. Mathematics, Science and Epistemology. Philosophical Papers Volume 2*. Cambridge University Press, Cambridge.

# Ethnomathematics and the philosophy of mathematics (education)

## Karen François and Bart Van Kerkhove*

Centre for Logic and Philosophy of Science, Vrije Universiteit Brussel, Pleinlaan 2, 1050 Brussel, Belgium

E-mail: karen.francois@vub.ac.be; bart.van.kerkhove@vub.ac.be

## 1  Introduction

This paper looks into the field of enquiry called ethnomathematics and the influence it has exerted on the philosophy of mathematics and mathematics education. Although a number of pointers to the most relevant and survey literature will be included, it is important to note that we shall not here be providing a systematic overview of empirical-anthropological studies in ethnomathematics, as originally conceived. Ethnomathematics is still being identified with this type of studies, although they have never gone under its name, being included instead in several domains, such as anthropology or history. Nowadays, the bulk of this research falls under the label of cognitive studies (cf., e.g., De Cruz et al., 2010, in this volume). What the reader may instead expect here is a general survey of theoretical work done within the research discipline of ethnomathematics during the last few decades, with a particular focus on Western science and society.

In Section 2, we first take a broad view, estimating the seemingly paradoxical place mathematics occupies in modern society, being as ubiquitous and important as it is invisible and loathed. In Section 3, we first elaborate on the shifted meaning of the concept 'ethnomathematics', and then explore a number of domains in which it has been applied. Until the early 1980s, only mathematical practices of 'nonliterate' peoples were studied, in an attempt to show that their mathematical ideas were as sophisticated as the modern, 'Western' ones. Then a broadening of the 'ethno' concept was proposed, to include all culturally identifiable groups. As a result, today, within the ethnomathematics discipline, scientists are collecting empirical data about the mathematical practices of culturally differentiated groups, literate or not. 'Ethno' should thus no longer be understood as referring to the exotic.

*Support for this paper was granted by project GOA049 of Vrije Universiteit Brussel (Belgium). The second author is also indebted to the Alexander von Humboldt Foundation (Germany), and to *Wissenschaftliches Netzwerk* PhiMSAMP funded by the *Deutsche Forschungsgemeinschaft* (MU 1816/5-1) for travel support to a number of very stimulating meetings. The authors would like to thank Dirk Schlimm and Rachel Rudolph for valuable comments on a previous version of the text.

Benedikt Löwe, Thomas Müller (*eds.*). *PhiMSAMP. Philosophy of Mathematics: Sociological Aspects and Mathematical Practice.* College Publications, London, 2010. Texts in Philosophy 11; pp. 121–154.

*Received by the editors:* 2 October 2009; 13 January 2010.
*Accepted for publication:* 25 February 2010.

This changed and enriched meaning of the concept 'ethnomathematics' has had an impact on both the philosophy of mathematics and mathematics education. Within the philosophy of mathematics, which we discuss in Section 4, it has lent weight to studies of mathematical practice as alternative or complementary to foundational studies. Within the field of mathematics education, ethnomathematics gained a more prominent role, since it became meaningful and relevant to explore various aspects of mathematical literacy in the context of Western curricula. In Section 5, we discuss a number of possibilities and dangers this has opened, and on the basis of this present ethnomathematics as an alternative, implicit philosophy of professional and school mathematical practices.

## 2   Mathematics in society

Mathematics pervades our everyday lives, sometimes obviously and sometimes on a more hidden or implicit level. The rate at which we have to, or at least are expected to, process numerical data is indeed stupefying.[1] But of course this superficial appearance is only a symptom of something deeply structural: through the ever growing impact of science and technology, our entire society has become thoroughly 'mathematized' (Resnikoff and Wells, 1973; Kline, 1990). Strangely enough, this evolution has brought about a rather perverse effect, as reported by Morris Kline:

> Just as a phrase either looses meaning or acquires an unintended meaning when removed from its context, so mathematics detached from its rich intellectual setting in the culture of our civilization and reduced to a series of techniques has been grossly distorted. Since the layman makes very little use of technical mathematics, he has objected to the naked and dry material usually presented. Consequently, a subject that is basic, vital, and elevating is neglected and even scorned by otherwise highly educated people. Indeed, ignorance of mathematics has attained the status of social grace. (Kline, 1990, p. 16)

In other words, the more we have become dependent upon mathematics, mostly in an indirect or invisible way, the less we are actually understanding its principles. While one needs to know little about the inside operations of a car or a personal computer in order to use them effectively, we are not only referring here to sophisticated applied mathematics, but also to basic skills, such as elementary probability theory. John Allen Paulos comments:

---

[1] E.g., Butterworth (1999) has estimated: "At a very, very rough guess, I would say that I process about 1,000 numbers an hour, about 16,000 numbers per waking day, nearly 6 million a year. People whose job entails working with numbers, in supermarkets, banks, betting shops, schools, dealing rooms, will process many more than this" (p. x).

> I'm distressed by a society which depends so completely on mathe-
> matics and science and yet seems so indifferent to the innumeracy and
> scientific illiteracy of so many of its citizens. [...] I'm pained as well
> at the sham romanticism inherent in the trite phrase 'coldly rational'
> [...] and at the belief that mathematics is an esoteric discipline with
> little relation or connection to the 'real' world. (Paulos, 1990, p. 134)

Despite there being arguably deeper reasons for perceiving mathematics
as an extremely difficult, unworldly and thus unsympathetic subject (see
below), this 'distorted picture' has often been identified with failing educational policy.[2] As performance in mathematics is, on the whole, fairly
poor in comparison with other subjects, it is widely argued that something
must be fundamentally wrong with its teaching. Morris Kline for example,
in connection with his previous remark about our loss of affinity for the
'backbone' of our civilization, made this link:[3]

> School courses and books have presented 'mathematics' as a series
> of apparently meaningless technical procedures. Such material is as
> representative of the subject as an account of the name, position, and
> function of every bone in the human skeleton is representative of the
> living, thinking, and emotional being called man. Just as a phrase
> loses meaning or acquires an unintended meaning when removed from
> its context, so mathematics detached from its rich intellectual setting
> in the culture of our civilization and reduced to a series of techniques
> has been grossly distorted. (Kline, 1990, p. 15–16).

Although this is an apt criticism, it cannot be the entire story. To begin
with, feedback mechanisms are in place between general performance on
the one hand, and the quantitative and qualitative supply of teachers (and
expert policy makers) on the other, so that it is a tricky affair to pinpoint at
which of these levels (most of) the problems start or are reinforced. International studies have indeed shown both that the level of recruitment of math
teachers is well below target and that the enrolment of students in higher
math education has been steadily declining over the past decades. This
could be partly due to a poor public image of mathematics, circumstantial
evidence shows. First, there are the cultural complaints ventilated by Kline
and Paulos above. Second, teachers prefer to lecture on other subjects, with
the consequence that similar staff problems are less common in these other
areas (Sam, 1999, p. 19–20).

What about the deeper causes of these unfortunate trends? Although
largely conjectural, the postmodern critical, skeptical and even hostile atti-

---

[2]Cf. (Sam, 1999, p. 21).

[3]Note that Kline also wrote separate monographs about the (then) contemporary
'debacles' of elementary (Kline, 1973) and undergraduate mathematical education (Kline,
1977).

tude towards science might be a good candidate. During the 1960s, there was a growing awareness in Western society of the inherent limits and— possibly detrimental—external effects of its reigning development model: one based on capitalism, science and technology, and that recognized no limit to economical success or to the mastery of nature. The era indeed witnessed the publication of a rising number of highly critical books commenting on this situation, including *Silent Spring* (1962) by Rachel Carson (on the ecological costs of pesticides), *The Population Bomb* (1968) and *Ecocatastrophe* (1969) by Paul Ehrlich, and of course *The Limits to Growth* (1972), the famous first report of the Club of Rome, by Dennis Meadows et al. The main message for science was that it is an activity that is intimately linked with, and that cannot be cut off from, the rest of society. This theme was picked up, among others, by Jerry Ravetz in his timely *Scientific Knowledge and its Social Problems* (1971). As Sardar explains:

> If science is seen as a craft [rather than as an autonomous fact-discovering machine], then 'truth' is replaced by the idea of 'quality' in the evaluation of scientific output. Quality firmly places both the social and ethical aspects of science, as well as scientific uncertainty, on the agenda. Ravetz showed that in the overall practice of contemporary science one could identify four categories that were seriously problematic: shoddy science, entrepreneurial science (where securing grants is the name of the game), reckless science, and dirty science; and they are all involved with runaway technology. He showed further that quality in science depended largely on the morale and commitment of working scientists and was reinforced by the moral acumen of the leadership of the scientific communities (Sardar, 2000, p. 38–39)

The general wariness of science that has since established itself is confirmed when it comes to mathematics. It is perceived as unimportant, dull, coldly calculating, thus potentially threatening and impoverishing (Boyle, 2000). One might indeed ask whether mathematics, the apex as well as one of the main instruments of hard science, should not be identified with the latter's excrescences. If we are to believe Ravetz, then an important role is to be played here by mathematicians themselves. Do they care about their reputations, and what are they prepared to invest in them? Let us briefly look into two issues connected with this question: ethics and popularization.[4]

It is rather surprising how little attention has been devoted to the ethical side of mathematics. And even when it is actually attended to, people remain reluctant to get too deeply involved. E.g., Reuben Hersh, one of the leading mathematical humanists,[5] has remarked that, contrary to moral

---

[4]Obviously, education—one of the central themes of this paper—is another one.

[5]In view of books such as (Davis and Hersh, 1981, 1990), but also of numerous articles published to date.

considerations about *any other* discipline, for mathematics, surely these are not "intrinsic to the actual practice of the particular profession" (Hersh, 1990, p. 13). Put simply, mathematicians—as opposed to, say, chemists—are not the fathers of any artifacts that put an actual burden on, or pose a direct threat to, society or environment. Hersh continues that "it's hard to see significant ethical content in improving the value of a constant in some formula or calculating something new" (ibidem). He further holds that any moral issues to be addressed are personal (not corporate) and academic (not global) in character, and he lists "five different categories of people to whom we [mathematicians] have duties: staff, students, colleagues, administrators, and ourselves" (ibidem). Cases involved may concern discrimination, innumeracy, loyalty, fraud, etc. The only more socially embedded topic referred to, "a little out of date, but interesting" [sic] (Hersh, 1990, p. 14), is that of the affinities between mathematics and the military.

The latter is indeed one of the few (politico-)ethical issues surrounding mathematics that have been studied in any depth. There is, e.g., the pathbreaking work of historian Jens Høyrup with mathematician Bernhelm Booss-Bavnbek, showing the intimate though not essential connection, past and present, between mathematics and the military (Høyrup, 1994, Chapter 8). Although the relation is considered to be one of interdependence, particular attention is devoted to warfare as an impetus for mathematical development. One might in this respect think of the active role of mathematicians in the furtherance of code making and breaking, ballistics, positioning, spying, or 'intelligent' bombing. A particular case in point is John von Neumann, who during the Second World War participated in the development project for the A-bomb, and was one of the respected advisors of the U.S. government. In the early stages of the Cold War, he also developed game theory, having on his mind, among other things, a rational justification for tougher behaviour towards Moscow.[6] Another famous example of ideological influence on mathematical development is the considerable damage to the German mathematical community, especially its Jewish members, in the interbellum period. Under the rule of Ludwig Bieberbach's programme of *Deutsche Mathematik*, important centres of mathematical research, such as the Göttingen school, suffered greatly, or even came to a virtual end.[7]

The previous issues, from education to international affairs, are naturally connected with that of popularization. Indeed well conducted popularization can (help) adjust negative images, whether justified or not. However, this is a specialty in itself that has to cope with a number of problems and limitations. To begin with, the field is deceptively diverse, as contri-

---

[6]Cf., e.g., (Strathern, 2001; Macrae, 1999).

[7]Cf. (Cornwell, 2003, Chapter 16) for a general account, and (Lindner, 1980; Mehrtens, 1985) for more detailed ones.

butions vary widely, depending on what elements of the subject are taken as points of entry: esoteric or merely funny puzzles, great mathematicians of all times, specific branches or programmes, cross-cultural differences etc. (Van Bendegem, 1996, pp. 216–23). Secondly, books about 'the real stuff', i.e., (Western) mathematical science, intended for an audience from outside the discipline—whether made up of other scientists willing to peep over (sub)disciplinary walls or of interested laymen—can only screen a tiny part of the gigantic field at a time. Even those written by excellent popularizers like Keith Devlin or Ian Stewart, can only provide some general considerations or mostly (have to) end up by illuminating a number of 'accessible' topics, which—unavoidably or not—fall well short of painting anything near a 'true' picture of the goings on within the field.[8] One of the possible means of countering this trend would be to devote more attention to a specific type of (hitherto neglected) vulgarization. In it, one turns away from working into the ground the internal histories (albeit from different angles), and instead focuses on the intimate connections between mathematics and daily life, more particularly through its numerous concrete *applications*. Examples include statistics, informatics, genetics, econometrics, etc. Note that the present paper aims to contribute on various fronts to this cause of 'bringing back' mathematics to where it arose, viz. the layman.

# 3  Cultural foundations recognized

## 3.1  The notion of 'ethno'

The prefix *ethno* originally refers to races, tribes, or groups of relatives. Correspondingly, *ethnomathematics* has been associated with the mathematical practices of particular tribes or indigenous, 'primitive' peoples, as well as those of a nation and/or human race.[9] In recent times, under the impulse of an encompassing research programme, the concept has received a much broader interpretation. Before we turn to that, however, a few words on the anthropological roots.

As a field of enquiry, ethnomathematics started in the 1960s, its subject being the mathematical practices of so-called illiterate (or, more politically correct: nonliterate) peoples, holding that—despite the fact that the category 'mathematics' is a strictly Western one—mathematical *ideas* are "pan-human", and are primarily "developed within cultures" (Ascher

---

[8]In a recent book, Devlin has himself confirmed this sorry circumstance. If one desires "to delve further into the world of mathematics, then there are a number of excellent books you can consult, written for a general audience. Most of them, however, are either written at a much more superficial level than the present one [*The Millennium Problems*], or else are focused on specific issues in mathematics" (Devlin, 2002, p. 229).

[9]Of course the latter concept, which was notoriously used to apply biological differences as a basis for segregation, has meanwhile been discredited as a result of advanced studies in genetics.

and Ascher, 1994, p. 1545). Since then, it has also become the (implicit or explicit) mission of many within the discipline to demonstrate that the practices in question and 'our' Western ones are equally complex.

Ubiratan D'Ambrosio (to whom we shall return in extenso below) has provided a more extended version of this idea, with his aim of encompassing the whole range of cultures. Thus he presupposes "a broader concept of 'ethno', to include all culturally identifiable groups with their jargons, codes, symbols, myths, and even specific ways of reasoning and inferring" (D'Ambrosio, 1997, p. 17). Here the word 'ethno' has ceased to refer to anachronistic concepts such as racial groups, primitive peoples or illiterates, and is instead given a comprehensive meaning, pointing to any group of people who share a cultural identity.

However, given the myriad of different interpretations of the concept of culture, the latter move is not uncontroversial. In static views that employ 'culture' to discriminate among groups, characteristics attributed to respective groups of people are considered essences of their culture. But pegging humans down to essences causes them to be characterized invariably and irreversibly, a way of identification that conflicts with the practices of human interaction and mutual influence. Cultural changes can thus only be appreciated exploiting a dynamic interpretation of cultural identity. Indeed identities are not homogeneous and eternal, but rather correspond to an area of tension between permanence and alteration, where—within given contexts—room is left for psycho-social growth processes. A similar dynamic interpretation of culture fully links up with D'Ambrosio's plea for educational reform: "More attention should be paid to students and teachers as human beings, and we have to realize that mathematics—the same is true with respect to other disciplines—are epistemological systems in their socio-cultural and historical perspective and not finished and static entities of results and rules" (D'Ambrosio, 1990, p. 374). We have thus come a long way from the initial meaning to the actual interpretation of 'ethnomathematics'. Similarly, ethnomusicology has evolved to comparative musicology, and more generally ethnology to ethnography, and later to cultural anthropology. The prefix 'ethno' has indeed experienced quite an evolution in its content, up to the moment where only the original term has been preserved.

The Brazilian mathematician and educationalist Ubiratan D'Ambrosio was the first, from the late 1980s, to propose a research program for ethnomathematics, based on the following analysis of the term:

> I call *mathema* the actions of explaining and understanding in order to survive. Throughout all our own life histories and throughout the history of mankind, *technés* (or tics) of *mathema* have been developed in very different and diversified cultural environments, i.e., in the diverse *ethnos*. So, in order to satisfy the drives towards survival and

> transcendence, human beings have developed and continue to develop, in every new experience and in diverse cultural environments, their *ethno-mathema-tics*. (D'Ambrosio, 1990, p. 369)

Following this, the research interests of the newly founded discipline pertain to the development, transmission and distribution of mathematical knowledge as dynamic processes *embedded in their socio-cultural context*. An important implication is that Western mathematics is also considered as having developed (and as continuing to develop) within a particular, contextual reality, *not* detached from it. What is currently known as 'academic' mathematics, though it originated in the Mediterranean area, later expanded to Northern Europe and then to other parts of the World, becoming so to say 'universal'. Nonetheless, it is difficult to deny that the codes and techniques which were developed (such as measuring, quantifying, inferring and abstract thinking) as strategies to express and communicate reflections on space, time, classifying, comparing, are contextual in their origin. Clearly, in other regions of the world, particular circumstances have given rise to different codes and techniques resulting from different perceptions of space, time, and different ways of classifying and comparing (D'Ambrosio, 2007a, p. 30). Reference to the socio-cultural roots of mathematical practices is common in the ethnomathematics literature.[10] This relates to the pragmatic grounds for the development of mathematics: the transmission and distribution of mathematical knowledge to help people cope with day-to-day reality. In the current description of ethnomathematics as given by the International Study Group on Ethnomathematics (ISGE) the narrow (anthropological) meaning of ethnomathematics as well as its broad (socio-cultural) meaning are combined as follows (from the ISGE webpage, July 2009):

> [Ethnomathematics] is sometimes used specifically for small-scale indigenous societies, but in its broadest sense the "ethno" prefix can refer to any group—national societies, labor communities, religious traditions, professional classes, and so on. Mathematical practices include symbolic systems, spatial designs, practical construction techniques, calculation methods, measurement in time and space, specific ways of reasoning and inferring, and other cognitive and material activities which can be translated to formal mathematical representation. The ISGE strives to increase our understanding of the cultural diversity of mathematical practices, and to apply this knowledge to education and development.

In the next section, we shall identify a number of (overlapping) ethnomathematics subdomains. These are identified according to the particular

---

[10]Cf., e.g., (Bishop, 1997, 1988; D'Ambrosio, 1989, 1990, 2007a,b; Gerdes, 1988, 1997; Pinxten, 1991; Zaslavsky, 1973, 1985, 1989).

subject that is their main target. We shall distinguish four: anthropology, history, philosophy and education.

## 3.2 Research domains influenced by ethnomathematics

### 3.2.1 Anthropology

A first important theme for ethnomathematicians is the description of non-Western mathematical practices that, in spite of colonization, have been retained and further developed by many so-called traditional cultures. From within cultural anthropology, the focus is on the description of these practices, and of the ideas underlying them (e.g., Ascher, 1998, 2002; Pinxten, 1991; Bazin et al., 2002). A considerable share of case-studies concerns the mathematical practices performed by professional groups such as fishermen, carpenters, carpet weavers, sugarcane farmers, salesmen and vendors (Vithal and Skovsmose, 1997, p. 134). The website of the International Study Group on Ethnomathematics (ISGE) gives a good overview of the ethnomathematical studies from the anthropological point of view. Studies of mathematical practices are listed by ethnicity/geography, e.g., African mathematics, Native American mathematics, Pacific Islander mathematics, African American mathematics, Asian mathematics, Math in European culture, Latino mathematics and Middle Eastern mathematics.

This study and description can also be coupled with a critical response to the superiority of Western mathematics, in an attempt to demonstrate the equivalent complexity of the ethnomathematics practices (Joseph, 1987, 1997; Bishop, 1995; D'Ambrosio, 2007a,b). The contrasts between the various mathematical practices are substantial but so are the similarities. Alan Bishop describes six mathematical competencies that every culture requires to be able to answer questions and respond to problems arising from the environment: counting, measuring, locating, designing, playing and explaining (Bishop, 1997, 1988, 1995). These six competencies are called 'mathematics' (with lower-case m) while 'Mathematics' (upper-case M) is used to refer to the Western or European version which is known world-wide. With this universal and intercultural interpretation of the six mathematical competencies Bishop emphasizes that a mathematical practice is a cultural product. In any culture mathematics is a symbolic technology that builds relationships between a person and his or her physical and social environment (Bishop, 1988, p. 147). Like D'Ambrosio, he points out that managing time and space, defining classifications and comparing are all human mathematical practices. To communicate and reflect on these universal practices, human beings develop codes and techniques that vary because of their development within a particular context (D'Ambrosio, 2007a, p. 30).

### 3.2.2  History

A second theme occupying ethnomathematicians is that of mathematical transmission and development. A critical stance is involved towards received views about the history of mathematics, by particularly examining the mechanisms of what is known as *Eurocentrism* about mathematics. While originally referring to a European attitude exclusively, Eurocentrism has gradually come to stand for *any* way of thinking that considers itself superior, thus justifying its own global distribution, thereby displaying an utter disregard for or even suppression of local practices elsewhere.[11] This self-sufficient way of thinking is also displayed in the way mathematical heritage is passed down. Typically, the history of mathematics indeed is and has been described as a *Western* affair, in which Arab, Chinese or Indian contributions figure as distant and exotic 'influences' at most—if at all—and certainly not as independent, let alone alternative developments.

Even prominent contemporary historians of mathematics, like Rouse Ball (early twentieth century) or Kline (middle twentieth century) have contributed to wiping out our collective memories of anything preceding and having inspired the 'Greek miracle', as Joseph laments at the outset of his book *The crest of the peacock* (1992). Chief among the intellectual debts one finds the Egyptian and Mesopotamian civilizations, acknowledged by the Greeks themselves, various clay tablets and papyri confirm. Further, the neglect of the vital Spanish-Arab contributions during the Middle Ages has to be counted with, viz., keeping record of the dormant Greek tradition before it was finally picked up again in the Renaissance. Moreover, nothing yet has been said about the (influence of the) rich Chinese and Indian cultures. Restivo (1992, Chapters 3–6) includes nice socio-historical readings on the mathematical traditions of China (relying on the unsurpassed Needham (1959)), the Arabic-Islamic world, India, and Japan. Clearly, something more complicated than the 'simple' diffusion model, with Europe at its centre, is needed. This is what Joseph (1992) sets out to provide; the success of this alternative will however not concern us here. Fortunately, in the meantime, an increasing number of historians have been going through great pains to show that the said influences were not so marginal after all, and instead profoundly shaped mathematics as we know it (or consitute viable alternatives to it).[12]

A fortiori, the story of mathematical practices that obviously did *not* contribute to the development of Western mathematics, gets little to no

---

[11] Joseph (1987, 1997) uses the notion 'Europe' for all regions that are being dominated by a population of European origin, such as the United States, Canada, Australia, New Zealand etc. Powell and Frankenstein (1997) use the phrase 'Europe and Europeanized areas' in this respect, which is clearly is no longer restricted to the European continent.

[12] Cf., e.g., Grattan-Guinness (1994, part one).

mention in the classical history books. This is for example the case for ancient cultures like the Inca, or those from Sub-Sahara Africa. Whenever they are referred to, immediately the label 'ethnomathematics' is attached to it, as an immunizing move as it were. With respect to the historiography of mathematics the obvious alternative would be to acknowledge that more than one history of mathematics can be made up. A parallel, less geographically dominated dimension of the prevailing history of mathematics, has been the pushing aside of the 'gender' dimension. It has been ages since one thought that histories could be described that in principle incorporate the story of every human being. This neglect of *herstory* next to history might very well connect to the 'great men'-syndrome that has been so typical for much historiography of science.[13]

Summarizing, in general, mathematics has been historically pictured as an almost exclusive product of European society, at the cradle of which stood ancient Greece. Although it should be added that there is nowadays, among historians, a growing tendency of subtly modifying this harsh picture, next to that, there is also the issue of past and present mathematical systems, or rather collections of ideas, "that did not feed into or effect this main mathematical stream" (Ascher, 2002, p. 1). The reason is they were developed by peripheral and minor indigenous people. In paying attention to voices unheard in the Western dominated mathematical debate, the anthropological branch of ethnomathematics "has the goal of broadening the history of mathematics to one that has a multicultural, global perspective" (Ascher, 1998, p. 188), so as to include the unknown or misunderstood.

### 3.2.3  Philosophy

Drawing heavily upon "the theories, knowledge, and methods of culture history, cognitive studies and, above all, linguistics and anthropology" (Ascher and Ascher, 1994, p. 1545), ethnomathematics research may very well be subsumed under social studies of mathematical practices; practices which, importantly, are *not* all of a kind. In studies of science in general, the introduction of a cultural category has made it possible to more firmly establish the epistemic connection between (external) context and (internal) content. Important consequences, in history as well as in philosophy, have included a partial drive away from the traditional focus on individuals (the 'great men' syndrome), and the acknowledgement, as Steve Woolgar put it, "that reason, logic and rules are *post hoc* rationalizations of scientific and mathematical practices, not their determining force" (Woolgar, 1993, p. 50). What could then be the consequences of this type of research for the philosophy of mathematics? Could it come to challenge the dominant views that mathematical truth is immutable, monolithic, universal, timeless?

---

[13]Cf. also Koblitz (1996).

In his plenary lecture as an invited speaker at the 1950 International Congress of Mathematicians in Cambridge Mass., Raymond Wilder observed that his fellows had been invariably led, at least to a large extent, by what was considered or supposed to be proper mathematics in their own culture, and that particularly in the history of mathematics (of which he was an amateur himself) this limitation ought no longer to be overlooked (Wilder, 1998). Below, in §4.2, we shall show how Wilder himself, near the end of his life, actually proposed a philosophical endorsement of this view.

It can be no surprise that the feeling expressed by Wilder is widely shared by (mathematical) anthropologists, those in the business of studying other cultures, usually from within a relativist framework. According to them, alternatives to prevailing Eurocentric historiography should not just pay *more* attention to other traditions, past and present, but also, importantly, be *different* in character: maximally unbiased and detached, i.e., conducted in a 'true' anthropological spirit, taking 'the other' for granted, rather than (unconsciously) moulding aspects of it into fictitious categories that stand in outright opposition to our 'regular' or 'normal' Western ones (for example, 'the exotic').[14] What will be of interest at the philosophical level is not so much these empirical studies as such, but the way they affect our image of (the nature of) mathematics, Western and non-Western alike. Below, in §4.1, we briefly look into an approach called ethnomethodology, an application of a similar 'unbiased' perspective to the mathematics of the Western tribe itself, as exemplified in the work of Eric Livingston. After that, in §4.2, we turn to Wilder's own epistemological interpretation.

### 3.2.4   Education

A fourth and final field suited for ethnomathematical approaches is education. Here, all previous subdisciplines converge. For 'peripheral' countries, the critical historical and philosophical tradition with regard to superior Western mathematics accommodates a favouring of local mathematical practices in the curriculum, rather than (implicitly) importing the Western one. To that purpose, a description of everyday mathematical practices is needed, a task which anthropologists can take to heart. In response to the challenge of incorporating living mathematics into curricula and school practices, D'Ambrosio has formulated a programme of 'learning in action' (cf. Figure 1), in which the teacher's part is that of a process manager.

In view of the set of instruments and the social context, with content being delivered from a variety of sources, obviously a highly interactive approach is required. Let us however point out that D'Ambrosio is still overlooking one particular and very obvious source: the pupil. The general

---

[14]Cf., e.g., Said (1979) for a renowned complaint against such (implicit) methods, as applied to 'the orient'.

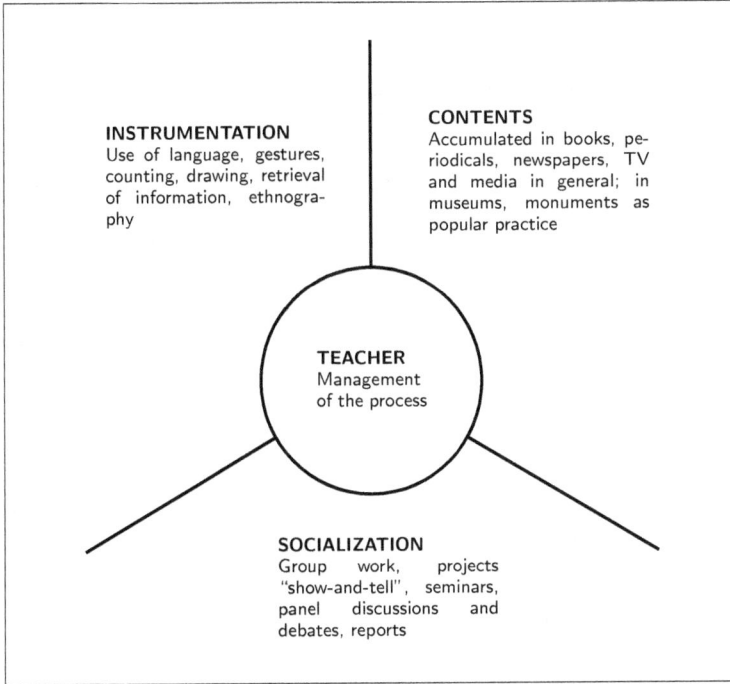

FIGURE 1. Interactive curriculum concept (D'Ambrosio, 1990, p. 376)

principle that the social environment of the person doing the learning must be taken into account provides a link between ethnomathematics and learning in diversity. In Section 5, we shall elaborate on the implementation of ethnomathematics in educational contexts.

## 4    Ethnomathematics and philosophy

### 4.1    Livingston

Among the more neutral definitions of ethnomathematics is one that refers to it as "the study of mathematical ideas and activities as embedded in their cultural context" (Gerdes, 2001, p. 12). As Paul Gerdes, the author of these words, himself indicates, this means that 'academic' mathematics is, in principle, also liable to the ethnomathematical approach. With his 1986 book *The ethnomethodological foundations of mathematics*, this is precisely what Eric Livingston has aspired to do: to 'descriptively analyse' what practices the "tribe of mathematical theorem-provers" are engaged in.[15] Livingston's

---

[15](Livingston, 1986). The source of the cited designation is (Livingston, 1999, p. 885).

central thesis is that of the local production of social order, in mathematics as in other scientific and non-scientific domains.[16] Social order in mathematics amounts to conveying and spreading a belief in the indubitable, 'transcendental' correctness of proofs. It is Livingston's contention that this is effectuated through, or supported by, nothing other than concrete displays of individual proofs (e.g., at the blackboard), despite –or rather thanks to– local contingencies. This approach has been criticized, by David Bloor among others, for being unable to explain how the processes described succeed in illuminating the nature of mathematical objects. Bloor contends that Livingston gets stuck in circularity between contingent practices and objective structures, as ethnomethodology "actually uses the very concepts whose questionable significance provided us with the philosophical problem with which we began. The whole problem was to illuminate what goes on when we 'realize' something in the course of a mathematical proof" (Bloor, 1987, p. 349). This way, says Bloor, Livingston does not in the least surpass the already sketchy accounts of both Lakatos and Wittgenstein, to which he owes substantial—though unacknowledged—debt. From Lakatos he takes, e.g., the idea that definitions are not laid down prior to the proof process, but emerge from that actual mathematical activity itself, or, more generally, that the phase of informal discovery intermingles with that of formal justification. With Wittgenstein, it shares the anti-foundational rationale for the primacy of concrete, low-level mathematical practice.[17]

It is fair to say that Livingston's ethnomethodology of mathematical proof remains theoretically very superfluous, resulting in a number of loose ends that are not attended to. From the circularity pointed out above flows a possibly unintended yet inescapable and strong sense of implicit Platonism, which does not square with its anthropological intentions.[18] There is another quasi-paradoxical point, in that the approach remains extremely

---

[16]Cf. the definition of Michael Lynch, where the approach to science in general is set out: "Ethnomethodology is commonly said to be the study of 'micro' social phenomena - the range of 'small' face-to-face interactions taking place on street corners and in families, shops, and offices" (Lynch, 1997, p. xii).

[17]Cf. "I go through the proof and then accept its result. – I mean: this is simply what we *do*. This is use and custom among us, or a fact of our natural history" (Wittgenstein, 1956, p. 61, i.e., RFM I–63).

[18]Cf. "Provers see through the representations of mathematical objects to the objects that are represented, and use those representations to inspect and discover properties of mathematical objects" (Livingston, 1999, p. 873). It should be added that in the paper just quoted from, it is additionally commented that ontological impartiality is implied here. However, something more seems to be needed in order to put into perspective persistent talk about mathematical objects and their representations. Although the bringing up of pebbles and diagrams seems to suggest that something like a (minimal) empiricist foundation of mathematical knowledge might be thought of here, these are not explicitly awarded more than illustrative power, and no (psychological) account is given of the passing from concrete to abstract practices (contrary to, e.g., Kitcher, 1983).

internalistic, with no attention paid to any of the small- or large-scale cultural surroundings (e.g., intellectual tradition) of the proofs 'lived through'. In sum, Livingston presents a genealogy of mathematical rigour that remains limited to a purely descriptive analysis of narrowly local practices. He takes us through specific proofs, e.g., by Euclid or Gödel, to show how they are supposed to convince (make certainty emerge). But the question of what actually constitutes their truth or falsity is *de facto* considered as totally irrelevant, and no normative criteria are proposed or (internal or external) explanatory forces identified. Therefore, let us now turn to a theoretically more elaborated anthropological account of mathematics, as Wilder attempted to give.

## 4.2   Wilder

When considering the body of original sources on 'humanistic' or 'non-foundational' approaches to mathematics (say, from the postwar era through the early 1980s), what is most striking is that nearly all of it appears to be by trained mathematicians, not philosophers. Raymond Wilder is exemplary of this tendency. As announced above, addressing the International Congress of Mathematicians as early as 1950, he urged his colleagues "to get outside mathematics, as it were, in the hope of attaining a new perspective, [...] [a] vantage point from which one can view such matters more dispassionately. [...] We 'civilized' people rarely think of how much we are dominated by our cultures—we take so much of our behavior as 'natural'" (Wilder, 1998, p. 186–187). One can hardly think of any philosopher of mathematics who would have (dared) put forward this type of externalist claim at the time. A full three decades later, in 1981, Wilder conceived the monograph *Mathematics as a cultural system*, in which he set out to systematically do what was only sketchily suggested in his earlier lecture: apply the anthropological approach to mathematics as a (Western) field of scientific inquiry. The following quote may function as a declaration of his programme.

> With the achievement of cultural status, such internal forces as hereditary stress, consolidation, selection, symbolization and abstraction have played an increasingly important part in the evolution of mathematical thought. To ignore these forces and the manner in which they have influenced characteristic patterns of evolution seems to deprive both the historian and the students who follow him of a fuller understanding of the historical process. Of course, mathematics is done by individuals, but these individuals share a common, albeit variable and diverse, mathematical culture, and along with the study of the achievements of a Gauss or Riemann, one should study the culture, both mathematical and environmental, in which they lived and worked, in order to achieve a fuller understanding and appreciation of what they accomplish. (Wilder, 1981, p. 162).

At the source of Wilder's work in the cultural anthropology of mathematics stands a general neglect—if not disdain—among mathematicians and philosophers alike for contemporary advances in the social sciences. Contrary to popular belief, the author puts forward the moderately naturalist thesis that mathematicians will greatly benefit, in terms of the deliberate choices they make between different problems and strategies, from considering and appreciating the cultural determinants of their work.[19] Wilder's approach is in fact adapted from that of Leslie White (1947), taking a cultural system as resulting from a dynamical interplay of 'vectorial' forces. In the case of mathematics, these forces are specified by Wilder as different types of 'stress' imposed on mathematical practices (see below). These influences vary, and the historical sequence of their different configurations constitutes mathematical evolution (indeed a largely neglected topic). Actually, this brings out specific affinities between Wilder's account and the operational part of Kitcher's approach, which we have elaborated on elsewhere (Van Bendegem and Van Kerkhove, 2004).

At a macro-level, a notable historical constant is that of the co-evolution (rise and decline) of mathematics and general culture (a factor also stressed by Marxists like Dirk Struik). For example, although the patterns that have followed the different specialties are very hard to discern, one may observe a general mathematical decline after the Greeks, and a subsequent resurrection, past the 'dark ages', from the seventeenth century onwards. This phenomenon Wilder calls *environmental stress*. "The environment suggested the invention of new concepts in mathematics, whose study resulted in mature techniques which were seized upon by environmental interests for the solution of their problems and advancement of their own theories" (Wilder, 1981, p. 55).[20] On top of this, note that developments have been regularly obeying the law of inertia. That is, even when in urgent need of accommodating present needs and worries, the time has to be ripe for launching new and possibly revolutionary ideas. For example, Saccheri failed to officially join the club of simultaneous developers of non-Euclidean geometry largely because of ideological reasons, since he was unable to free himself of the dominant notion of 'absolute' (physical or Platonic) truth. Similarly, Gauss felt way ahead of his time when it came to this topic, and for genuine fear of unfavourable public reaction left his work on it unpublished.

A form of cultural influence more limited in scope is *hereditary stress*. Individual mathematicians inherit a reigning culture of specific concepts and tools from their predecessors, and from within that context try to ac-

---

[19] A metamathematical example of such potential influence, suggested by Wilder himself, is that exerted by the development (also in other cultures!) of alternative logics, upon how to conceive of logicism.

[20] There is of course a strong link with the issue of applicability here.

complish set goals and contribute to the further development of their field and of mathematics generally. Wilder considers the following components connected with the mathematical domains scholars happen to be engaged in, and thus with the hereditary stress exerted on one freshly 'raised' in it: capacity, significance, challenge, status, and body of concepts (particularly its consistency). *Capacity*, or what remains to be done, while essentially indeterminate, is pretty easy for the trained mathematician to (roughly) assess. "If the capacity is large, he may either be attracted to do research in it, or direct others, such as younger colleagues or students, to consider helping to develop the field" (Wilder, 1981, p. 69). A domain with ample capacity is thus likely to attract talented people. And this, in its turn, might be a good way, although no guarantee, to reinforcing its further blossoming. A good example hereof is twentieth century Cantorian set theory, a counter-example the slow or even non-start of projective geometry in the seventeenth century. The element of *significance* is closely tied to that of capacity. Take computability theory, particularly in the light of the ever growing importance of the computer, for mathematics as for society in general. Exemplifying the move in the reverse direction is Euclidean geometry, or even geometry in general, having been increasingly marginalized during the modern process of rigorization and formalization. Also notice that these shifts are not necessarily 'internal' to mathematics, as is illustrated by the example of computer science, but also, e.g., that of additional fields connected with the military (such as cryptology or ballistics).

A third factor of hereditary stress, in its turn rather intimately connected with the first and second, is *challenge*, or the degree of difficulty of problems facing mathematicians. Assessing this dimension, one can refer to Hilbert-like commented lists of outstanding problems.[21] One of the most intriguing challenges recently brought to a good end was no doubt Andrew Wiles's feat, solving Fermat's Last Theorem. Famous challenging problems still available as we write this include Goldbach's Conjecture and the Riemann Hypothesis. A fourth element, again of a considerable cultural dependence, is that of the present *status* of the field in question. It is based on philosophical esteem (e.g., on whether the analytic or synthetic method is dominant),[22] but also on the services paid by mathematics to other fields in which it is applied, inside and outside science.

One of the regularities or patterns that marks the history of mathematics is that of 'multiples' or (quasi-)simultaneous discoveries, such as those of the calculus, non-Euclidean geometries or informational complex-

---

[21] A couple of recent such lists can be found in Smale (1998); Griffiths (2000); Devlin (2002).

[22] Cf. (Otte and Panza, 1997).

ity theory.[23] This phenomenon, Wilder conjectures, is the uncommon result of severe *conceptual stress* (the large fifth hereditary category) exerted on mathematicians to avoid or resolve theoretical tensions, whether alleged or resulting from truly paradoxical situations. Threats to the internal consistency of mathematics boosts development, it is claimed, because they pose a special challenge. It is especially paradoxes connected with the infinite that have served this purpose very well: incommensurables and Zeno paradoxes in the Greek period; the continuum, Cantorian set theory, and the Russell antinomy in modern times. In the face of these, given culturally accepted axiom systems and proof methods, as well as uniform distribution of mathematical capacities, it is not so surprising that (quasi-)identical theorems are derived around the same time by different scholars. In general, conceptual stress is defined as the urge to devise a satisfactory conceptual system with suitable and properly interpreted symbolism. More examples of conceptual stress include the conception of the zeroth digit, of the imaginary numbers, or of the completed infinity as a number (by Cantor). All of these instruments, while workable, nevertheless did force a whole process of conceptual clarification (as did the Leibnizian $\varepsilon$-$\delta$-symbolism for the calculus). But work on (non-paradoxical) specific problems might also require and thus induce conceptual innovation. E.g., solving general algebraic equations required group and Galois theory, and the foundations of analysis required set theory. An additional motive for conceptual innovation is the creation of order or unification. Examples include the case of Klein, whose "analysis of existing geometries led him to realization of the importance of transformation groups and geometrical invariants as their distinguishing features" (Wilder, 1981, p. 76), which supposedly brought order in the field of geometry, the introduction and gradual development of the group concept, and that of category theory.

In view of our topic of socio-history, it is interesting to eventually see Wilder himself propose, like Crowe, a number of developmental 'laws' for mathematics (cf. François and Van Bendegem, 2010, in this volume). These capture most of the dimensions covered and elaborated in the rest of his book in an explicit developmental pattern, inspired by and showing a substantial overlap with Crowe's laws, as there are: the role of hereditary, environmental and conceptual stress (and the interpretative flexibility in the face of the latter), the ubiquity of multiples, the decisive influence of status and beneficiary cultural climate, the limitations imposed by prevailing culture, the importance of tacit aspects or hidden assumptions (to be unearthed and made explicit), the cultural relativism and ultimately intu-

---

[23]A general sociological treatment of this phenomenon, at times—though in mathematics rarely—followed by harsh priority disputes, can be found in (Merton, 1973, §16 and §17). For a specific such analysis of the second case, cf. (Restivo, 1983, app. D).

itive basis of mathematical knowledge, diffusion between (sub)cultures, the driving force of conceptual fruitfulness (problem-solving capacity and aesthetic appeal), the absence of revolutions in the core of mathematics (cf. François and Van Bendegem, 2010, in this volume), the ongoing increase in rigour and abstraction, the occasional (and often erroneous) feeling that entire fields have been 'worked out'.

The whole of Wilder's 'anthropological' account might then be summarized thus: cultural evolution of mathematics follows its own laws, depending on various specific types of environmental (external) and—especially— hereditary (internal) stress, and as a process of change certainly has to be distinguished from the mere chronology of contingent historical facts.

## 5 Ethnomathematics and education

With D'Ambrosio as its intellectual father, the application of an ethnomathematical approach to education has Brazilian roots, but has since become common practice all over the world. In 1975, D'Ambrosio founded a masters programme in Teaching Sciences and Mathematics at the State University of Campinas (UNICAMP), to which, as we write, he is still affiliated as an emeritus. By doing this, and through taking part in the International Congresses on Mathematical Education (ICME), he established international contacts with other pioneers of mathematics education, such as Luis Santaló, Hans Freudenthal and Edward G. Begle. At the 1976 ICME-3 congress in Karlsruhe, Germany, participating in the panel discussing *Why teach mathematics?*, he was one of the first to assert that mathematics education is nested in a socio-cultural context. Since 1984, after the first explicit international articulation of his ideas regarding ethnomathematics during his plenary lecture "Socio-Cultural Bases for Mathematical Education" at the fifth ICME conference in Adelaide, Australia, D'Ambrosio has developed his ideas in full detail through numerous articles. In 1985, he was the co-founder of the International Study Group on Ethnomathematics (ISGEm).

Meanwhile ethnomathematics has developed into both an education practice and a separate discipline within academics. From the start, D'Ambrosio has emphasized that with the notion 'ethnomathematics' he aspired to cover a much larger area than in its traditionally ethnocentric or exotic interpretation, and that the approach should be applicable to any learning context. He holds on to this distinction through what he calls ethnic-mathematics as opposed to ethnomathematics (D'Ambrosio, 2007a, p. 30).

Even though ethnomathematics is a critical research program and practice regarding mathematics education, it should still be considered different from so-called Critical Mathematics Education (Vithal and Skovsmose, 1997, pp. 132–133). That is, although both research programs reside within the much larger Learning in Diversity research community, ethnomathematics originated in post-colonies that have opposed themselves to importing a Western curriculum, instead developing their own mathematical practices that could serve as an instructive basis for education; while Critical Mathematics Education originated from within Western high-tech societies criticizing the idea of linear progress, and defining a number of suppression types, for example those based on class and gender (Atweh et al., 2001; Burton, 2003).

Within ethnomathematical studies, one can distinguish two major, but interrelated points of attention. On the one hand, as an application of human and childrens' rights in general, the importance of mathematical education and literacy for all, that is, in all cultures and for each and every one within those cultures. This is the theme of §5.2 and §5.3. On the other hand, the diversity of mathematical practices across cultures, a diversity which, in order to challenge and supersede a monolithical and universalist image of mathematics, should be represented in the math curricula. This is the theme of §5.1. As said, both themes are connected. Culturally adjusted math education does no longer import Western curricula without further ado, but actively examines and uses local practices as an entry to the subject matter. As a result, both a better access to math education and a recognition of mathematical diversity are served.

## 5.1   Cultural diversity

As already explained above (§3.1), the meaning of the 'ethno' concept has been extended since the beginning of its application, now designating ethnical groups, national groups, racial groups, professional groups, groups with a philosophical or ideological basis, socio-cultural groups and groups that are based on gender or sexual identity respectively (Powell, 2002, p. 19). Although this list is in principle incomplete, and some distinctions are or have been relevant only in specific contexts, it is clear that throughout the history of the concept 'ethno', elements of diversity have always been aimed at. Accordingly, with respect to the field of mathematics, 'ethnomathematics' can be characterized as being engaged with cultural diversity within the whole of mathematics. Comparative cultural studies of mathematics, while describing various mathematical practices, not only reveal their diversity, but also emphasize the complexity of each system, by showing how these practices arise and are used within specific socio-cultural and historical contexts. Finally, this approach is translated to mathematics education, challenging the instructor to introduce a diversity of mathematical practices in the curriculum.

The latter application exceeds the mere introduction in class of the study of other cultures or—to put it more dynamically—new culture fields. Indeed, this was done at first, namely looking into practices related to mathematics (and adjacent sciences) as part of the enquiry of all kinds of exotic traditions: music from Brazil, card games played in Madagascar, the arithmetic system of the Incas or the Egyptians, basket or carpet weaving, the Mayan calendar, the production of dyes out of natural substances, tea drinking habits or terrace cultivation in China, water collection in the Kalahari desert, the construction of Indian arrows, house construction or sand drawing in Africa, etc. (Bazin et al., 2002). Notwithstanding the good intentions of these and similar projects, we would like to emphasize, with reference to (Powell and Frankenstein, 1997), that these initiatives, while originally intended to offer intercultural education, run an easy risk of turning into some kind of folklore. It should be clear that we ourselves are not advocating such a simplistic curricular use of other people's ethnomathematical knowledge. Quite to the contrary, many examples show that foreign mathematical practices may be used in class not as some kind of exotic diversion next to 'the real stuff', but as truly mathematical in their own right. For instance, interesting cases about the number concept or certain geometrical aspects have been elaborated (Zaslavsky, 1985; Katz, 1994). Now pupils not only learn about how people elsewhere count or have originated the decimal, vigesimal or duodecimal systems; they also get to know how the various mathematical practices were (and continue to be) answers to the circumstances in which they arose.

This is of more than local importance. For example, while most European languages use the decimal system, they still make lots of references to the vigesimal (e.g., "quatre-vingts" in French) and the duodecimal systems (e.g., "half a dozen of eggs", or the watch), systems which are closely related to the structure of the human body: fingers, toes, finger bones, ... In Sierra Leone's *Mena* language, twenty stands for a complete person, which derives from groups of twenty representing all fingers and toes. Counting all finger bones (excluding the thumb, which is the instrument used to count with) results in the figure of twelve. Taking the fingers of the other hand to register the number of times that we reach twelve with the one, the result will be 5 times 12 or 60. Another example pertains to the geometrical concepts of area, perimeter and their ratio. Based on the construction of houses all over the world, students try to work out the most economical form of building when materials are restricted (Zaslavsky, 1985, 1989; Pinxten, 1994). By comparing the ratio between area and perimeter of various surfaces (circle, square, rectangle, triangle, ...), students should conclude circular constructions are most economical.

In (Lengnink and Schlimm, 2010, in this volume), an additional illustration is provided, analyzing the common difficulties in learning arithmetic in the decimal place-value system, and discussing pedagogical strategies for overcoming the difficulties. By comparing the algorithms for operations in the decimal place-value system with the additive, Roman numeral system, they have been able to show that lots of the mistakes students make when learning the decimal place-value system would not occur within the additive system. This analysis of the semantic aspects of number representation is an appropriate example of how techniques other than the common mathematical operations may be successfully used in class.

Let us give one more example, viz. about ethnocomputing. Indeed, ethnomathematics has recently evolved so dramatically that even digital applications now exist. It was actually but a small step for Eglash et al. (2006) to develop a software package that places mathematical ideas and practices in a cultural context. The Culturally Situated Design Tools (CSDTs) software more particularly gives students the possibility of reproducing art of different cultures, ranging from Indian strings of beads or Afro-American hair braiding to metropolitan graffiti. This can be done in an independent and creative way, exploiting underlying mathematical principles. With their program, the designers explicitly want to counter multicultural mathematics, which they regard as mostly holding an exotic as well as a static vision on culture. Their contrasting ethnomathematical view is based on four principles. First, mathematical designs are by no means autonomous, but are linked to other cultural expressions such as cosmology, spirituality, medicine,. . . Second, the complexity of mathematical designs refutes their alleged primitiveness. Third, not only are the designs interpreted from a Western perspective, for example looking for symmetries, but they are also analyzed with an eye on discovering underlying models that reappear in other designs. Fourth, they herald a dynamical view of culture.

In the following section, we look into the philosophy of egalitarianism and emancipation that helped bring about the changes that allowed for the implementation of the classroom practices just discussed.

## 5.2   Egalitarianism

The ethical-theoretical foundations on which egalitarian projects within education are based, assume that equality is measured at the end of the educational process. For as pointed out in the justice theories of Rawls (1999) and Sen (1992), pupils' starting positions can be dissimilar in such a way that strictly equal deals will prove insufficient to achieve equality. This implies that a meritocratic approach—which measures equality at the start of the process—cannot fully guarantee equal chances. Egalitarian approaches, therefore, start from a certain pedagogic optimism, taking into account the diversity of those learning in order to give them the greatest chance of equality.

By extending the notion 'ethno' to include diversity, it has been claimed, the distinction between mathematics and ethnomathematics might be disappearing. Hence the question has been raised whether thereby the achievements of ethnomathematics will not also get lost. We think not, as the said distinction can and will only disappear once ethnomathematics' achievements have been acknowledged and implemented in mathematics education (Pinxten, 1994; Setati, 2002). Nevertheless, although ethnomathematics was originally mostly just critical of (the dominance of) Western mathematics, now the time seems to have come to also raise critical questions in and about the programme itself. What indeed exactly distinguishes ethnomathematics from mathematics, if all of mathematics is constituted by mathematical practices performed by cultural groups that identify themselves on the basis of philosophical and ideological perspectives (Setati, 2002, p. 31)?

To make this point more specific with respect to education: Every maths teacher is supposed to apply a series of standards that come with the profession. These standards are philosophical, ideological, and argumentative in nature, pertaining to ways of being, ways of perceiving and ways of expressing respectively. Hence both mathematics and ethnomathematics turn out to be embedded in a normative framework, which casts doubts over whether the values of mathematics and ethnomathematics are really that distinctive.

What does distinguish ethnomathematics as a programme, is its emancipatory and critical spirit, promoting the equality of groups hitherto discriminated against (Powell and Frankenstein, 1997). To use Gilligan's (1982) terms, ethnomathematics gives air to the *different voices* in mathematics education. Carrying this objective into education in general, institutions organizing education mostly do so from an emancipatory perspective. This very egalitarian idea of respecting diversity is also to be found in UNESCO's educational mission. This connects mathematics education closely with wealth, endurance and peace, to converge in a general human rights discourse. Indeed, UNESCO firmly believes that education is the key to sustainable social and economic development, as the following quote from the UNESCO mission statement shows:

> The mission of the UNESCO Education Sector is to: Provide international leadership for creating learning societies with educational opportunities for all populations; Provide expertise and foster partnerships to strengthen national educational leadership and the capacity of countries to offer quality education for all.

Having established that general educational objectives clearly relevant to the issue of equal opportunities, which is central to the ethnomathematics programme, let us now turn to the philosophical question of the value-ladenness of mathematics education.

## 5.3   Values and equal opportunities

D'Ambrosio has situated mathematics education within a social, cultural and historical context, and also explicitly linked mathematics education with politics (D'Ambrosio, 1997). According to him, mathematics education is a lever for the development of the individual, national and global well-being (D'Ambrosio, 2007a,b). In other words, teaching mathematics has a political foundation. D'Ambrosio has advanced that mathematics education should hence be available to all; a proposition which has been registered in the OECD/PISA report, the basis of the PISA-2003 continuation enquiry.

> Mathematical literacy is an individual's capacity to identify and understand the role that mathematics plays in the world, to make well-founded judgements and to use and engage with mathematics in ways that meet the needs of that individual's life as a constructive, concerned and reflective citizen. (Watanabe and McGaw, 2004, p. 37)

This specification of mathematical literacy clearly implies that it has now explicitly been acknowledged as a basic right for every child, allowing him or her to participate in society in a constructive, relevant and thoughtful way. The theme recurs in the essays of Alan Bishop, demonstrating the link between mathematics, ethnomathematics, values and politics (cf., e.g., Bishop, 2002; Bishop et al., 2006). Because mathematics is traditionally perceived as being non-normative *par excellence*, the mere mentioning of mathematics education and education of values in the same breath sounds ambiguous. However Bishop comments:

> It is a widespread misunderstanding that mathematics is the most value-free of all school subjects, not just among teachers but also among parents, university mathematicians and employers. In reality, mathematics is just as much human and cultural knowledge as any other field of knowledge, teachers inevitably teach values... (Bishop, 2002, p. 228)

Be that as it may, mathematics education is a field in which a multicultural approach, dealing with diversity or cooperative learning is very hard to find. Indeed, mathematics does not appear to easily lend itself to integrating the so-called 'cross-curricular' learning objectives, such as consumer, health, environmental, safety or leisure education. Bishop (1997) observes that the curricula, on the contrary, have a particularly narrow, technical orientation, in which there is (almost) no room for historical considerations or philosophical exercise. We have managed to confirm this on the basis of our own research into the role of philosophy in the mathematics curriculum of secondary education in Flanders–Belgium (François and Van Bendegem, 2007).

Bishop was one of the first to investigate values in relation to mathematics and mathematics education (Bishop, 1997). To facilitate their study, he identified a cluster of six values, presumably used by mathematics and science teachers (Bishop et al., 2006). There are three groups of two values each. The first group touches on the ideological dimension: rationalism and objectivism. The second group connects with the attitudinal dimension: control (prediction, checks and rules) and progress (cumulative knowledge). The third group introduces the sociological dimension: openness (knowledge as a human activity and product) and mystery (the 'unreasonable' origins and uses of knowledge).

In line with Bishop, Paul Ernest also focuses on the analysis of values in the context of mathematics and mathematics education. He holds that philosophy of mathematics is currently at the heart of a Kuhnian revolution. As he has it, mathematics has been dominated by an absolutist paradigm for more than two thousand years: it has been considered to be a foolproof system consisting of objective truth, hence steering clear of values. In the past decades, however, the viability of studying mathematics just as any other branch of human knowledge has become increasingly obvious. In this spirit, where one does not exclude fallibility, change or cultural context, mathematics is conceived as a knowledge system that is the result of human investments. Ernest (1991) has analyzed different ideological groups and the values that they adhere to within mathematics education. As for the inherent values of mathematics, he has documented that 'abstract' is esteemed higher than 'concrete', 'formal' higher than 'informal', 'objective' higher than 'subjective', 'justification' higher than 'discovery', 'rationality' over 'intuition', 'reason' over 'emotion', 'general' over 'particular', 'theory' over 'practice', 'intellectual' over 'manual', etc. (Ernest, 1991, p. 259). Prediger (2006) has added some more values to this list, by showing that 'coherence' and 'consensus' are held in particularly great respect by the community of mathematicians, and that big efforts are made to remove instances of inconsistency or dissensus, whenever they occur.

It is particularly within ethnomathematics research that the link between mathematics education and values is further extended into the political domain. According to D'Ambrosio, too many people are still convinced that mathematics education and politics have nothing in common. In recent work (D'Ambrosio, 2007a,b), he brings to attention the Universal Declaration of Human Rights, articles 26 and 27 of which highlight the right to education and the right to share in scientific advancements and their benefits. These were further developed and confirmed in UNESCO's World Declaration on Education for All (1990), ratified by 155 countries. The principles have also been applied in the OECD/PISA declaration on mathematical literacy (2003). D'Ambrosio deeply regrets that these manifestos are not at all well-known by maths teachers, given that it is precisely they who play a key role in the emancipatory process at which the declarations aim.

To stress the actual importance of the matter, D'Ambrosio once more points out that mathematics is at the basis of many significant developments in technology, industry, the military, economy and politics, fields in which many problems faced by mankind are embedded. D'Ambrosio then argues that mathematics and its education are valuable tools to help solve these problems. At this point, he attributes a key role to the study of mathematical practices showing mathematics to be a creative force.

> I see my role as an educator and my discipline, mathematics, as complementary instruments to fulfill commitments to mankind. To make good use of these instruments, I must master them, but also need to have a critical view of their potentialities and of the risk involved in misusing them. This is my professional commitment. (D'Ambrosio, 2007a, p. 27)

According to D'Ambrosio, a critical way of teaching mathematics does not get enough chances. Like Bishop (1997) he criticizes predominantly formally-oriented curricula with their emphasis on technical drill, giving hardly any place to history, philosophy or reflection in general. D'Ambrosio has proposed three concepts to focus on in a new curriculum, if one wants to take the international (UNESCO) emancipatory objectives to heart: literacy, matheracy and technoracy. Literacy pertains to communicative skills in containing and using information. Both spoken and written language are concerned, but so are symbols, meanings, codes and numbers, so mathematical literacy is undoubtedly part of it. Matheracy denotes the necessary qualities of a scientific attitude: being capable of developing hypotheses, deducing and drawing conclusions from data. Technoracy refers to opportunities to familiarize oneself with technology, such that every pupil at least has the chance to get to know the basic principles, possibilities and risks of technological artifacts.

To conclude this section, let us summarize that the ethnomathematics programme is a value-driven view on mathematical practice and education. In sharp contrast to most Western mathematical research and educational traditions, in ethnomathematics practices are not decoupled from their cultural origins and contexts. More strongly than that, education is explicitly associated with social justice and even politics.

## 6   Conclusion

In this paper, we have aimed to give a general overview of how the field of ethnomathematics has developed from its origins in anthropological studies of mathematical practices outside the 'Western' tradition. It has been observed that, because of their implicit or explicit criticism of Eurocentrism, these studies have inspired a growing interest in the cultural embedding of mathematical practices, including 'Western' ones. Within the philosophy of mathematics, this has added to the array of studies, arising since the middle of the twentieth century, that collectively propose themselves as 'alternative' to foundational studies, by bringing back the nature of mathematical knowledge to where it seems to belong: in the practices of concrete, limited and fallible human beings.

With respect to mathematics education, a number of crucial evolutions should be noted as well. Less developed countries have increasingly ceased merely to import the 'superior' Western mathematical curricula, and instead have begun implementing local practices in education. In the developed world itself, the interest has moved (or widened) from exotism to cultural diversity: Ethnomathematics in school no longer means a five minute break from technical drill, but has become one of the instruments for better learning to deal with cross-cultural differences within or without the immediate environment. The philosophies of egalitarianism and emancipation have influenced these changes in education, to the point that they have also been written into chapters of global political programs (UNESCO, OECD, PISA) that strive to ensure every human's right to mathematical literacy.

## Bibliography

Ascher, M. (1998). *Ethnomathematics. A Multicultural View of Mathematical Ideas.* Hapman and Hall / CRC, New York NY.

Ascher, M. (2002). *Mathematics Elsewhere. An Exploration of Ideas Across Cultures.* Princeton University Press, Princeton NJ.

Ascher, M. and Ascher, R. (1994). Ethnomathematics. In Grattan-Guinness, I., editor, *Companion Encyclopedia of the History and Philosophy of the Mathematical Sciences*, pages 1545–1554. Routledge, London.

Atweh, B., Forgasz, H., and Nebres, B., editors (2001). *Sociocultural Research on Mathematics Education: An International Perspective.* Lawrence Erlbaum Associates Inc., Mahwah NJ.

Bazin, M., Tamez, M., and the Exploratorium Teacher Institute (2002). *Math and Science Across Cultures. Activities and Investigations from the Exploratorium.* The New Press, New York NY.

Bishop, A. J. (1988). The interactions of mathematics education with culture. *Cultural Dynamics*, 1(2):145–157.

Bishop, A. J. (1995). Western mathematics. The secret weapon of cultural imperialism. In Ashcroft, B., Griffiths, G., and Tiffin, H., editors, *The Post-Colonial Studies Reader*, pages 71–76. Routledge, London.

Bishop, A. J. (1997). *Mathematical Enculturation, A Cultural Perspective on Mathematics Education. Mathematics Education Library*, volume 6. Kluwer Academic Publishers, Dordrecht.

Bishop, A. J. (2002). Research policy and practice: The case of values. In Valero, P. and Skovsmose, O., editors, *Proceedings of the 3rd International Mathematics Education and Society Conference*, pages 227–233. Centre for Research in Learning Mathematics, Copenhagen.

Bishop, A. J., Clarke, B., Corrigan, D., and Gunstone, D. (2006). Values in mathematics and science education: Researchers' and teachers' views on the similarities and differences. *For the Learning of Mathematics*, 26(1):7–11.

Bloor, D. (1987). The living foundations of mathematics. *Social Studies of Science*, 17:337–358.

Boyle, D. (2000). *The Tyranny of Numbers. Why Counting Can't Make Us Happy*. HarperCollins, London.

Brummelen, G. V. and Kinyon, M., editors (2005). *Mathematics and the Historian's Craft*. Springer, New York NY.

Burton, L., editor (2003). *Which Way Social Justice in Mathematics Education?* Praeger Publishers, Santa Barbara CA.

Butterworth, B. (1999). *The Mathematical Brain*. Macmillan, London.

Carson, R. (1962). *Silent Spring*. Houghton Mifflin, Boston MA.

Cornwell, J. (2003). *Hitler's Scientists. Science, War and the Devil's Pact*. London, Viking.

D'Ambrosio, U. (1989). A research program and course in the history of mathematics: Ethnomathematics. *Historia Mathematica*, 16(3):285–287.

D'Ambrosio, U. (1990). The history of mathematics and ethnomathematics. how a native culture intervenes in the process of learning science. *Impact of Science on Society*, 40(4):369–377.

D'Ambrosio, U. (1997). Ethnomathematics and its address in the history and pedagogy of mathematics. In Powell, A. B. and Frankenstein, M., editors, *Ethnomathematics. Challenging Eurocentrism in Mathematics Education*, pages 13–24. State University of New York Press, Albany NY.

D'Ambrosio, U. (2007a). Peace, social justice and ethnomathematics. *The Montana Mathematics Enthusiast*, 1:25–34.

D'Ambrosio, U. (2007b). Political issues in mathematics education. *The Montana Mathematics Enthusiast*, 3:51–56.

Davis, P. J. and Hersh, R. (1981). *The Mathematical Experience*. Birkhauser, Boston MA.

Davis, P. J. and Hersh, R. (1990). *Descartes' Dream. The World According to Mathematics*. Penguin, London.

De Cruz, H., Hansjörg, N., and Schlimm, D. (2010). The cognitive basis of arithmetic. In Löwe, B. and Müller, T., editors, *PhiMSAMP. Philosophy of Mathematics: Sociological Aspects and Mathematical Practice*, volume 11 of *Texts in Philosophy*, pages 59–106, London. College Publications.

Devlin, K. (2002). *The Millennium Problems. The Seven Greatest Unsolved Mathematical Puzzles of Our Time*. Basic Books, New York NY.

Eglash, R., Bennett, A., O'Donnell, C., Jennings, S., and Cintorino, M. (2006). Culturally situated design tools: Ethnocomputing from the field site to classroom. *American Anthropologist*, 108(2):347–362.

Ehrlich, P. R. (1968). *The Population Bomb*. Ballantine Books, New York NY.

Ehrlich, P. R. (1969). Ecocatastrophe. *Ramparts*, 8:24–28.

Ernest, P. (1991). *The Philosophy of Mathematics Education. Studies in Mathematics Education*. Routledge-Falmer, London.

François, K. and Van Bendegem, J. P. (2007). The untouchable and frightening status of mathematics. In François, K. and Van Bendegem, J. P., editors, *Philosophical Dimensions in Mathematics Education*, pages 13–39. Springer, New York NY.

François, K. and Van Bendegem, J. P. (2010). Revolutions in mathematics. More than thirty years after Crowe's "Ten Laws". A new interpretation. In Löwe, B. and Müller, T., editors, *PhiMSAMP. Philosophy of Mathematics: Sociological Aspects and Mathematical Practice*, volume 11 of *Texts in Philosophy*, pages 107–120, London. College Publications.

Gerdes, P. (1988). On culture, geometrical thinking and mathematics education. In Bishop, A. J., editor, *Mathematics Education and Culture*, pages 137–162. Kluwer Academic Publishers, Dordrecht.

Gerdes, P. (1997). Survey of current work on ethnomathematics. In Powell, A. B. and Frankenstein, M., editors, *Ethnomathematics. Challenging Eurocentrism in Mathematics Education*, pages 331–371. State University of New York Press, Albany NY.

Gerdes, P. (2001). Ethnomathematics as a new research field, illustraties by studies of mathematical ideas in african history. In Saldaña, J. J., editor, *Science and Cultural Diversity. Filling a Gap in the History of Science*, pages 11–36. Cuadernos de Quipu, México.

Gilligan, C. (1982). *In a Different Voice. Psychological Theory and Women's Development*. Harvard University Press, Cambridge MA.

Grattan-Guinness, I., editor (1994). *Companion Encyclopedia of the History and Philosophy of the Mathematical Sciences*. Routledge, London.

Griffiths, P. A. (2000). Mathematics at the turn of the millennium. *American Mathematical Monthly*, 107(1):1–14.

Hersh, R. (1990). Mathematics and ethics. *The Mathematical Intelligencer*, 12(3):12–15.

Høyrup, J. (1994). *In Measure, Number, and Weight: Studies in Mathematics and Culture*. State University of New York Press, Albany NY.

Joseph, G. G. (1987). Foundations of Eurocentrism in mathematics. *Race and Class*, 28(3):13–28.

Joseph, G. G. (1992). *The Crest of the Peacock*. Penguin, London, 2nd edition.

Joseph, G. G. (1997). Foundations of Eurocentrism in mathematics (revised). In Powell, A. B. and Frankenstein, M., editors, *Ethnomathematics. Challenging Eurocentrism in Mathematics Education*, pages 61–81. State University of New York Press, Albany NY.

Katz, V. (1994). Ethnomathematics in the classroom. *For the Learning of Mathematics*, 14(2):26–30.

Kitcher, P. (1983). *The Nature of Mathematical Knowledge*. Oxford University Press, New York NY.

Kline, M. (1973). *Why Johnny Can't Add. The Failure of the New Math*. St Martin's Press, New York NY.

Kline, M. (1977). *Why the Professor Can't Teach. Mathematics and the Dilemma of University Education*. St Martin's Press, New York NY.

Kline, M. (1990). *Mathematics in Western Culture*. Penguin, London.

Koblitz, A. H. (1996). Mathematics and gender: Some cross-cultural observations. In Hanna, G., editor, *Towards Gender Equity in Mathematics Education*, pages 93–109. Kluwer, Dordrecht. Page numbers refer to the reprint in (Brummelen and Kinyon, 2005, pp. 329–345).

Lengnink, K. and Schlimm, D. (2010). Learning and understanding numeral systems: Semantic aspects of number representations from an educational perspective. In Löwe, B. and Müller, T., editors, *PhiMSAMP. Philosophy of Mathematics: Sociological Aspects and Mathematical Practice*, volume 11 of *Texts in Philosophy*, pages 235–264, London. College Publications.

Lindner, H. (1980). 'Deutsche' und 'gegentypische' Mathematik. Zur Begründung einer 'arteigenen' Mathematik im 'Dritten Reich' durch Ludwig Bieberbach. In Mehrtens, H. and Richter, S., editors, *Naturwissenschaft, Technik und NS-Ideologie*, pages 88–115. Suhrkamp, Frankfurt a.M.

Livingston, E. (1986). *The Ethnomethodological Foundations of Mathematics*. Routledge & Kegan Paul, London.

Livingston, E. (1999). Cultures of proving. *Social Studies of Science*, 29(6):867–888.

Lynch, M. (1997). *Scientific Practice and Ordinary Action. Ethnomethodology and Social Studies of Science*. Cambridge University Press, Cambridge.

Macrae, N. (1999). *John von Neumann. The Scientific Genius Who Pioneered the Modern Computer, Game Theory, Nuclear Deterrence, and Much More*. American Mathematical Association, Providence RI.

Meadows, D., Meadows, D., Randers, J., and Behrens III, W. (1972). *The Limits to Growth*. Potomac Associates, Washington DC.

Mehrtens, H. (1985). The social system of mathematics and national socialism. *Sociological Inquiry*, 57(2):159–182. Page numbers refer to a reprint in (Restivo et al., 1993, pp. 219–246).

Merton, R. K. (1973). *The Sociology of Science. Theoretical and Empirical Investigations*. The University of Chicago Press, Chicago IL.

Needham, J. (1959). *Science and Civilization in China. Vol. III: Mathematics and the Sciences of the Heavens and the Earth*. Cambridge University Press, Cambridge.

Otte, M. and Panza, M. (1997). *Analysis and Synthesis in Mathematics: History and Philosophy.* Kluwer Academic Publishers, Dordrecht.

Paulos, J. A. (1990). *Beyond Numeracy. Mathematical Illiteracy and Its Consequences.* Penguin, London.

Pinxten, R. (1991). Some remarks on Navajo geometry and Piagetian genetic theory. *Infancia y Aprendizaje*, 54:41–52.

Pinxten, R. (1994). Anthropology in the mathematics classroom? In Lerman, S., editor, *Cultural Perspectives on the Mathematics Classroom*, pages 85–97. Kluwer Academic Publishers, Dordrecht.

Powell, A. B. (2002). Ethnomathematics and the challenges of racism in mathematics education. In Valero, P. and Skovsmose, O., editors, *Proceedings of the 3rd International Mathematics Education and Society Conference*, pages 17–30. Centre for Research in Learning Mathematics, Copenhagen.

Powell, A. B. and Frankenstein, M. (1997). Ethnomathematical praxis in the curriculum. In Powell, A. B. and Frankenstein, M., editors, *Ethnomathematics, Challenging Eurocentrism in Mathematics Education*, pages 249–259. State University of New York Press, Albany NY.

Prediger, S. (2006). Mathematics: Cultural product or epistemic exception? In Löwe, B., Peckhaus, V., and Räsch, T., editors, *Foundations of the Formal Sciences IV. The History of the Concept of the Formal Sciences*, volume 3 of *Studies in Logic*, pages 271–232. College Publications, London.

Ravetz, J. (1971). *Scientific Knowledge and its Social Problems.* Oxford University Press, Oxford.

Rawls, J. (1999). *A Theory of Justice.* Harvard University Press, Cambridge MA. Revised Edition.

Resnikoff, H. L. and Wells, R. O. (1973). *Mathematics in Civilization.* Holt, Rinehart and Winston, New York NY.

Restivo, S. (1983). *The Social Relations of Physics, Mysticism, and Mathematics. Studies in Social Structure, Interests, and Ideas.* D. Reidel Publishing Company, Dordrecht.

Restivo, S. (1992). *Mathematics in Society and History. Sociological Inquiries.* Kluwer Academic Publishers, Dordrecht.

Restivo, S., Van Bendegem, J. P., and Fischer, R., editors (1993). *Math Worlds. New Directions in the Social Studies and Philosophy of Mathematics.* State University New York Press, Albany NY.

Said, E. W. (1979). *Orientalism. Western Conceptions of the Orient.* Penguin, London.

Sam, L. C. (1999). *Public Images of Mathematics.* PhD thesis, University of Exeter.

Sardar, Z. (2000). *Thomas Kuhn and the Science Wars.* Icon Books / Totem Books, Duxford.

Sen, A. K. (1992). *Inequality Reexamined.* Harvard University Press, Cambridge MA.

Setati, M. (2002). Is ethnomathematics = mathematics = antiracism? In Valero, P. and Skovsmose, O., editors, *Proceedings of the 3rd International Mathematics Education and Society Conference*, pages 31–33. Centre for Research in Learning Mathematics, Copenhagen.

Smale, S. (1998). Mathematical problems for the next century. *Mathematical Intelligencer*, 20(2):7–15.

Strathern, P. (2001). *Dr. Strangelove's Game.* Penguin, London.

Van Bendegem, J. P. (1996). The popularization of mathematics or the pop-music of the spheres. *Communication & Cognition*, 29(2):215–238.

Van Bendegem, J. P. and Van Kerkhove, B. (2004). The unreasonable richness of mathematics. *Journal of Cognition and Culture*, 4(3):525–549.

Vithal, R. and Skovsmose, O. (1997). The end of innocence: A critique of ethnomathematics. *Educational Studies in Mathematics*, 34(2):131–157.

Watanabe, R. and McGaw, B., editors (2004). *Learning for Tomorrow's World. First Results from PISA 2003.* Organisation for Economic Co-operation and Development, Paris.

White, L. A. (1947). The locus of mathematical reality: An anthropological footnote. *Philosophy of Science*, 17:189–203.

Wilder, R. L. (1981). *Mathematics as a Cultural System.* Pergamon, New York NY.

Wilder, R. L. (1998). The cultural basis of mathematics. In Tymoczko, T., editor, *New Directions in the Philosophy of Mathematics. An Anthology*, pages 185–199. Princeton University Press, Princeton NJ.

Wittgenstein, L. (1956). *Remarks on the Foundations of Mathematics. Bemerkungen über die Grundlagen der Mathematik*. Basil Blackwell, Oxford, 3rd edition. Edited by G. H. von Wright, R. Rhees and G. E. M. Anscombe; translated by G. E. M. Anscombe; page numbers refer to the 1978 edition.

Woolgar, S. (1993). *Science: The Very Idea*. Routledge, London.

Zaslavsky, C. (1973). *Africa Counts*. Prindle, Weber and Schmidt, Boston MA.

Zaslavsky, C. (1985). Bringing the world into the math class. *Curriculum Review*, 24(3):62–65.

Zaslavsky, C. (1989). People who live in round houses. *Arithmetic Teacher*, 37(1):18–21.

# Peer review and knowledge by testimony in mathematics

Christian Geist[1], Benedikt Löwe[1,2,3] and Bart Van Kerkhove[4,*]

[1] Institute for Logic, Language and Computation, Universiteit van Amsterdam, Postbus 94242, 1090 GE Amsterdam, The Netherlands

[2] Department Mathematik, Universität Hamburg, Bundesstrasse 55, 20146 Hamburg, Germany

[3] Mathematisches Institut, Rheinische Friedrich-Wilhelms-Universität Bonn, Endenicher Allee 60, 53115 Bonn, Germany

[4] Centre for Logic and Philosophy of Science, Vrije Universiteit Brussel, Pleinlaan 2, 1050 Brussel, Belgium

E-mail: cgeist@gmx.net; bloewe@science.uva.nl; bart.van.kerkhove@vub.ac.be

## 1  Introduction

Mathematics has been called an "epistemic exception" with a type of knowledge being categorically more secure than that of other sciences (Heintz, 2000; Prediger, 2006). At the other end of the epistemological spectrum, we have the whimsical "knowledge by testimony", disputed by some (cf. § 2) to be knowledge at all.

In this paper, we shall discuss two closely related question fields spanning the gap between these two epistemological extremes:

1. If mathematical knowledge is categorically different from other types of knowledge, and if the published mathematical research papers are a part of the written codification of this knowledge, then the level of certainty of claims made in the published literature should be higher than in other scientific disciplines. Is this true? And if so, how is this higher level of certainty achieved? A piece of mathematical text becomes part of the published literature by means of going through the process of peer review. Is mathematical peer review different from peer review in other disciplines?

2. Mathematicians refer to the published literature, sometimes without checking the proofs themselves. This is a form of knowledge by testimony; so how can the epistemic exception of mathematics survive if some of the proofs rely on pointers to the literature?

*The second and third author should like to thank the *Wissenschaftliches Netzwerk* PhiMSAMP funded by the *Deutsche Forschungsgemeinschaft* (MU 1816/5-1) for travel support.

Benedikt Löwe, Thomas Müller (*eds.*). *PhiMSAMP. Philosophy of Mathematics: Sociological Aspects and Mathematical Practice*. College Publications, London, 2010. Texts in Philosophy 11; pp. 155–178.

*Received by the editors*: 2 October 2009; 27 December 2009.
*Accepted for publication*: 11 January 2010.

A simple and naïve answer to both questions would be that the deductive nature of mathematics allows referees to check correctness of the proofs of published papers with absolute certainty, and thus the written codification of mathematical knowledge is certain knowledge, relieving us of any qualms about referring to it. However this is very far from the truth; in his opinion piece published in the *Notices of the American Mathematical Society*, Nathanson (2008) paints a dark picture of the mathematical refereeing process:

> Many (I think most) papers in most refereed journals are not refereed. There is a presumptive referee who looks at the paper, reads the introduction and the statement of the results, glances at the proofs, and, if everything seems okay, recommends publication. Some referees check proofs line-by-line, but many do not. When I read a journal article, I often find mistakes. Whether I can fix them is irrelevant. The literature is unreliable.

Given that mathematical correctness of a paper is so important for the decision of whether a paper should be published or not, it might come as a surprise that there have been no studies of the mathematical refereeing process. In other fields, in particular in the medical sciences, the refereeing process is heavily scrutinized; research on the effect of peer review on the quality of papers (Goodman et al., 1994; Pierie et al., 1996; Roberts et al., 1994), on indicators for good referees (Evans et al., 1993; Black et al., 1998; Callaham et al., 1998; Nylenna et al., 1994), on referee bias (Link, 1998), on instruments that help to improve the quality of the refereeing process (Das Sinha et al., 1999; Garfunkel et al., 1990; Feurer et al., 1994), and on the question of blinding author identities and referee identities in the process (Walsh et al., 2000; Justice et al., 1998; Cleary and Alexander, 1988; Katz et al., 2002; McNutt et al., 1990; Fisher et al., 1994) abound in the medical and biological literature. In these fields, we find much more explicit rules for what is expected of the refereeing process and the individual referees than in mathematics (cf. Footnote 3).

An important *caveat* is in place here: measuring the quality of the refereeing process requires definitions of quality criteria for papers, referee reports, and referees; e.g., if you want to know whether the refereeing process improves the quality of papers, you first need to give a gauge for this quality. Let us give one such example: in (Abby et al., 1994), the authors conclude that the peer review process is successful in monitoring quality as "rejected manuscripts often were not published in other indexed medical journals". Definitions like this are circular: instead of measuring properties of the rejected papers, they rather measure whether the process is homogeneous across journals, using the refereeing process of other journals as a gauge for the quality of the refereeing process at a given journal. These methodological issues are discussed in the meta-study (Jefferson et al., 2002); we shall not discuss them in depth in the given paper, but they will form the background of our discussions with empirical data in the later sections.

In this paper, we shall give a description of the mathematical refereeing process and its role in ascertaining that only correct results are published. In § 2, we give a brief overview of the discussion of testimony in epistemology before moving on to discussing the uses of trust and reference without checking details of proof in mathematical research in § 3.

In § 4, we give a schematic description of the mathematical refereeing process based on personal experience and a number of text sources. The description in this section is not based on any empirical research, but collects anecdotal data. For the next two sections, §§ 5 and 6, we then move to empirical data: in § 5, we give the results of a questionnaire that we sent to editors of mathematical journals with questions about the refereeing process; in § 6, we compare the results of a study on referee agreement in the neurosciences and information sciences to similar results for more mathematical conferences (actually, most of our examples are from theoretical computer science). Our empirical results can only be a very first step for understanding the mathematical refereeing process. The results of § 6 seem to indicate that there is in fact a higher degree of referee agreement in mathematical fields than in others. Further approaches and future work are discussed in our concluding § 7.

## 2   Knowledge by testimony

In daily life as in science, we heavily depend on reports by others. Inescapable as this may be, it nevertheless opens a deep epistemological problem. In principle, holding beliefs on the mere assertion by someone else is a precarious affair; by choosing to rely on the authority of other agents, we put ourselves at their epistemic mercy. Taking into account its ubiquity, it might surprise that testimony has only recently become a topic of major philosophical scrutiny. As a consequence, debates are still wide open, e.g., in the existing literature, there is no consensus over what exactly is to count as a proper instance of testimony and what not. Further, there are a number of complicating factors hampering a good assessment, e.g., what is the nature of the relationship between speaker and hearer, possibly illuminating or obscuring the latter's judgement. For an overview of the current debate, cf. (Adler, 2008).

Even when limiting ourselves to cases "of simple informational exchange over easily known matters, where there is little or no motivation to deceive" (Adler, 2008, § 1), serious epistemological questions remain. As the essence of the matter before us is that checking propositions for oneself is impossible, other elements of the testimonial setting have to supply us with evidence for the trustworthiness of what is asserted.

A *default rule* for dealing with testimony is to accept assertions unless one has a special reason not to:

> Otherwise put, as long as there is no available evidence *against* accepting a speaker's report, the hearer has no positive epistemic work to do in order to justifiedly accept the testimony in question. (Lackey and Sosa, 2006, p. 4)

Limiting ourselves to the simple conversational exchanges mentioned above, in 'normal conditions', mostly no such defeaters apply. On the basis of past experience, we thus accept claims of others in these cases. This results in a large degree of uniformity in (justly) relying on testimony for situations of daily life. But also in science, having left behind 'gentlemen's culture', with peer review and reproducibility demands now seemingly constituting a system of organized skepticism rather than trust, things are at most different at the surface.[1] In this paper, we are illustrating this point with a particular view towards mathematics.

There are two major philosophical positions with respect to knowledge by testimony, the *anti-reductionist* or *non-inferentialist* stance and the *reductionist* or *inferentialist* stance. The anti-reductionists treat testimony as a fundamental source of knowledge, requiring no further justification beyond its apparent success. In this, testimony would be akin to, e.g., perception or memory. Contemporary discussions trace back to Reid (1969) and Coady (1973). Reductionists on the contrary deny that testimonial knowledge can be basic in the sense just specified, as it epistemically depends upon —and thus should always be inductively derived from— other resources, most notably sense perception and memory. In other words, reductionists demand positive reasons, not just the lack of defeaters, for accepting testimonial reports: *nullius in verba*, as the motto of the Royal Society reads. The prototype of such a thinker is David Hume.

## 3   Mathematical research based on trust

How much of mathematical research practice is based on testimony? Traditionalists will claim that if indeed mathematical research has proceeded on the basis of testimony, this reliance was and is *in principle* removable, i.e., all mathematicians can go through any proof in question and do it for themselves.[2] We know a substantial number of mathematicians who want

---

[1] Cf. (Lipton, 1998, p. 1): "Science is no refuge from the ubiquity of testimony. At least most of the theories that a scientist accepts, she accepts because of what others say".

[2] As an illustrative anecdote, let us report that an American mathematical logician teaches his graduate students to read mathematical papers as follows: read the statement

to understand all proofs that form a part of their papers and who will re-prove even classical statements to be completely sure of their own results based on them; but we also know that many mathematicians are not as meticulous and accept results from the published literature as black boxes in their own research. Many mathematicians tend to trust the experts and (in Auslander's words) "[t]his is the case even if we haven't read the proof, or more frequently when we don't have the background to follow the proof." (Auslander, 2008, p. 64)

The fact that written mathematical proofs are not complete formal derivations is acknowledged by many authors dealing with the epistemology of mathematics. Fallis (2003) discusses gaps in mathematical proofs; some of them are *enthymematic gaps* where the author has checked all details and omits them from the published paper for reasons of style or brevity (cf. (Heintz, 2000, p. 170), where the author compares the original set of notes written by Hirzebruch with the much terser final publication); others are what Fallis calls an *untraversed gap*:

> A mathematician has left an untraversed gap whenever he has not tried to verify directly that each proposition in the sequence of propo-sitions that he has in mind (as being a proof) follows from previous propositions in the sequence by a basic mathematical inference. (Fal-lis, 2003, pp. 56–57)

Fallis notes that "there are [...] cases where it is considered acceptable for a mathematician to leave an untraversed gap" (Fallis, 2003, p. 58). In general, research mathematicians agree with Fallis's observation. Referring to Almgren's proof that establishes the regularity of minimizing rectifiable currents up to codimension two, Hales writes:

> The preprint is 1728 pages long. Each line is a chore. He spent over a decade writing it in the 1970s and early 1980s. It was not published until 2000. Yet the theorem is fundamental. [...] How am I to develop enough confidence in the proof that I am willing to cite it in my own research? Do the stellar reputations of the author and editors suffice, or should I try to understand the details of the proof? I would consider myself very fortunate if I could work through the proof in a year. (Hales, 2008, pp. 1370–1371)

Nathanson (in the cited opinion piece in the *Notices of the American Mathematical Society*) is more critical of this described practice:

> Many great and important theorems don't actually have proofs. They have sketches of proofs, outlines of arguments, hints and intuitions

of the theorem, cover its proof with a sheet of paper, and then try and prove the theorem yourself.

that were obvious to the author (at least, at the time of writing) and that, hopefully, are understood and believed by some part of the mathematical community. But the community itself is tiny. In most fields of mathematics there are few experts. [...] In every field, there are 'bosses' who proclaim the correctness or incorrectness of a new result, and its importance or unimportance. Sometimes they disagree, like gang leaders fighting over turf. In any case, there is a web of semi-proved theorems throughout mathematics. (Nathanson, 2008)

## 4   The mathematical refereeing process

We hope to have convinced the reader in §3 that there are at least some relevant instances of references to testimony in mathematical research: even if some mathematicians check meticulously whether all of the theorems that their results depend on are correct, not all do it, and so, if a research mathematician uses a theorem from the literature, the correctness of the result depends not only on the accuracy of the refereeing process of the paper he or she uses, but also on the refereeing processes of the papers used by that paper, and so on.

So, if mathematicians are relying on these iterated refereeing processes, how much security does the refereeing process in mathematics actually generate? In this section, we shall describe the mathematical refereeing process. There is hardly any (systematic) discussion about this topic[3] and therefore, our description here is largely based on the guidelines given in the excellent books by Steven Krantz (1997, 2005) on mathematical writing and publishing and the personal experience of the second author of the present paper as an author, referee, journal editor, and book editor of mathematical papers.

In mathematics, papers are mostly published in journals; conferences and their proceedings volumes play a subordinate role. In the case of journal

---

[3]Cf. (Auslander, 2008, p. 65): "The issue of the refereeing process —real and ideal— in mathematics is fascinating and largely unexplored. Gossip on this topic abounds but I know of no systematic study."

We should mention that there are meta-discussions about the refereeing process as part of the discussions about major changes of the mathematical publishing process: some mathematicians do not like the role of commercial publishers in the publishing process, and would like to replace the current process with a web-based alternative, leading to changes also to the refereeing process; cf. (Birman, 2000; Jackson, 2002, 2003; Borwein et al., 2008). However, these discussions rarely touch the epistemological issues relevant here, and so we shall not discuss them further.

We should also like to mention the highly interesting case study (Weintraub and Gayer, 2001) in which the authors analyse the refereeing process of a particular result from mathematical economics with access to the reports and the paper (Thompson, 1983) in which an author displays the history of the rejection of one of his papers, raising "questions about the role of the referee in the professional development of a mathematician" (Thompson, 1983, p. 661).

publishing, the journal editor typically asks a single referee to write a report on a given submission; these referees are experts and have often published on material very closely related to the material in the submission.[4] Often, the referee is personally known to the editor, allowing the editor to read between the lines of the report. It is not standard practice to give referees many instructions apart from a deadline:[5] it is understood that referees know what is expected of them. Somewhat in contrast to our empirical findings of § 5, it is accepted by authors that the refereeing process takes more than six months. Many authors consider it inappropriate to remind an editor before six months after submission have passed, and some even do not ask about the status of their papers before a year has passed.

Ideally, referee reports "should address Littlewood's three precepts: (1) Is it new? (2) Is it correct? (3) Is it surprising?" (Krantz, 1997, p. 125). The level of detail of referee reports varies a lot: many reports are very short (less than one page of text), but some can be very long, sometimes longer than the submission itself. Typically, even for longer reports, the core of the report (the statement of the recommendation and the argument for this recommendation) is rather short, and the bulk of the report consists of detailed comments to be considered for revisions. Reports recommending rejection tend to be much shorter, sometimes only a few lines.[6] In general, it seems fair to say that the default decision for mathematical journals is *reject*: in order for an editor to accept a paper for a mathematical journal, the referee has to give arguments supporting acceptance. Editors will only very rarely overrule a referee's recommendation to reject.[7]

Mathematicians disagree about the amount of detail checking that has to be done by the referees. While some (few) mathematicians think that checking the correctness of the proofs is the main task of the referee, others disagree with this and consider mathematical correctness the problem of the author rather than that of the referee.[8] Methodologically, this is an

---

[4] "There are several parameters to consider [to find a good referee]: (i) the referee should be an expert in the subject area, (ii) the referee should be dependable, (iii) the referee should not be prejudiced, (iv) the referee should be someone who can get the job done." (Krantz, 2005, p. 119)

[5] "[The] guidelines [...] may certainly suggest a time frame for the refereeing process. [...] It is not often that the instructions to the referee will give detailed advice on what points to address." (Krantz, 2005, p. 121)

[6] Krantz writes that "[a] typical referee's report is anywhere from one to five pages (or, in rare instances, even more)." (Krantz, 1997, p. 125).

[7] This may be a special situation in the mathematical review process. In his account of the peer review process, Gross reports that "editors generally assume that the rejection of a paper depends on a clear negative decision on the part of both referees; a split decision ordinarily favors the authors" (Gross, 1990, p. 134). However, in an endnote commenting on this statement, Gross says: "It is on this point that journals in the humanities deviate most; in the case of split decisions they are inclined to reject." (Gross, 1990, p. 215)

[8] Cf. (Auslander, 2008, p. 65): "[S]tandards of refereeing vary widely. Some papers [...]

important issue: the anonymity of the referee means that the reader of a paper does not know which type of refereeing treatment the paper has received before acceptance.

To conclude this section with illustrating statements of mathematicians, let us give two excerpts from an interview study performed by Eva Müller-Hill (2010). This study provided a qualitative extension of the quantitative work reported on in (Müller-Hill, 2009; Löwe et al., 2010) in which mathematicians were supposed to assess the knowledge of protagonists in a fictitious story about mathematical proofs. The interviews were not primarily concerned with the mathematical refereeing process; we give two excerpts (extensively rewritten and sometimes reworded in grammatical English from the original transcript of spoken English) that are relevant here and corroborate the positions of Auslander and Krantz:[9]

> Let's say a famous mathematician comes up with a paper, and I have to referee it. Then I am preoccupied with the fact that he is a very well known mathematician, and so that it probably will be ok. And then, you say "yes, this really seems plausible, but I'm not really sure if it's true" and you end up with the question "is this because I don't have enough knowledge?" And then there's time pressure and you have other things to do when they ask you to referee this 50 pages paper. Then you have a tendency of believing that it is correct, and you think "he's publishing it, not I, so it's his responsibility that it is correct".

The same interviewee had to comment on the story about a fictitious world-famous expert named Jones who proved a result, submitted it, published it after a refereeing process, only to find out a few years later that the main result is wrong:

> It depends on a lot of things. Firstly, Jones is a world famous expert, so this means she's teaching at a university like Harvard. Then the paper was sent to a mathematical journal of high reputation, so,

---

concern famous problems, and thus have received intense scrutiny. Other papers receive more routine treatment". Auslander also reports that "referees are generally told that it is not their job to determine whether a paper is correct — this is the responsibility of the author — although the referee should be reasonably convinced. The referee is typically asked to determine whether the paper is worthwhile" (Auslander, 2008, p. 65). This is reflected in Krantz's recommendation to the referee: "While you may not have checked every detail in the paper, you should at least be confident of your opinion as to the paper's correctness and importance" (Krantz, 1997, p. 125). Cf. also our survey results in § 5.2.

[9]We should like to thank Eva Müller-Hill for the permission to include these examples. We reworded the texts from the transcripts to give the quotations the flow necessary for written texts while preserving the content and style of the original utterances. The literal quotes will be contained in Müller-Hill's dissertation (2010).

say, *Acta Mathematica*; this tells us something about the size of the mathematical community involved. You cannot be a world famous expert on something that nobody else does. That it went to a good journal means that the journal thought of looking for good referees, so it was established more surely than that it would have been sent to the journal of a tiny mathematical society with very few members. This puts the scenario in a framework which makes it very likely that the result is correct. And of course it happens that things are not right.

## 5    A survey of mathematical journal editors

The description of the mathematical refereeing process given in §4 was based on personal experience and the descriptions in (Krantz, 1997, 2005; Auslander, 2008). In the spirit of *Empirical Philosophy of Mathematics*, we aim to corroborate this personal account with empirical data.

In March 2009, we selected 27 editors of mathematical journals: 9 editors each of (what we estimated to be) top, mid or lower level journals. During the month after that, we received 13 answers (4 from top journal editors, 4 from mid level journal editors and 5 from lower level journal editors). Our modest questionnaire aimed at getting opinions of what the refereeing process is about; the questionnaire can be found in Figure 1.

### 5.1    Question 1. Importance of referee's tasks.

Overall, the editors agreed that Littlewood's precepts (cf. p. 161) are important. On the scale from 1 to 5, novelty gets an average score of **4.7**, correctness a score of **4.5**, and interest a score of **4.3**. The importance of whether the paper is well-written only gets an average score of **3.5**. The general tendency is that editors of higher-ranked journals give higher scores to all of the categories than editors of lower-ranked journals. The biggest difference between top-ranked journals and lower-ranked journals was in the category "is it interesting". One editor sums up our averaged findings as follows:

> [T]o be published an article must be novel (but new, enlightening proofs of older central results are publishable in exceptional cases), correct and interesting. In the process of refereeing, we try to improve the writing.

### 5.2    Question 2. Checking of proofs.

We received eleven answers to this question: six editors thought that the referee should check all proofs in detail; five thought that the referee should check some proofs in detail. Option (c) was not selected by anyone. As a realistic side remark, one of the editors who checked (a) wrote "but to be

---

1. Rate the importance of the following tasks of a referee, from 1 (not at all) to 5 (most certainly):

    (a) checking the correctness of results

    (b) estimating the novelty of results

    (c) judging whether the paper is interesting

    (d) judging whether the paper is well-written

2. Pick one answer out of the following:

    (a) I think the referee should check all proofs in detail.

    (b) I think the referee should check some proofs in detail.

    (c) I think the referee should check none of the proofs in detail.

3. How many weeks approximately do you grant a referee to write a report for an average 20-page research paper?

4. How many hours approximately do you expect a referee to spend on checking the correctness of a paper's claims?

5. What percentage of referees approximately do a good job checking the correctness of a paper's claims?

---

FIGURE 1. The questionnaire sent to 27 editors of mathematical journals in March 2009.

reasonable, I am happy when I find a referee doing (b)." One of the editors who did not provide an answer described a slightly non-standard procedure:

> We actually work with several referees for a given article; first an overview referee checking novelty, interest, [and] correct references. [Then, after that] if we [...] feel the paper is interesting and new, a[nother] referee [who] checks [the proofs] for correctness.

### 5.3   Question 3. Overall time for refereeing.

The average amount of time that our editors give their referees is about 14 weeks. The period of 3 months seems to be a standard expected length of the refereeing process (six out of thirteen responses), but two, four, and six months also occurred as answers.

### 5.4   Question 4. Time spent on checking correctness.

The answers to this question varied widely (from 5 to 80 hours), and a large number of the respondents refused to answer it on grounds that it depends too much on the individual paper.

## 5.5   Question 5. Quality of referees.

The overall average to the question how many referees do a good job checking the correctness is **52.3%**. There is a marked difference along the lines
of the ranking of the journals: among the editors of top-level journals, the
estimate was 61.3%, among the editors of mid-level journals, the estimate
was 42.5%, and the answers in the lower-ranked journals differed too much
to give a meaningful average.[10]

## 5.6   Some additional results.

In addition to the answers to the five questions, we received some comments
that confirm parts of the description from §4. One of the editors, when
asked about the quality of the job of the referees, answered

> After [many years], I know many colleagues convenient for refereeing
> papers. I am sending, if possible, papers to be refereed only to those
> who are responsible and I can believe they do a good job. Exception-
> ally, I must send a paper to a [different] person.

Also the following quotation from an editor reinforces our statements from
§3 that mathematics is largely built on trust:

> There are situations where almost nothing needs be checked (e.g., the
> results come from a seminar where the results were checked, or I see
> the paper is not too good and then it is useless to check details, or the
> author is well-known and it is his concern to submit a correct paper).
> There are situations when I insist to check all the procedures (e.g.,
> when it concerns good results from a less known author).

It is interesting to see the criteria according to which this editor decides that
"almost nothing needs to be checked": it is the reputation of the author that
drives decisions about how much detail has to be checked.

# 6   Quantitative data for conference refereeing in theoretical computer science

In §1, we mentioned that there is a large body of research on the refereeing
process in the natural and medical sciences, for instance on the question
of reliability. One indicator for reliability or objectivity of the process is
whether referees agree in their judgments of refereed papers. This was
investigated, e.g., by Rothwell and Martyn (2000) who considered the question whether the agreement of referees is greater than it would be by mere
chance on the basis of data from journals in the neurosciences, and a similar

---

[10]It is interesting to compare this to the quote from Müller-Hill's interview study
(already mentioned on p. 163): "That it went to a good journal means that the journal
thought of looking for good referees, so it was established more surely than that it would
have been sent to the journal of a tiny mathematical society with very few members."

study by Wood, Roberts and Howell (2004) in the information sciences. In this section, we aim at providing the same analysis for the mathematical refereeing process.

In our description of the mathematical refereeing process in § 4 (cf. also § 5.6), we stressed that the mathematical peer review is largely a communication between an editor and one referee based on trust due to a close personal relationship. The question of referee agreement does not make sense in this situation. Therefore, we had to leave the immediate area of journal refereeing in mathematics and move to the cognate area of conference refereeing in theoretical computer science.

Theoretical computer science, as mathematics, is largely based on the deductive method, and its main results are mathematical theorems. Therefore, the epistemic character of results in theoretical computer science is comparable to that of results in mathematics. On the other hand, computer science has a rather different publication culture from mathematics. While journal publications in theoretical computer science follow the mathematical refereeing process described in § 4, computer science developed a distinctive culture of refereed conference publications. The highest-ranking conferences clearly outrank some of the good journals of the field in terms of reputation. Due to the time pressure of the production schedule, the refereeing process here is markedly different from the mathematical refereeing process.

The following description is not meant to be an empirically verified description of the refereeing process in theoretical computer science, but a generalized personal account of the second author, based on his experience in programme committees for conferences in this field. Some of the claims (e.g., the ones on the tendencies about the length and depth of the referee reports) certainly will require an empirical and methodologically clean analysis in the future.

A computer science conference has a programme committee that is responsible for the selection of papers. Papers are submitted to a conference about half a year before the conference; after the submission deadline for the conference, the chair of the programme committee assigns papers to members of the programme committee, often after a phase of bidding during which the members can announce their preferences for being assigned certain papers. Typically, each paper is assigned to three or four members of the programme committee who then take a role similar to that of an editor in the mathematical peer review process. An assigned member of the programme committee can either decide to write a referee report herself or find a so-called *subreferee* who will write a referee report. Typically, the referee has three to four weeks to complete the report. Reports come with a numerical score on a scale fixed in advance by the chairs of the programme committee and tend to be shorter than reports in the mathematical journal refereeing process described in § 4: some can be as detailed as in the journal peer review, but others can just consist of a couple of lines. While it is preferred that the referees check the mathematical details, it is acceptable to submit a referee report stating "I did not check the details", "I did not have the time to check the details" or even "I am not an expert in the area and didn't follow the proofs".

After the referee reports are in, the chairs initiate the so-called *PC Session*, a time period of about a week during which the members discuss the papers and referee reports electronically. During this period the chairs moderate the discussion of the members, propose to accept certain papers and reject others, and in some cases announce votes on particular decisions. During the session, it is not uncommon to ask subreferees for more clarification and further comments on their reports.

In past years, the refereeing process for computer science conferences has been uniformized strongly by the use of conference submission software. An important system used is the EasyChair system of Andrei Voronkov which was used for 1313 conferences in 2008 and for 2186 conferences in 2009.

## 6.1   Available data

From the papers (Rothwell and Martyn, 2000; Wood et al., 2004) serving as our comparative data set from non-mathematical areas of science, we obtained the data on two journals in the neurosciences (neuro1 and neuro2) and two conferences in the information sciences (infor1 and infor2). In these four cases, the majority of refereed papers had two referees, and the studies (Rothwell and Martyn, 2000; Wood et al., 2004) reduced their data set to the subset of those submissions. The journals neuro1 and neuro2 used the categories "accept", "accept after revision", "reject"; the conferences infor1 and infor2 used "accept", "accept with minor revisions", "accept with major revisions", and "reject".[11]

We obtained anonymized data for eight conferences for which the refereeing was organized via the EasyChair system. One of these conferences was purely mathematical (math1), the others were in theoretical computer science (comp1 to comp7). In the following, we shall call the original data from (Rothwell and Martyn, 2000; Wood et al., 2004) *non-mathematical conferences* and our new data *mathematical conferences*.

Each of the conferences used a slighly different grading scales, and we decided to use a three-category scale 1 ("accept"), 0 ("accept with revisions" or "borderline"), −1 ("reject") as the most natural common grading scale.

---

[11] Note that while the data of (Wood et al., 2004) had four categories, the analysis was done in terms of two categories.

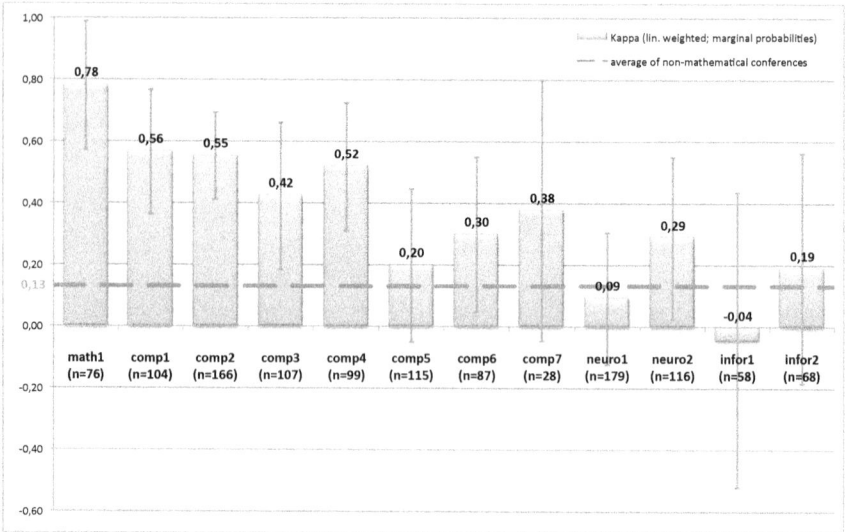

FIGURE 2. Comparison of Kappa values (partly averaged; linearly weighted; using marginal probabilities) to the average value of the non-mathematical conferences. The error bars show the induced 95% confidence intervals.

We needed to translate the various scales used into this three-category set-up.

The major difference between the data used by (Rothwell and Martyn, 2000; Wood et al., 2004) and our new data was that the mathematical conferences used variable numbers of referees (the minimum number was two and the maximum number was seven). We shall discuss how we dealt with this in § 6.2.

The data sets used for this calculation (i.e., the data translated to our three-category scale as described) can be found in Appendix B..

## 6.2 Method

Following (Rothwell and Martyn, 2000; Wood et al., 2004), we did our analysis by means of Kappa statistics. Cohen's Kappa is a standard measure for the analysis of inter-rater reliability (two raters) proposed in (Cohen, 1960). The Kappa value is scaled such that 0 represents the level of agreement that would have been expected by chance, and 1 represents perfect agreement between the raters. Various suggestions for improvement of the method of Kappa statistics have been made (Brennan and Prediger, 1981; Cohen, 1960; Sim and Wright, 2005): e.g., it is strongly recommended by Sim and Wright (2005) to use a weighted Kappa in the case of ordinal categories.

We decided to use a linearly weighted Kappa (making disagreement by one category count as half an agreement).[12] We did not use the results of the analysis from (Rothwell and Martyn, 2000; Wood et al., 2004) directly, but instead used their data and recalculated the Kappa values to make sure that we used exactly the same method for all conferences investigated.

As mentioned, the new conferences had a variable number of referees (between two and seven). Such a situation cannot be handled by Kappa analysis[13] and would make the comparison between conferences rather difficult. In order to carry out this type of analysis to compare it to the result obtained from the data in (Rothwell and Martyn, 2000; Wood et al., 2004), we had to choose two referees per article. We let a computer pick two different referees per paper uniformly at random and iterated this procedure 100 times; every time computing Kappa value as well as its 95% confidence interval as induced by the confidence intervals of the observed level of agreement, which were computed using the standard formula for a proportion on a 95% confidence level and applying the usual continuity correction. The values we report here are the arithmetic means of these 100 computations.[14]

### 6.3 Results

We have visualized the results of our analysis in Figure 2, which exhibits the Kappa values obtained (with 95% confidence intervals) for each conference (as bars) and compares them to the average Kappa of the non-mathematical conferences (dashed line at 0.13).

The values for Kappa range from -0.04 (infor1) up to 0.78 (math1) and all mathematical conferences have Kappa values that are strictly higher than the threshold of 0.13 (average Kappa of the non-mathematical conferences). Furthermore, with one exception, all mathematical conferences have strictly higher Kappa values than all four non-mathematical conferences.

## 7   Conclusion and future work

We started this paper with the question: If mathematics is an epistemic exception, shouldn't the mathematical literature be more reliable than that of other fields; and if so, how does the refereeing process contribute to this?

The answer given in this paper is somewhat ambiguous: we argued that a lot of mathematical research uses black boxes from the literature without checking the proofs, and claimed that there are serious issues with the reliability of the literature. Looking into the mathematical refereeing process, we saw that it is not universally expected that referees check the correctness

[12]The definition of the Kappa we used is given in Appendix A..

[13]Fleiss (1971) introduced a new version of Kappa in that is able to treat fixed larger numbers of raters (Fleiss' Kappa), but not variable numbers of raters.

[14]We attempted to avoid the randomization by choosing the extreme two raters in any given set, but this did not generate realistic results.

of all claims in the papers, and this was corroborated by our survey study in §5. But the survey study also showed that editors have a lot of trust in their referees. In our empirical study in §6, we saw a different facet of this: compared to other fields where referees do not come to the same conclusion much more often than they would by pure chance, the agreement between referees in fields based on the deductive method is higher, indicating (but not proving) that the degree of objectivity is higher.

We also saw that the empirical methods employed in §6 do not really fit the core of the mathematical refereeing business: a more qualitative study is necessary, analogous to the field work of Greiffenhagen (2008) on the process of graduate student supervision in mathematics. A possible source of data could be the new and unusual journal *Rejecta Mathematica*, a journal that

> publishes only papers that have been rejected from peer-reviewed journals in the mathematical sciences [...] [together with] an open letter from its authors discussing the paper's original review process, disclosing any known flaws in the paper, and stating the case for the paper's value to the community. (Wakin et al., 2009, p. 1)

It remains to be seen how useful *Rejecta Mathematica* will be as a source for studying the mathematical refereeing process (the published letters from the authors in its inaugural issue do not give much insight in the original refereeing process). We should also like to mention the interesting paper (Weintraub and Gayer, 2001) (cf. Footnote 3): written by economists, it reports on how a mathematical result (the Arrow-Debreu theorem) became accepted by the whole community within a very short time, even though the refereeing process was anything but unproblematic. One of the reviewers, the mathematician Cecil G. Phipps, originally selected by the associate editor to "thoroughly check the mathematics of the argument",[15] strongly objected to the publication of the paper. But the editors decided to accept the paper on the basis of one very short referee report of just a few lines and four typographical corrections (Weintraub and Gayer, 2001, p. 430) and a detailed comment of the associate editor who, however, had decided not to do a "thorough checking of the mathematics".[16] Weintraub and Gayer raise interesting questions like "Did the persuasion occur before Arrow and Debreu submitted the article for publication at *Econometrica*?" (Weintraub and Gayer, 2001, p. 440). Case studies like this, either dealing with historical cases like the Arrow-Debreu theorem or with a refereeing process accompanied as it is going on, would be the natural next step.[17]

---

[15]From a letter by associate editor Nicholas Georgescu-Roegen to the managing editor Robert Strotz, dated 8 October 1953, quoted after (Weintraub and Gayer, 2001, p. 431).

[16]From the same letter; quoted after (Weintraub and Gayer, 2001, p. 434).

[17]Note that also Gross's study of the peer review process in his (Gross, 1990, §9) is based on textual data from actual referee/editor and author/editor exchanges.

# Bibliography

Abby, M., Massey, M. D., Galandiuk, S., and Polk Jr., H. C. (1994). Peer review is an effective screening process to evaluate medical manuscripts. *Journal of the American Medical Association*, 272:105–107.

Adler, J. (2008). Epistemological problems of testimony. In Zalta, E. N., editor, *Stanford Encyclopedia of Philosophy*. Fall 2008 Edition.

Auslander, J. (2008). On the roles of proof in mathematics. In Gold, B. and Simons, R. A., editors, *Proof & Other Dilemmas: Mathematics and Philosophy*, pages 61–77. Mathematical Association of America, Washington DC.

Birman, J. S. (2000). Scientific publishing: A mathematician's viewpoint. *Notices of the American Mathematical Society*, 47(7):770–774.

Black, N., van Rooyen, S., Goglee, F., Smith, R., and Evans, S. (1998). What makes a good reviewer and a good review for a general medical journal? *Journal of the American Medical Association*, 280:231–233.

Borwein, J. M., Rocha, E. M., and Rodrigues, J. F., editors (2008). *Communicating Mathematics in the Digital Era*. AK Peters, Natick MA.

Brennan, R. L. and Prediger, D. J. (1981). Coefficient $\kappa$: Some uses, misuses, and alternatives. *Educational and Psychological Measurement*, 41:687–699.

Callaham, M. L., Wears, R. L., and Waeckerle, J. F. (1998). Effect of attendance at a training session on peer reviewer quality and performance. *Annals of Emergency Medicine*, 32:318–322.

Cleary, J. D. and Alexander, B. (1988). Blind versus nonblind review: Survey of selected medical journals. *Drug Intelligence and Clinical Pharmacy*, 22:601–602.

Coady, C. A. J. (1973). Testimony and observation. *American Philosophical Quarterly*, 10:149–155.

Cohen, J. (1960). A coefficient of agreement for nominal scales. *Educational and Psychological Measurement*, 20:37–46.

Das Sinha, S., Sahni, P., and Nundy, S. (1999). Does exchanging comments of Indian and non-Indian reviewers improve the quality of manuscript reviews? *National Medical Journal of India*, 12:210–213.

Evans, A. T., McNutt, R. A., Fletcher, S. W., and Fletcher, R. H. (1993). The characteristics of peer reviewers who produce good-quality reviews. *Journal of General Internal Medicine*, 8:422–428.

Fallis, D. (2003). Intentional gaps in mathematical proofs. *Synthese*, 134:45–69.

Feurer, I. D., J., B. G., Picus, D., Ramirez, E., Darcy, M. D., and Hicks, M. E. H. (1994). Evaluating peer reviews pilot testing of a grading instrument. *Journal of the American Medical Association*, 272:98–100.

Fisher, M., Friedman, S. B., and Strauss, B. (1994). The effects of blinding on acceptance of research papers by peer review. *Journal of the American Medical Association*, 272:143–146.

Fleiss, J. L. (1971). Measuring nominal scale agreement among many raters. *Psychological Bulletin*, 76(5):378–382.

Garfunkel, J. M., Ulshen, M. H., Hamrick, H. J., and Lawson, E. E. (1990). Problems identified by secondary review of accepted manuscripts. *Journal of the American Medical Association*, 263:1369–1371.

Goodman, S. N., Berlin, J., Fletcher, S. W., and Fletcher, R. H. (1994). Manuscript quality before and after peer review and editing at annals of internal medicine. *Annals of Internal Medicine*, 121:11–21.

Greiffenhagen, C. (2008). Video analysis of mathematical practice? Different attempts to 'open up' mathematics for sociological investigation. *Forum: Qualitative Social Research*, 9(3):art. 32.

Gross, A. G. (1990). *The Rhetoric of Science*. Harvard University Press, Cambridge MA.

Hales, T. C. (2008). Formal proof. *Notices of the American Mathematical Society*, 55(11):1370–1380.

Heintz, B. (2000). *Die Innenwelt der Mathematik. Zur Kultur und Praxis einer beweisenden Disziplin*. Springer, Vienna.

Jackson, A. (2002). From preprints to e-prints: The rise of electronic preprint servers in mathematics. *Notices of the American Mathematical Society*, 49(1):23–31.

Jackson, A. (2003). The digital mathematics library. *Notices of the American Mathematical Society*, 50(8):918–923.

Jefferson, T., Wager, E., and Davidoff, F. (2002). Measuring the quality of peer review. *Journal of the American Medical Association*, 287:2786–2790.

Justice, A. C., Cho, M. K., Winker, M. A., Berlin, J. A., and Rennie, D. (1998). Does masking author identity improve peer review quality? A randomized controlled trial. *Journal of the American Medical Association*, 280:240–242.

Katz, D. S., Proto, A. V., and Olmsted, W. W. (2002). Incidence and nature of unblinding by authors: Our experience at two radiology journals with double-blinded peer review policies. *American Journal of Roentgenology*, 179:1415–1417.

Krantz, S. G. (1997). *A Primer of Mathematical Writing. Being a Disquisition on having your ideas recorded, typeset, published, read, and appreciated.* American Mathematical Society, Providence RI.

Krantz, S. G. (2005). *Mathematical Publishing. A Guidebook.* American Mathematical Society, Providence RI.

Lackey, J. and Sosa, E., editors (2006). *The Epistemology of Testimony.* Oxford University Press, Oxford.

Link, A. M. (1998). US and non-US submissions: An analysis of reviewer bias. *Journal of the American Medical Association*, 280:246–247.

Lipton, P. (1998). The epistemology of testimony. *Studies in History and Philosophy of Science A*, 29(1):1–31.

Löwe, B., Müller, T., and Müller-Hill, E. (2010). Mathematical knowledge: A case study in empirical philosophy of mathematics. In Van Kerkhove, B., De Vuyst, J., and Van Bendegem, J. P., editors, *Philosophical Perspectives on Mathematical Practice*, volume 12 of *Texts in Philosophy*, pages 185–203. College Publications, London.

McNutt, R. A., Evans, A. T., Fletcher, R. H., and Fletcher, S. W. (1990). The effects of blinding on the quality of peer review. A randomized trial. *Journal of the American Medical Association*, 263:1371–1376.

Müller-Hill, E. (2009). Formalizability and knowledge ascriptions in mathematical practice. *Philosophia Scientiae*, 13(2):21–43.

Müller-Hill, E. (2010). *Die epistemische Rolle formalisierbarer mathematischer Beweise. Formalisierbarkeitsbasierte Konzeptionen mathematischen Wissens und mathematischer Rechtfertigung innerhalb einer sozioempirisch informierten Erkenntnistheorie der Mathematik.* PhD thesis, Rheinische Friedrich-Wilhelms-Universität Bonn.

Nathanson, M. B. (2008). Desperately seeking mathematical truth. *Notices of the American Mathematical Society*, 55(7):773.

Nylenna, M., Riis, P., and Karlsson, Y. (1994). Multiple blinded reviews of the same two manuscripts. Effects of referee characteristics and publication language. *Journal of the American Medical Association*, 272:149–151.

Pierie, J., Walvoort, H., and Overbeke, A. (1996). Readers' evaluation of effect of peer review and editing on quality of articles in the Nederlands Tijdschrift voor Geneeskunde. *Lancet*, 348:1480–1483.

Prediger, S. (2006). Mathematics: Cultural product or epistemic exception? In Löwe, B., Peckhaus, V., and Räsch, T., editors, *Foundations of the Formal Sciences IV. The History of the Concept of the Formal Sciences*, volume 3 of *Studies in Logic*, pages 271–232. College Publications, London.

Reid, T. (1969). *Essays on the Intellectual Powers of Man*. MIT Press, Cambridge MA.

Roberts, J. C., Fletcher, R. H., and Fletcher, S. W. (1994). Effects of peer review and editing on the readability of articles published in Annals of Internal Medicine. *Journal of the American Medical Association*, 272:119–121.

Rothwell, P. M. and Martyn, C. N. (2000). Reproducibility of peer review in clinical neuroscience: Is agreement between reviewers any greater than would be expected by chance alone? *Brain*, 123(9):1964–1969.

Sim, J. and Wright, C. C. (2005). The kappa statistic in reliability studies: Use, interpretation, and sample size requirements. *Physical Therapy*, 85(3):257–268.

Thompson, R. C. (1983). Author vs. referee: A case history for middle level mathematicians. *American Mathematical Monthly*, 90(10):661–668.

Wakin, M., Rozell, C., Davenport, M., and Laska, J. (2009). Letter from the editors. *Rejecta Mathematica*, 1(1):1–3.

Walsh, E., Rooney, M., Appleby, L., and Wilkinson, G. (2000). Open peer review: A randomized controlled trial. *British Journal of Psychiatry*, 176:47–51.

Weintraub, E. R. and Gayer, T. (2001). Equilibrium proofmaking. *Journal of the History of Economic Thought*, 23(4):421–442.

Wood, M., Roberts, M., and Howell, B. (2004). The reliability of peer reviews of papers on information systems. *Journal of Information Science*, 30:2–11.

# Appendix A.   Calculation of the Kappa value

The data sets we use are assigning two reviewer recommendations for each of a set of $N$ papers; derived from this, we consider functions $\mathbf{d} : N \rightarrow \{-1, 0, 1\}$ picking out the first and the second reviewers recommendation from the data set. To simplify notation, we introduce the following functions on $\{-1, 0, 1\} \times \{-1, 0, 1\}$, representing "equality" and "distance 1":

$$\delta_{x,y} := \begin{cases} 1 & x = y \\ 0 & \text{otherwise} \end{cases} \quad \text{and} \quad \xi_{x,y} := \begin{cases} 1 & |x - y| = 1 \\ 0 & \text{otherwise.} \end{cases}$$

We fix $\mathbf{a}, \mathbf{b} : N \rightarrow \{-1, 0, 1\}$ as given from our data set and write $\mathbf{c}_i$ for the constant function with value $i \in \{-1, 0, 1\}$. We then define

$$\Delta_{\mathbf{a},\mathbf{b}} := \sum_{i=0}^{N-1} \delta_{\mathbf{a}_i, \mathbf{b}_i}, \ \Xi_{\mathbf{a},\mathbf{b}} := \sum_{i=0}^{N-1} \xi_{\mathbf{a}_i, \mathbf{b}_i}, \text{ and}$$

$$\eta_i := (\Delta_{\mathbf{a},\mathbf{c}_i} + \Delta_{\mathbf{b},\mathbf{c}_i})/2N.$$

Then the *observed agreement* is defined as

$$O_{\mathbf{a},\mathbf{b}} := \frac{1}{N}\Delta_{\mathbf{a},\mathbf{b}} + \frac{1}{2N}\Xi_{\mathbf{a},\mathbf{b}},$$

and the *expected agreement* as

$$E_{\mathbf{a},\mathbf{b}} := (\eta_{-1})^2 + (\eta_0)^2 + (\eta_1)^2 + \eta_1\eta_0 + \eta_0\eta_{-1}.$$

Finally, the Kappa value is given by

$$\kappa_{\mathbf{a},\mathbf{b}} := \frac{O_{\mathbf{a},\mathbf{b}} - E_{\mathbf{a},\mathbf{b}}}{1 - E_{\mathbf{a},\mathbf{b}}}.$$

# Appendix B.   The data used in § 6

A semicolon separates the different papers, whereas a comma separates the recommendations for a paper by different reviewers.

**math1** $(n = 76)$

1, 1; 1, 1; 1, 1; 1, 1; 1, 1; 1, 1; 1, 1; 1, 1; 1, 1; 1, 0; 1, 0; 1, 0; 1, 0; 0, 0; 0, 0; 0, 0; 0, 0; 0, 0; 0, 0, 0; 0, 0; 0, 0; 0, 0; 0, 0; 0, 0; 0, 0; 0, 0; 0, 0; 0, 0; 0, 0; 0, 0; 0, 0; 0, 0; 0, 0; 0, 0; 0, 0; 0, 0; 0, 0; 0, 0; 0, 0; 0, 0; 0, 0; 0, 0; 0, 0; 0, 0; 0, 0; 0, 0; 0, 0; 0, 0; 0, 0; 0, 0; 0, 0; 1, −1; 1, −1; 1, −1; −1, −1; −1, −1; −1, −1; −1, −1; −1, −1; −1, −1; −1, −1; −1, −1; −1, −1; −1, −1; −1, −1; −1, −1; −1, −1; −1, −1; −1, −1

**comp1** $(n = 104)$

1, 1; 1, 1; 1, 1; 1, 1; 1, 1; 1, 1; 1, 1; 1, 1; 1, 1; 1, 1; 1, 1; 1, 1; 1, 1; 1, 1; 1, 1; 1, 1; 1, 1; 1, 1; 1, 1; 1, 1; 1, 1; 1, 1; 1, 1; 1, 1; 1, 1; 1, 1; 1, 1; 1, 1; 1, 1; 1, 1; 1, 1; 1, 1; 1, 1; 1, 1; 1, 1; 1, 1; 1, 0; 1, 1; 1, 1; 1, 1; 1, 1; 1, 1; 1, 0; 1, 1; 1, 1; 1, 1; 1, 1; 1, 0; 1, 1; 1, 1; 1, 1; 1, 0; 1, 0; 1, 1; 1, 0; 0, 1; 1, 0; 1, 1; 1, 1; 0, 1; 1, 0; 1, 1; 1, 0; 1, 1; 1, 1; 0, 1; 0, 1; 1, 0; 1, 0; 1, 1; 1, 1; 1, 0, 1; 1, 1; 0, 1; 1, 0; 0, 1; 1, 0; 1, 0; 0, 1; 0, 1, 0, 0; 1, 1, 0; 1, 0, 1; 1, 1, −1, 1; 0, 1, 0, 0; 0, 0, 1; 0, 0, 1, 0; 0, 1, 0; 0, 1, 0; 1, 1, 0, 0; 1, 1, 0, 0; 0, 0; −1, 1, 1; 0, 1, 0, 0; 0, 0, 1, 1; −1, 1, 1; 0, 1, 0; 0, 0, 0; 0, 0, 1; 0, 0, 1; 0, 0, 1; 0, 0, 0; 1, 0, 0; 0, 0, 0; 0, 0, 0; 0, 1, 0; 1, 0, 0; −1, 0, 1, 0; −1, 1, 0; 0, 0, 0; 0, 0, −1; 0, 0, −1; −1, 0, −1, 1; −1, 0, −1, 0; −1, 0, −1, 0; −1, 0, 0; −1, 0, −1; −1, −1; −1, 0, −1; −1, −1; −1, −1, −1, −1; −1, −1, −1, −1; −1, −1, −1, −1; −1, −1, −1, −1; −1, −1, −1, −1; −1, −1; −1, −1, −1, −1; −1, −1, −1; −1, −1, −1

## comp2 $(n = 166)$

1, 1, 1; 1, 1, 1; 1, 1, 1; 1, 1, 1; 0, 1, 1, 1; 1, 1, 1; 1, 1, 1; 1, 1, 1; 1, 0, 1; 1, 1, 0; 1, 1, 1; 1, 1,
0; 1, 1, 1; 1, 1, 1; 1, 1, 1; 0, 1, 1; 0, 1, 0, 1; 1, 1, 1; 1, 1, 1; 0, 0, 1, 1; 1, 1, 0; 1, 1, 0; 1, 1, 1; 1,
1, 0; 1, 1; 1, 1, 1; 1, 1, 1; 1, 1, 0; 1, 1, 1; 1, 1, 0; 1, 0, 1; 0, 0, 1, 1; 1, 1, 0, 1, 1; 1, 1, 1, 0; 1, 1,
0, 1; 1, 1, 1, 1, 0, 0; 0, 1, 1; 1, 0, 1; 0, 1, 1; 1, 0, 1; 0, 1, 1; 1, 0, 0; 1, 1, 0; 1, 1, 0; 0, 1, 1; 0, 1,
1; 1, 1, 0, 0; 1, 1, 0, 1; 1, 0; 0, 0, 1; 1, 0, 1, 0; 0, 1, 0; 1, 0, 1, 0, 1; 0, 1, 0; 1, 1, 1, 1, 0, 0; 1,
0, 0; 1, 0, 0, 0, 0, 1; 0, 0, 0; 0, 1, 0; 0, 1, 0, 0; 0, 0, 1; 1, 0, 0; 1, -1, 1, 0; 1, -1, 1, 0, 0, 0;
1, 0, 0; 0, 0, 1; 1, 0, 0; 0, 1, 0, 0; 1, 0, 1, -1, 0; 0, 0, 0; -1, 1, 0; 0, 0, 0; 0, 0, 1; 0, 0, 0; 0,
1, 0, -1; -1, 0, 0; 0, 0; 0, 0, 1; -1, 0, 1; 0, -1, 1, 0; -1, 0, 1; 0, 0, -1, 1; 0, 0; 0, 0; 0, -1, 0;
-1, 1, -1; 0, -1, 0; 0, -1, 0; 0, 0, 0; 1, -1, -1; 1, -1, 0; 0, -1, 0; -1; -1, 0; 0, 0, 0; -1, 0,
0, -1, -1; 1, -1, -1; -1, -1, 0; -1, -1, 0; 1, -1, -1; -1, 0, 0; 0, 0, -1, -1; -1, -1, -1,
-1; -1, -1, 0, -1; -1, 0, -1; 0, -1, -1; -1, 0, -1; -1, 0, -1; -1, 0, -1; -1, -1, 1; -1, 0;
-1, 0, -1; -1, -1, -1; -1, -1, -1; 0, -1, -1; -1, -1, -1; 0, -1, -1; -1, -1, -1; -1, -1,
-1; -1, 0, -1; 0, -1, -1; -1, -1; -1, -1, -1; -1, -1, -1; -1, -1, -1; -1, -1, -1,
-1, -1; -1, -1, -1; -1, -1, -1; -1, -1, -1; -1, -1, -1; -1, -1, -1; -1, -1, -1; -1, -1, -1,
-1; -1, -1, -1; -1, -1, -1; -1, -1, -1; -1, -1, -1; -1, -1, -1; -1, -1, -1; -1, -1, -1;
-1, -1; -1, -1, -1; -1, -1, -1; -1, -1, -1; -1, -1, -1; -1, -1, -1; -1, -1, -1; -1, -1, -1;
-1; -1, -1, -1; -1, -1, -1; -1, -1, -1; -1, -1, -1; -1, -1, -1; -1, -1

## comp3 $(n = 107)$

1, 1, 1, 1, 1; 1, 1, 1, 1; 1, 1, 1; 1, 1, 1; 1, 1, 1; 1, 1, 1; 0, 1, 1, 1; 1, 1, 0; 0, 1, 1; 1, 1, 1; 0, 1, 1, 1;
1, 1, 1; 0, 1, 1, 1, 1; 1, 1, 1, 1; 1, 1, 1, 0, 1; 0, 1, 1, 0, 1; 1, 1, 1, 1; 1, 0, 1, 1; 1, 1, 1, 0; 0, 1, 1,
0; 1, 0, 1, 1; 1, 1, 0; 1, 0, 1; 0, 1, 1; 0, 1, 0, 1, 1, 0; 1, 0, 0; 1, 0, 1; 1, 1, 1, 0, 0; 1, 0, 0, 1; 0, 0,
1, 1, 1, 0, 0, 1, 1, 1; 1, 1, 0, 1; 0, 1, 1, 0, 1; 1, 0, 1; 0, 1, 0, 0; 0, 0, 0; 0, 1, 1; 0, 1, 1, 1; 0, 0, 1, 1;
1; 1, 0, 1; 1, 0, 0; 1, 0, 0; 1, 0, 0, 0; 0, 0, 1, 0; 0, 1, 1; 0, 0, 1, 1, 0; 1, 1, -1, 1, 0; 0, 0, 1, 1, 1;
0, 0, 1; 0, 1, 1, 0; 0, 1, 0; 0, 1, 1; 0, 0, 0, 1, 0; 0, 0, 0, 1; 0, 1, 1, 0; 0, 1, 0; 0, 0, 1; 1, 0, 0, 0, 1; 1,
0, 0, 0; 0, 0, 0, 0, 1; 0, 1, 1, 0, 0; 0, 1, 0; 0, 0, 0, 0, 1, 0; 0, 1, 0; 0, 0, 0, 0, 0, 1; -1, 0, 1, 0, 1,
1; 0, 0, 0; 1, 0, 0, 0, 0; 0, 0, 0, 0; 0, 0, 0, 0; 0, 0, -1, 1; 1, -1, 0; 1, -1, 0; 0, 0, 1, 0, 1;
-1, 0; 0, 0, 0; -1, 0, 0, 1; 0, 0, -1, 0; 0, -1, 0; 0, 0, 0, -1, 0; 0, 0, 0; 0, 0, 0, 0, 0; -1, 1, 0, -1,
0; 0, -1, 0, 0, 0; 0, 0, -1, 0, 1, -1; -1, 0, 1; 1, 0, -1; 0, 0, 0, 0; 0, -1, 1; -1, 0, 0, -1, 0, 1;
-1, 0, 0, 0; 0, 1, -1, 0; -1, 1, 0, 0; 0, 0, 0, -1, -1, 0; 0, 0, -1; 0, -1, 0; 0, 0, 0; 0, 0, 0, -1,
-1, 0; 0, -1, 0, 0, 0, 0; 0, -1, 0; 0, 0, 0, 0, -1; -1, 1, -1, 0; 0, -1, 0, -1, 0, 0; -1, 0, 0, -1;
0, -1, -1; 0, 0, 0, -1, -1, 0; 0, -1, 0, 0, -1, -1; 0, -1, -1; -1, 0, -1; 0, -1, -1, 0; -1, 0,
-1; -1, -1, 0; -1, -1, 0, -1; -1, -1, -1

## comp4 $(n = 99)$

1, 1, 1; 1, 1, 1, 1; 1, 1, 1; 1, 1, 1, 1; 1, 1, 1; 1, 1, 1; 1, 1, 1; 1, 1, 1; 1, 1, 1; 1, 0, 1; 1, 1, 1; 1, 1,
1, 1; 1, 1, 1, 1; 1, 1, 1, 1, 0; 1, 1, 0; 0, 1, 1; 1, 1, 1, 1; 1, 1, 1, 1, 0; 1, 1, 1; 1, 1, 1, 1; 1, 1, 1; 1;
1, 0, 1; 0, 1, 1, 1, 1; 1, 1, 1, 1; 1, 1, 1, 1; 1, 1, 1; 0, 1, 1, 1, 0; 0, 1, 1, 0; 0, 1, 1, 0; 1, 0, 1, 0, 1;
1; 1, 1, 0; 0, 1, 1; 1, 0, 0; 0, 1, 1; 1, 1, 1, 0; 1, 1, 0; 1, 1, 0; 1, 1, 0; 1, 1, 0, 0; 1, 0, 0, 1; 0, 1, 1, 0;
0, 0, 1; 1, 0, 0; 1, 0, 0; 0, 1, 0; 1, 0, 0; 1, 0, 0; 0, 0, 1; 1, 0, 0; 0, 1, 0; 0, 1, 0; 1, 0, 0; 0, 0, 1; 0,
0, 1; 0, 1, 0; 0, 0, 1, 0; 0, 1, 0, 1; 0, 0, 0, 1, 0; 0, 0, 1; 0, 0, 0, 1; 0, 0, 0, 1, 0; 0, 0, 1, 0; 0, 0, 1;
1, 0, 0; 0, 0, 0, 1; 0, 0, 0, 1; 0, 0, 0; 0, 0, 0, 0; 0, 0, 0; 0, 0, 0, 1; -1, 1, 0; 0, 0, 0, 0; 0, 0, 0; 0, 0,
0; 0, 0, 0, 0; 0, 0, 0, 0; 0, 0; 0, 0, 0, 0; 0, 0, 0, 0; 0, 0, 0; 0, 0, 0, 0; 0, 0, 0; 0, 0, 0; 0, 0, 0; 0, 0,
-1; 0, 0, -1; 0, -1, 0; 0, 0, -1; 0, -1, 0; 0, 0, -1; 0, 0, -1; -1, 0, -1; 0, -1, -1; -1, 0, -1;
-1, 0, -1; -1, -1, 0; -1, 0, -1; -1, 0, -1; -1, 0, -1; -1, -1, -1; -1, -1, -1; -1, -1, -1

## comp5 $(n = 115)$

1, 1, 1; 1, 1, 1, 1; 1, 1, 1, 1; 1, 1, 1, 1; 1, 1, 1, 1; 1, 1, 1, 1; 1, 1, 1, 1; 1, 1, 1, 1; 1, 1, 1, 1, 1; 1,
1, 1, 0, 1; 1, 1, 1, 1; 1, 1, 1, 1; 1, 1, 1, 1; 1, 1, 0, 1; 1, 1, 0, 1; 1, 1, 1, 1; 1, 1, 1, 1; 1, 1, 0; 1,
1, 1, 1; 1, 1, 1, 1; 1, 1, 1; 1, 1, 1, 1; 1, 1, 1, 1; 1, 1, 1, 1; 1, 1, 1, 1; 1, 1, 0; 1, 1, 1, 1; 1, 1, 1, 0,
1; 1, 1, 1, 1; 1, 0, 1, 1; 1, 1, 0; 1, 0, 1; 0, 1, 1, 0; 1; 1, 0, 1; 1, 1, 1, 0; 1, 1, 0, 0; 1, 0, 0; 0, 1, 1;
1, 0, 1; 0, 1, 1, 0; 1; 0, 1, 1, 0, 1; 0, 1, 0, 1; 0, 0, 1, 1; 1, 0, 0; 1, 0, 1; 0, 1, 0, 1, 0; 1, 0, 0;
0, 0, 1; 0, 0, 0, 1; 0, 0, 0, 1; 1, 0, 0, 0; 1, 0, 0, 0; 0, -1, 1, 1, 1; 0, 1, 0, 0; 0, 1, 0, 0, 0, 0, 1;
0, 0, 0; 1, 1, 0, 0; 0, 0, 0, 0; 0, 0, 1; 0, 0, 0, 0; 0, 0, 0, 0; 0, 0, 0; 0, 1, 0, 0; 0, 1, -1, 1;
1, 0, 0, -1; 0, 0, 0, -1; 0, 0, -1, 0; 0, 0, -1, 0; 0, 0, 0, -1; 0, -1, 0, 0, 0; 0, -1, -1, 0; 0, 0,
0, 0, 0; 0, -1, 0, 0; 0, -1, 0, 0; 0, 1, 0; -1, 0, 0, 0; 0, 1, -1, 0, 0, 0; -1, -1; -1, 0, 0, 0;
-1, 0, 0, 0; 0, -1, 1; 0, -1, 0, -1, -1; -1, -1, 0, 0; -1, -1, 0, 0; 0, -1, -1; 0, -1, -1, -1,
-1; 0, -1, -1; -1, -1, -1; -1, -1, -1; -1, -1, -1; -1, -1, -1; -1, -1, -1, -1; -1,
-1, -1, -1; -1, -1, -1, -1; -1, -1, -1, -1; -1, -1, -1, -1

## comp6 $(n = 87)$

1, 1, 1; 1, 1, 1; 1, 1, 1; 1, 1, 1; 1, 1, 1; 1, 1, 1; 1, 0, 1; 1, 1, 1; 0, 1, 1; 1, 1, 1; 1, 1, 1; 1, 1, 1;
0, 1, 1; 1, 1, 1, 0, 1, 1; 1, 0, 1; 1, 1, 1; −1, 1; 1, 1, −1, 1; 1, 1, 1, 0; 0, 1, 1; 0, 1, 1; 1, 1, 1;
1, 1, 1; 1, 1, 1, 1; 1, 0, 0; 1, 0, 1; 0, 0, 1, 1, 1, 1; 1, 0, 1; 1, 0, 0; 0, 1, 0, 0; 0, 1, −1, 1; 0, 1, 0; 0,
1, 1, 0; 0, 0, 1, 1; 0, 0, 1; 1, 0, 0; 1, 0, 1, 1; 0, 1, 0; −1, 0, 0, 1; −1, 1, 1; 1, 0, 0; 0, 1, 0, 1; 0,
0, 1, 0, 0, 0; 0, 1, 0; 1, 0, 0; 0, 0, 1, 1, 0, 0; 0, 1, 0, 0, 0; −1, −1, 1, 1; 1, −1, 1, 0; 0, 0, 0;
0, 1, 0; −1, 0, 0; 0, 1, 0; 0, 0, 0; 0, 0, 1; 0, 0, 0; 0, 1, −1; 0, 0, −1; 1, −1, 0; 0, −1, 1; −1, 0, 0;
0, 1, −1; 0, 0, 0; 1, −1, 0; 0, 0, 0; 0, 0, 0; −1, −1, 1; −1, 0, 0; −1, 0, −1; 1, −1, −1; −1, −1,
0; 0, −1, 0; −1, 0, −1; 0, 0, −1; 0, −1, −1; −1, −1, −1; −1, −1, −1; −1, −1, −1; −1, 0, −1;
−1, −1, 0; 0, 0, −1; −1, 0, 0; −1, −1; 0, −1, −1

## comp7 $(n = 28)$

1, 1; 1, 1; 1, 1; 1, 1; 1, 0; 1, 1; 0, 1; 0, 0; 0, 1; 0, 1; 1, 0; 1, 0; 0, 0; 1, −1; 0, 0; 0, 0; −1, 0; −1,
1; −1, 0; 0, 0; −1, 0; 0, −1; −1, −1; −1, 0; −1, −1; −1, −1; −1, −1; −1, −1

## neuro1 $(n = 179)$

1, 1; 1, 1; 1, 1; 1, 0; 1, 0; 1, 0; 1, 0; 1, 0; 1, 0; 1, 0; 1, 0; 1, 0; 1, 0; 1, 0; 1, 0; 1, 0; 1, 0; 1,
−1; 1, −1; 1, −1; 1, −1; 1, −1; 1, −1; 0, 1; 0, 1; 0, 1; 0, 0; 0, 0; 0, 0; 0, 0; 0, 0; 0, 0; 0, 0;
0, 0; 0, 0; 0, 0; 0, 0; 0, 0; 0, 0; 0, 0; 0, 0; 0, 0; 0, 0; 0, 0; 0, 0; 0, 0; 0, 0; 0, 0; 0, 0; 0, 0;
0, 0; 0, 0; 0, 0; 0, 0; 0, 0; 0, 0; 0, 0; 0, 0; 0, 0; 0, 0; 0, 0; 0, 0; 0, 0; 0, 0; 0, 0; 0, 0; 0, 0;
0, 0; 0, 0; 0, 0; 0, 0; 0, 0; 0, −1; 0, −1; 0, −1; 0, −1; 0, −1; 0, −1; 0, −1; 0, −1; 0, −1; 0,
−1; 0, −1; 0, −1; 0, −1; 0, −1; 0, −1; 0, −1; 0, −1; 0, −1; 0, −1; 0, −1; 0, −1; 0, −1;
0, −1; 0, −1; 0, −1; 0, −1; 0, −1; 0, −1; 0, −1; 0, −1; 0, −1; 0, −1; 0, −1; 0, −1; 0, −1;
1; −1, 1; −1, 1; −1, 1; −1, 1; −1, 0; −1, 0; −1, 0; −1, 0; −1, 0; −1, 0; −1, 0; −1, 0; −1, 0;
−1, 0; −1, 0; −1, 0; −1, 0; −1, 0; −1, 0; −1, 0; −1, 0; −1, 0; −1, 0; −1, 0; −1, 0; −1, 0; −1,
0; −1, 0; −1, 0; −1, 0; −1, 0; −1, 0; −1, 0; −1, −1; −1, −1; −1, −1; −1, −1; −1, −1;
−1, −1; −1, −1; −1, −1; −1, −1; −1, −1; −1, −1; −1, −1; −1, −1; −1, −1; −1, −1; −1, −1;
−1, −1; −1, −1; −1, −1; −1, −1; −1, −1; −1, −1; −1, −1; −1, −1; −1, −1; −1, −1; −1, −1;
−1, −1; −1, −1; −1, −1; −1, −1

## neuro2 $(n = 116)$

1, 1; 1, 0; 1, 0; 1, 0; 1, 0; 1, 0; 1, 0; 1, 0; 1, 0; 1, 0; 1, −1; 1, −1; 1, −1; 1, −1; 0, 1; 0, 1; 0, 1; 0, 0;
0, 0; 0, 0; 0, 0; 0, 0; 0, 0; 0, 0; 0, 0; 0, 0; 0, 0; 0, 0; 0, 0; 0, 0; 0, 0; 0, 0; 0, 0; 0, 0; 0, 0; 0, 0;
0, 0; 0, 0; 0, 0; 0, 0; 0, 0; 0, 0; 0, 0; 0, 0; 0, 0; 0, 0; 0, 0; 0, 0; 0, 0; 0, 0; 0, 0; 0, 0; 0, 0; 0, 0;
0, 0; 0, 0; 0, 0; 0, 0; 0, 0; 0, 0; 0, 0; 0, −1; 0, −1; 0, −1; 0, −1; 0, −1; 0, −1; 0, −1; 0, −1; 0,
−1; 0, −1; 0, −1; 0, −1; −1, 1; −1, 0; −1, 0; −1, 0; −1, 0; −1, 0; −1, 0; −1, 0; −1, 0; −1, 0;
−1, 0; −1, 0; −1, 0; −1, 0; −1, 0; −1, 0; −1, 0; −1, 0; −1, −1; −1, −1; −1, −1; −1, −1; −1,
−1, −1; −1, −1; −1, −1; −1, −1; −1, −1; −1, −1; −1, −1; −1, −1; −1, −1; −1, −1; −1, −1;
−1, −1; −1, −1; −1, −1; −1, −1; −1, −1; −1, −1; −1, −1; −1, −1

## infor1 $(n = 58)$

1, 1; 1, 0; 1, 0; 1, 0; 1, 0; 1, 0; 1, 0; 1, 0; 1, 0; 1, 0; 1, 0; 1, 0; 1, 0; 1, −1; 1, −1; 1, −1; 1, −1; 0, 0;
0, 0; 0, 0; 0, 0; 0, 0; 0, 0; 0, 0; 0, 0; 0, 0; 0, 0; 0, 0; 0, 0; 0, 0; 0, 0; 0, 0; 0, 0; 0, 0; 0, 0; 0, 0;
0, 0; 0, 0; 0, 0; 0, 0; 0, 0; 0, 0; 0, 0; 0, −1; 0, −1; 0, −1; 0, −1; 0, −1; 0, −1; 0,
−1; 0, −1; 0, −1; 0, −1; 0, −1; 0, −1; −1, −1

## infor2 $(n = 68)$

1, 1; 1, 1; 1, 1; 1, 1; 1, 0; 1, 0; 1, 0; 1, 0; 1, 0; 1, 0; 1, 0; 1, 0; 1, 0; 1, 0; 1, 0; 1, 0; 1, 0; 1, 0; 1,
0; 1, 0; 1, 0; 1, 0; 0, 0; 0, 0; 0, 0; 0, 0; 0, 0; 0, 0; 0, 0; 0, 0; 0, 0; 0, 0; 0, 0; 0, 0; 0, 0; 0, 0; 0, 0;
0, 0; 0, 0; 0, 0; 0, 0; 0, 0; 0, 0; 0, 0; 0, 0; 0, 0; 0, 0; 0, 0; 0, 0; 0, 0; 0, −1; 0, −1; 0, −1; 0, −1;
0, −1; 0, −1; 0, −1; 0, −1; 0, −1; 0, −1; 0, −1; 0, −1; 0, −1; 0, −1; −1, −1; −1, −1; −1, −1;
−1, −1

# Embodied strategies in mathematical cognition

Mikkel Willum Johansen

Center for Naturfilosofi og Videnskabsstudier, Københavns Universitet, Blegdamsvej 17,
2100 Copenhagen, Denmark

E-mail: willum@nbi.dk

## 1 Introduction

Most traditional theories of cognition, such as the computational theory of
mind favored by cognitive science during the last half of the 20th century,
imagine cognitive content to be located inside the head of the thinking
agent. The head is, in short, conceived as a container isolating the cognitive
content inside from the physical world on the outside. Furthermore, sensing,
planning, and acting are supposed to be three clearly distinct activities, and
they are supposed to be performed in the mentioned order: First you sense,
then you use your internal cognitive resources to form a plan, and finally
you enact your plan.

This general picture of thinking is also visible in some theories of mathematical cognition, such as the 'abstract code model' where a tripartition
between comprehension (of the mathematical problem), calculation, and response (e.g., in the form of a written number) is hypothesized (Campbell
and Epp, 2005).

During the 1980s this conception of cognition was met with considerably
opposition from many different fronts (such as studies in animal vision,
robot engineering, philosophy of consciousness, neuroscience *etc.*). I will
not review all of the arguments here, but instead focus on a single line
of criticism. This line of criticism simply points out that the container
metaphor is inadequate. Parts of human cognition can indeed be described
as contemplation taking place inside the head, as the container metaphor
suggests, but not all of it. Much of human cognition can only be understood
as interactive processes involving the brain, the body, and the surrounding
environment (both social and physical).

The interactive nature of human cognition is evidenced by our use of a
number of cognitive tools including:

1. Epistemic actions, i.e., actions taken in order to gain knowledge and
   not in order to achieve a practical end.

2. Cognitive artifacts, i.e., artifacts developed in order to facilitate thinking. These can be either physical or conceptual.

Benedikt Löwe, Thomas Müller (*eds.*). *PhiMSAMP. Philosophy of Mathematics: Sociological
Aspects and Mathematical Practice*. College Publications, London, 2010. Texts in Philosophy
11; pp. 179–196.

*Received by the editors:* 31 August 2008; 14 April 2009; 14 June 2009.
*Accepted for publication:* 1 July 2009.

3. Conceptual metaphors, i.e., metaphors where a—typically abstract—phenomenon or idea is expressed and understood using the terms and knowledge from another—typically more basic—conceptual domain.

It should be noted however, that these tools are not neutral instruments that merely enhance our cognitive powers. The tools might shape our cognitive style and they are in some instances constitutive of particular types of reasoning.

In what follows, I will briefly present the cognitive tools and exemplify their use in mathematics. I will demonstrate how the three tools complement each other in the mathematical practice, and discuss to what extent they have shaped what mathematics is.

## 2   Epistemic actions and distributed cognition

Our first example of interactive cognitive processes that involves brain, body and environment is *epistemic actions*. They were identified and defined by Kirsh and Maglio (1994). Where pragmatic actions are actions performed with a pragmatic goal, such as peeling potatoes, an epistemic action is defined as:

> "[A] physical action whose primary function is to improve cognition by:
>
> 1. Reducing the memory involved in mental computation, i.e., space complexity;
>
> 2. Reducing the number of steps involved in mental computation, i.e., time complexity;
>
> 3. Reducing the probability of error of mental computation, i.e., unreliability." (Kirsh and Maglio, 1994, p. 514)

Kirsh and Maglio give several examples of epistemic actions. If you have a tendency to forget your key, you might for instance leave it in your shoe. Then you are sure to be reminded of it when you put on your shoes before you leave your apartment. By putting the key in the shoe, the key becomes a cognitive device that reduces both the probability of error and the demands on internal memory. Hence, the act of putting your key in your shoe is an epistemic action.

The concept of epistemic actions fits well in the general theory of cognition called *distributed cognition*. In contrast to traditional theories of cognition, this approach does not identify cognition with information processing going on inside the human skull, but allows cognition to be distributed across individual human minds, social groups, and resources in the external environment (Zhang, 1997; Holland et al., 2000).

In the example with the key given above a cognitive task was solved by exploiting already existing environmental resources. Distributed cognition however, also includes another class of cases where cognitive tasks are solved by the use of specially created *cognitive artifacts*.

The midwife's wheel is a good example of such an artifact. The wheel consists of two discs, one marked with the months of the year and the other with the weeks 1 to 42 of a normal pregnancy. Given the date of conception, the discs are aligned in accordance with a simple algorithm, and both the due day and the current duration of the pregnancy can be read off from the artifact. This artifact allows the user to substitute complicated mental calculations with epistemic actions; she simply manipulates the disks and reads off the result. Notice, that the artifact in this example does not enhance or strengthen cognition. Instead, mental calculations are substituted with an entirely different task: the correct alignment of discs.

Artifacts might also change or influence the total outcome of a task. To use an example most academics might have experienced, a talk delivered by heart is different form a talk read from a manuscript. And a talk given by the help of Power Point is something different again. Paper and Power Point-presentations serve as excellent external substitutes for memory, but they do more than that. The two artifacts offer different possibilities, they make different things easy and hard, and that might change not only the style of the talk, but also the content. In other words, cognitive artifacts are not always neutral tools that simply make it easier for us to do, what we have always done. They might in a more profound way change us and influence the nature and content of the cognitive tasks we perform.

## 3  Cognitive artifacts in mathematics

It is not hard to find examples of cognitive artifacts used in mathematics. The very basic operation of counting is commonly supported by external scaffoldings in the form of tally marks, pebbles, fingers or other objects at hand. Multiplication and other arithmetic operations can be supported by a number of different devices such as written tables, counting boards, abaci, computers, and calculation machines. All of these devices allow mental calculations to be substituted by different types of epistemic actions; using written tables, calculations are substituted by perceptual and search processes, using abaci and counting boards, calculations are substituted by manipulation of physical counters in accordance with given algorithms, and using computers, calculations are substituted by keyboard operations.

Written tables have been in use at least since the second millennium BCE (where tables of multiplication, reciprocals, square- and cube roots were used by the Babylonians (Kline, 1990, pp. 5)). The origin of the abacus and counting boards are more uncertain. The oldest known counting board is the Salamis tablet from the fourth century BCE (Menninger, 1992, p. 299), but the invention of such devices is probably much older; according to linguistic evidence, primitive tallying boards were already used by the Sumerian civilization in the fourth millennium BCE (Nissen et al., 1993, p. 134). So it seems as if the use of physical artifacts and the embodied strategy of substituting mental calculations with epistemic actions is an integral part of mathematical cognition and has been more or less so from the dawn of calculation.

The widespread use of such artifacts raise a number of interesting philosophical questions, not least the question of impact, i.e., has the introduction of such artifacts altered the practice and content of mathematics?

This question is especially pressing in current years as the use of one type of physical artifact—electronic computers—might alter the mathematical practice in a number of ways with the introduction of computer assisted proofs, experimental mathematics, and new visualization techniques. The mathematical community has not yet decided exactly what stand to take on these issues, but it seems likely that an acceptance of one or more of these new techniques will have an impact on both the epistemic standards and the content of mathematics (Tymoczko, 1979).

### 3.1

Apart from physical calculating devices such as those mentioned above, the most important artifacts in modern mathematics are written symbols. In order to shed some light on the cognitive significance of the use of symbols, I will begin with an in-depth cognitive analysis of the Hindu-Arabic numerals. The Hindu-Arabic numerals are no doubt one of the most successful systems of mathematical notation, and by gaining a better understanding of the cognitive role played by the symbols used in the system we might get some understanding of the role played by mathematical symbolism in general.

In one such analysis, Zhang and Norman (1995) compare the performance of Hindu-Arabic, Greek alphabetic, and Egyptian hieroglyphic numerals on multiplication tasks. Zhang and Norman conclude that the superiority of the Hindu-Arabic numerals can (at least in part) be explained by the fact that, compared to the other systems, they allow most of the steps of the multiplication algorithm to be externalized; i.e., performed as epistemic actions using pen, paper, and the numerals of the system in question.

Unfortunately, this analysis suffers from several weaknesses. Most importantly, all three systems of numerals are compared on the same polynomial algorithm for multiplication.[1] I find this questionable, as it is unlikely

---

[1]In a numeral system with base $x$, a number $a$ can be represented in polynomial form as $\sum a_i x^i$. In this representation the algebraic structure of polynomial multiplication of two numbers $a$ and $b$ is: $a \cdot b = \sum a_i x^i \cdot \sum b_j x^j = \sum \sum a_i x^i b_j x^j = \sum \sum a_i b_j x^{i+j}$

that either the Greek or the Egyptians actually used this algorithm. We do not know much about how the Greeks did their calculations, but they probably performed the actual calculation on abaci or counting boards, and only used the alphabetic numerals as a way to record the result (Menninger, 1992, pp. 299). For this reason, the proper unit of cognitive analysis in this case is the numerals and the counting board in combination.

The Egyptians on the other hand used a binary algorithm and not a polynomial algorithm for multiplication.[2] Using the binary algorithm with Egyptian hieroglyphic numerals, multiplication can quite easily—and to a large extend externally—be performed as a series of doublings and reductions. Thus, looking only at the ratio between internal and external workload cannot tell us why or if it is easier to use the Hindu-Arabic than the Egyptian hieroglyphic numerals. Other factors, such as the total number of operations performed, must be considered as well.

In order to get a better understanding of the unique qualities of the Hindu-Arabic numerals, we might instead use the typology of numeral systems developed by Stephen Chrisomalis (2004, cf. our Table 1).

In this typology, the Hindu-Arabic numerals are characterized as positional and ciphered. We shall analyze these two characteristics one by one starting with ciphering.

In comparison with cumulative systems, the advantage of ciphering is the possibility of a much more compact way of writing numbers. The number eight for instance can be written with a single symbol in the Hindu-Arabic system: '8', whereas it takes eight symbols to represent the same number in the Egyptian hieroglyphic system: 'IIIIIIII', and four symbols using Roman numerals: 'VIII'.

Due to the compactness of the script, one would expect calculations in general to take fewer operations in a ciphered than in a cumulative system. Unfortunately very little empirical work has been done in this area, but the hypotheses is backed up by at least one study (Schlimm and Neth, 2008), where the ciphered Hindu-Arabic system is compared with the cumulative Roman system. Using virtual agents to perform a large number of addition and multiplication tasks in ways similar to human agents, Schlimm and Neth found that the number of basic operations, such as perceptual steps, attention shifts, and motor actions, was considerably more numerous when using Roman numerals than when using Hindu-Arabic numerals.

The compactness of ciphered numerals however, comes at a price. In a cumulative system, the value of each power of the base is represented by a repetition of a specific symbol. So for instance, in Egyptian hieroglyphic

---

[2]Cf. (Katz, 1998, pp. 8); in the binary algorithm, the multiplier $a$ is decomposed into its binary representation $\sum a_i 2^i$ and the multiplicand $b$ is multiplied with each term: $a \cdot b = \left( \sum a_i 2^i \right) \cdot b = \sum a_i 2^i b$.

|  | **Additive (sign-value)** The sum of the value of all the numerals gives the total value of the whole number | **Positional** The position of each numeral decides which power of the base the numeral is to be multiplied with. |
|---|---|---|
| **Cumulative** Many signs per power of the base. These are added to obtain the total value of that power. | Egyptian hieroglyphic Roman | Babylonian sexagesimal cuneiform |
| **Ciphered** Only one sign per power of the base. This sign alone represents the total value of that power. | Greek alphabetic | Hindu-Arabic |
| **Multiplicative** Two components per power, unit-sign(s) and power-signs. These are multiplied to give the total value of the power. | Chinese traditional | LOGICALLY EXCLUDED |

TABLE 1. Typology of numerical notation systems (redrawn with small adjustments from Chrisomalis, 2004, p. 42).

system eight tokens of the symbol 'I' means eight, and eight tokens of the symbol '∩' means eight tens (i.e., eighty) and so forth. In other words, in a cumulative system there is an iconic likeness between the value of a power and the number of signs used to represent this value. This is not so in ciphered systems. The sign '8' gives no clue to the fact that its value is eight. In a ciphered system the numerals are conventional symbols, and their values must be remembered. Or differently put: The numerals are meaningless symbols until interpreted.

In sum, from a cognitive point of view the choice of a ciphered over a cumulative system is in fact a trade-off, where a reduction in the number of operations is obtained by increased demands for internal cognitive work (cf. also Schlimm and Neth, 2008).

Turning to the positional character of the Hindu-Arabic system Zhang and Norman (1995) might be right in pointing out that a positional system allows for an easy separation of the power and base dimensions; the power of each numeral of a number is represented by its position, and the base value by its shape. Due to this fact, calculations can easily be broken down into two simpler tasks: 1) calculations involving only the numerals 0 through 9 and 2) the recording of the result of such calculations in the right positions on the paper.

In ciphered, additive systems such as the Greek alphabetic both base and power values are represented by the shape of the individual numeral. In the Greek system eighty is represented by a single symbol 'π', and the reader will have to infer both the base value—eight—and the power value—tens—from the shape of the sign. Calculating using such a sign, you will either have to separate the base and the power dimension in order to reduce the calculation to simpler facts, or you will simply have to memorize the necessary tables for all the numerals used in the system. As such systems need unique numerals for each unit of each decimal order, the tables get very big, and consequently memorizing them pose a considerable challenge to long-term memory.[3] Separating the dimensions on the other hand greatly increases the demands on internal, mental work-load (cf. Zhang and Norman, 1995, for a detailed analysis of the last). Either way, the written numerals of ciphered, additive systems does not seem to offer much support for calculations.

In conclusion, the Hindu-Arabic numerals are a very special kind of symbols. Unlike the symbols of ciphered, multiplicative systems, the Hindu-Arabic numerals allow calculations to be performed largely externally as series of epistemic actions, and unlike the numerals of cumulative systems, they are conventional, i.e., abstract symbols that have no iconic likeness with that, which they represent. Due to these characteristics, the Hindu-Arabic numerals serve as a very effective cognitive tool. They allow calculations to be (largely) externalized and performed as purely formal manipulations of the symbols; you do not need to worry about the meaning of the symbol as you calculate, you only need to remember the correct algorithms and transformation tables, and then the symbols will take care of the rest. The numerals, in other words, allow you to perform calculations as epistemic actions in ways similar to those allowed by physical artifacts such as the abacus or the midwife's wheel. Only, when you use the numerals you have to write down the transformations of the stings of symbols as you go along instead of simply manipulating preexisting physical tokens.

This cognitive characteristic of the Hindu-Arabic numerals is particularly interesting, as the very same characteristic applies to the symbols used in modern mathematics. The symbols are (mostly) conventional signs that have no likeness to that, which they represent. This allows computations to be performed as series of epistemic actions, where the symbols are treated as physical objects and manipulated according to strict, formal rules until

---

[3] The Greek system for instance has 27 different numerals, resulting in a multiplication table with 729 entries

a result can simply be read off from the paper. The modern symbols are a cognitive artifact, allowing mathematical computation to be externalized as a formal and meaningless game with uninterpreted symbols functioning as physical tokens (De Cruz, 2005).

The power and enormous influence of this cognitive tool is witnessed by the fact, that the formalist movement simply identified mathematics with "manipulation of signs according to rules" (Hilbert, 1926, p. 381). From my point of view, this is a mistake. The use of external cognitive artifacts is simply a tool used to do mathematics, and one should not take the tools for the trade. Although this particular cognitive strategy is at the heart of modern mathematics, it does not exhaust what mathematics is. As we shall see in the next section, other cognitive strategies are equally important.

## 3.2

With a few exceptions (such as Diophantus's notation for powers Kline, 1990, pp. 138), abstract symbols apart from numerals were only introduced to mathematics in the late 16th and (mainly) 17th century by mathematicians such as François Viète, René Descartes, William Oughtred and John Wallis.

The introduction of this new cognitive artifact has had a very clear impact on both the epistemic standards and the content of mathematics.

As an example of the first, I will mention the use of formalizability as a criterion on the acceptability of a proof. Although proofs are rarely given as rigorous formal deductions, most mathematicians will only accept a proof if it is somehow made probable that it could be formalized and given as a series of purely formal transformations in a formal theory (cf. Tymoczko, 1979, pp. 60). Such a criterion quite clearly only makes sense in a praxis involving abstract symbols. Furthermore, it rules out many of the proofs created before the introduction of abstract symbols, especially all proofs relying on diagrams in a non-trivial way. In sum it seems clear, that the use of abstract symbols have had an enormous impact on the epistemic standards of mathematics, i.e., the standards used to judge whether a knowledge claim is acceptable or not.

As an example of an impact on content, I will turn to geometry, where the idea of substituting calculations with manipulation of abstract symbols primarily was carried out within the paradigm of analytic geometry, founded by Descartes and Pierre de Fermat in the 17th century. This new paradigm quickly led to new and very powerful ways to solve problems which were difficult or even impossible to solve by traditional synthetic means. Take the advent of the calculus as an example; here the extremely difficult problems of constructing tangents and determining areas were replaced by (more or less) mechanical syntactic transformations of symbols. However, the paradigm of analytic geometry also led to a change in the conception of which objects were accepted in geometry. In analytic geometry, any curve which can be given an equation is accepted. This is clearly an expansion in comparison to the traditional Euclidian framework, where only objects which can be constructed using straight lines and circles were admitted. Ultimately, the adoption of analytic methods led to the discovery of several curves (such as space filling curves and continuous, but nowhere differentiable curves) which were strictly impossible from a geometrical point of view. However, because these objects could be expressed analytically they were accepted anyway (and the geometrical intuition was rejected). So all in all, the introduction of symbols as a cognitive artifact has had a clear impact on the content of mathematics, i.e., which objects to accept.

A number of other types of impact could also be discussed. De Cruz (2005) suggests, for instance, that abstract symbols might serve as a way to 'anchor' semantically opaque concepts (such as square roots of negative numbers) in concrete external representations. This idea is highly interesting, but it is unfortunately slightly problematic. Abstract symbols are only one of many ways to represent mathematical content. Square roots of negative numbers can for instance be represented symbolically (as "$i$"), rhetorically (as "the square root of minus one") or even diagrammatically using the complex plane. When it comes to computations, abstract symbols are clearly superior to rhetoric forms of representation because symbols as we saw above allow computations to be externalized and performed as epistemic actions. However, it is not clear that symbols are similarly privileged when it comes to anchoring opaque content. In fact, complex numbers were at first discovered and handled using purely rhetoric means (in Gerolamo Cardano's *Ars Magna* from 1545). So, De Cruz might be right in noting that abstract symbols are in fact used to anchor opaque content, but it is not clear that they are necessary means of doing so. Perhaps content could be anchored just as well by purely rhetoric or other means.

## 4   Conceptual metaphor

The use of epistemic actions and physical cognitive artifacts implies that human cognition is embodied in a very concrete sense. A disembodied mind cannot put a key in her shoe or operate a tool such as the midwife's wheel. Our physical body offers possibilities and has limitations for interacting with the surrounding world, and these constrains on our bodily interactions condition which artifacts we can and cannot use.

But that is not all. Our body and basic bodily experiences also influence our cognitive life in a much more profound way. As it turns out, we seem to use basic life-world experiences as a way to structure abstract thinking. This structuring is revealed by our heavy use of metaphors taking basic life-world experiences as their source-domain in our conceptualization of abstract phenomena (Lakoff and Johnson, 1980).

Examples of such metaphors are easily found in everyday language. Take for instance the expressions: "I couldn't quite grasp what he was saying", "Everything he said just flew over my head", and "Did you get it?" In all of these examples ideas are conceptualized as physical objects, understanding is conceptualized as grasping or holding such objects, and an exchange of ideas is understood as an exchange of objects. Consequently, the situation of attending a lecture can be understood as a situation, where the lecturer is throwing objects to the audience. When objects are thrown from one person to another, it might be difficult for the receiver to catch the objects, the thrower should be careful to aim his throw at the receiver and so on. In other words, something as abstract as learning and understanding can be structured and understood using the basic bodily experience of someone throwing things at us.

As pointed out in Lakoff and Johnson (1980, pp. 46) ideas can be conceptualized using a wide range of other metaphors. Ideas can be seen as living organisms; they can be born, mature, get old and die, and they can come to fruition or be planted in someone's mind. Ideas can be understood as food; they can be hard to digest, half-baked, rotten, fresh, or hard to swallow. Or ideas can be seen as cutting instruments or weapons: They can be sharp, dull or cut right to the heart of matters.

All of these metaphors help us understanding and structuring the abstract phenomena of ideas using well-known and concrete everyday experiences. The various metaphors highlight different aspects of the target-domain and offer guidance in different situations; When giving a lecture, you should be careful to aim what you are saying at the audience in order for them to catch your ideas, and when going to a debate (which is commonly conceptualized in terms of warfare!) it is wise to bring ideas at least as sharp as those of your opponent.

The cognitive approach to metaphor was introduced in the late 1970's, most influentially by Lakoff and Johnson (1980). According to this approach, the structuring of a concept in terms of concrete experience is not something exceptional or rare. In fact: "[...] metaphor is pervasive in everyday life, not just in language but in thought and action. Our ordinary conceptual system, in terms of which we both think and act, is fundamentally metaphorical in nature" (Lakoff and Johnson, 1980, p. 3).

A point of debate is the exact cognitive significance of such metaphors. When we talk about ideas as objects using the metaphorical expression "I couldn't quite grasp what he was saying", do we also *think* of ideas as objects, or is the metaphor merely a linguistic phenomenon? It is not very hard to find examples of 'dead metaphors', i.e., metaphors which might once have had cognitive significance, but clearly do not have so any more. Take for instance the expression: "I have examined 14 students today". The word 'examine' originates in the Latin word *examen*, which literally means 'tongue of a balance'. So, examining student is—or was—originally a metaphor, where the process of judging the knowledge of a student was described by comparing it to the process of weighing goods at the marketplace. Today however, most English speakers do not know the original meaning of the word 'examine', and they do not think of balances or processes of weighing when they use it. The word has simply obtained a new meaning, and consequently the original metaphor is dead and has ceased doing any cognitive work.

So, how do we know that not all of the metaphors discussed above are dead? This is a very important question that needs to be answered before the cognitive approach to metaphor and language can carry any philosophical weight.

The cognitive approach argues their case by putting forth different types of evidence. One type of evidence is studies in linguistics. Most people would probably not accept expressions such as: "That idea is hard to swallow, I fear it won't grow at all". The two parts of the sentence express the same phenomena of disliking an idea, but still, the sentence seems to be somehow inconsistent. The reason for this inconsistency is the fact that the sentence contains two metaphors exploiting two different source-domains; edibles and living organisms respectively. And although ideas can be understood as both edibles and organisms, they cannot be done so at the same time. The metaphors used to conceptualize ideas must in other words be applied in a coherent way. This suggests, that the analogies to basic experiences expressed in the metaphors still have cognitive significance and structure not only the way we talk about ideas, but also the way we think about ideas.

Similarly, a sentence such as "that idea is full of vitamins" will probably be understood immediately and effortlessly by most people familiar with English, even if it is the very first time they hear the expression. This suggests that the analogy exploited in the ideas-are-food-metaphor is still active and allows us to use knowledge of nutritional facts to understand new aspects of the abstract domain of ideas.

Apart from such linguistic evidence, the basic claim of the cognitive approach is backed up by empirical evidence from neural science. A full review of this evidence is unfortunately beyond the scope of this paper, so I will restrict myself to a more thorough review of the evidence put forth in connection to the metaphors of mathematics (below). The reader is referred to Lakoff and Johnson (1999, pp. 36) for a comprehensive list of the different types of evidence.

## 5  Metaphors of mathematics

Imagine teaching school children the fact that a negative number multiplied by a negative number gives a positive number. The children might simply learn the rule: this is how it is, this is how we play the game of arithmetic. But simply learning the rule does not give the children any comprehension of why it might be so.

If the children are to understand why a negative number multiplied with a negative number gives a positive number, you might start telling them about the number line and explain multiplication in terms of movement and locations in space. It might go something like this: Negative numbers are opposite numbers. Where positive numbers are located to the right of 0, negative numbers are located at similar distances, but to the left of 0. Multiplying any number $b$ with a positive number $a$ simply means walking the distance from 0 to $a$, and then keep on walking until this distance has been repeated a total of $b$ times. Then the number located at the point where you end up is the result of the multiplication. As negative numbers are opposite numbers, multiplying any number $a$ with a negative number $b$, you will have to walk exactly the same distance as before, only in the opposite direction. So, if $a$ is a negative number located to the left of 0, multiplying it with another negative number $b$, you will end up walking to the right, and consequently you end up at a location inhabited by a positive number. Hence, a negative multiplied by a negative gives a positive.

This story might explain why negative numbers multiplied by negative numbers give positive numbers, but none of it is literal. Numbers are not locations on a line, and multiplication is not movement. The explanation making the formal operations meaningful is made up of metaphors, or, to be more precise, it is an inference from one basic metaphor where arithmetic is conceptualized as motion along a path and numbers as locations on the path.

As pointed out by George Lakoff and Rafael Núñez, this metaphor is only one out of three basic metaphors used to conceptualize arithmetic (Lakoff and Núñez, 2000, pp. 54). In the other two, numbers are respectively conceptualized as collections of objects and as objects constructed by other objects. These conceptualizations are visible in expressions such as: "If you add three and four, you get seven," "If you put two and two together, you get four," "Five is made up of two and three," and "If you take three from seven, how much do you have left?" (examples from Lakoff and Núñez, 2000, pp. 54). Lakoff and Núñez also add a fourth metaphor, the measuring stick metaphor, where numbers are conceived as locations on a measuring stick. Based on the linguistic evidence, however, it is hard to distinguish this metaphor from the more general metaphor, where numbers are conceived as locations on a line or path.

Lakoff and Núñez claim the basic laws of arithmetic, such as commutativity, associativity, and even closure of the natural numbers to be derived from the structure of the source-domains of these metaphors. To take one example, we know from basic experience that adding a collection of objects to another collection of objects always results in a collection of objects. When numbers are metaphorically conceptualized as collections of objects, we can infer that the addition of two numbers must always result in a number, and consequently ℕ must be closed under addition (Lakoff and Núñez, 2000, p. 60).

The exact status of this claim is unfortunately somewhat unclear. Is it a genealogical claim, i.e., a claim about how the laws of arithmetic were once established, or is it a psychological claim about how each individual person understands mathematics? In either case more evidence is needed in order to support it.

The use of metaphors in mathematics is not restricted to basic arithmetics. Metaphors are frequently used in more advanced subjects as well. Calculus for instance, draws heavily on metaphors of physical movement. Although all central concepts such as function, continuity, differentiability etc. are now defined in terms of sets, i.e., discrete and motionless entities, in textbooks functions still oscillate, approach, tend to etc. (cf. Núñez, 2004, for an interesting treatment of several examples from textbooks).

In the examples given so far all metaphors take life-world experiences as their source-domain, but metaphors taking already accepted theories as their source-domain also play an important part in both the expansion and the unification of mathematics. As an example of the first, William R. Hamilton's use of mappings between geometry and algebra in his discovery of the quarternions deserves mentioning (Pickering, 1995, cf.), for an interesting analysis of this case). As an example of the second—unification—the modern reduction of virtually all mathematical entities to sets is an obvious case. Numbers are not literarily, only metaphorically sets, lines are not literarily infinite sets of points, and so on.

Turning to the question of the cognitive significance of the metaphors, we shall continue the discussion opened above by reviewing some of the empirical evidence used to argue, that the metaphors of mathematics have real cognitive significance.

One line of evidence comes from the study of gestures. In short, there seems to be a close link between gestures and speech, and it is believed that gestures and speech both reflect the same cognitive processes. A study of college professors teaching calculus suggests that the metaphors of movement used by the professors are active, i.e., when the teachers use words of movement they also think in terms of movement (Núñez, 2004, pp. 68). In a revealing example, a professor describes how the values in an infinite sequence oscillate between two fixed values. While he gives this description, he waves his hand from side to side and presses his thumb and index fingers together as if he was holding a small object. The professor seems to be thinking about the two bounding values as fixed locations in space, and about the values in the sequence as a tiny object which is moving—literally oscillating—between the two bounding locations.

The study of gestures gives strong support to the thesis, that the metaphors used by the professors actually reflect how they think—at least in that particular situation. But, the teaching situation is a very special situation, and quite often teachers use colorful metaphors for purely didactical reasons; the abstract subject must be related to something well-known by the students. This however, does not give us any proof that the professors still think in terms of movement when they return to their offices and start doing mathematical research. The study of gestures cannot give us the final answer to the fundamental question: To whom and when are the metaphors active?

Another line of evidence comes from experimental psychology and the study of the human brain. A number of experiments suggest that at least the natural numbers seems to be encoded in the form of some sort of magnitude. Reaction time experiments for instance, reveal that the time it takes a subject to judge whether a number represented in Hindu-Arabic digits is larger or smaller than a given target depends on how close the given number is to the target; the lesser the distance between target and given number, the longer the reaction time (Dehaene et al., 1990). Furthermore, studies of patients who have lost part of their mathematical capabilities due to injuries of the brain suggest a close connection between basic arithmetic skills, body maps, and spatial maps (Dehaene, 1997, pp. 189). This has been used to support the view that at least basic arithmetic is closely connected to basic life-world experiences of the body and physical space (Lakoff and Núñez, 2000, pp. 23).

This type of evidence should be treated with much care. The evidence is still somewhat inconclusive, and in addition it is questionable to what extend studies of the physical brain can tell us anything about mental phenomena such as thinking and understanding. Does the (supposed) fact that arithmetic is encoded in the same region of the brain as spatial- and body maps really prove, that we understand arithmetic using bodily and spatial experience? Even granted such an intimate relation between the physical brain and understanding, the evidence only supports a limited connection between life-world experiences and mathematics. Only arithmetic is (apparently) located in the brain area in question, while other mathematical capacities are not. This was, for instance, the case in a patient suffering from acalculia (Hittmair-Delazer et al., 1995). Due to the effects of cancer treatment the patient had lost the ability to solve even basic arithmetic problems such as $2 + 3$ and $3 \cdot 4$. Nevertheless, he was still able to solve abstract algebraic problems such as recognizing that $(d/c)+a$ is not in general equal to $(d+a)/(c+a)$. Does this mean that arithmetic is closely connected to life-world experiences, but that algebra is not? And does it prove that all metaphors connecting algebra to space, the body, or physical objects are dead, while metaphors of arithmetic are always alive?

The thesis that the metaphors of mathematics are always active and determines the way mathematicians think, seems to me to be much too strong to be defended. Furthermore, it focuses solely on one cognitive tool. The above review of the cognitive tools used by mathematicians suggest another more modest hypothesis to me: perhaps the cognitive tools, I have reviewed, compliments each other. Mathematics can be done as a formal game, where external symbols are treated and manipulated as meaningless tokens of a cognitive artifact, or mathematics can be done as something meaningful, where the meaning—at least in part—comes from cognitive metaphors adding flesh and blood to the naked symbols.

As we have seen, the Hindu-Arabic numerals are conventional, abstract symbols that have no resemblance of collections or any other objects the numbers might signify. This allows for easy calculations: you do not need to understand anything. As long as you know the formal rules, pen, paper, and the written symbols will take care of the rest. But on the other hand, the symbols and formal rules can be treated as something meaningful, when we metaphorically conceive of numbers as collections, as locations in space and so on. Similarly, calculus is formalized in set theory, which allows for powerful formal calculations, but still, college professors use metaphors of motion when they teach calculus in order to ground the students' understanding of the abstract mathematical theory in bodily experiences.

The reader should keep in mind, that the above is only a description of the cognitive behavior of mathematicians and students. The normative aspect, i.e., how one ought to behave, falls outside of the scope of this paper. The description of the cognitive behavior inspires at least one normative question: Mathematicians seem to use metaphors, but should they keep doing that or was it better if mathematicians practiced mathematics as a meaningless, formal game?

## 6   Is mathematics special?

Finally, we might return to the overall theme of the conference PhiMSAMP-3: is mathematics special? This is a very general question which could be understood in many different ways. In this paper, we have seen that mathematics is not special from an epistemic point of view. Mathematical knowledge is obtained using the very same cognitive tools as many other kinds of human knowledge.

This might lead us to wonder whether mathematical knowledge consists of necessary and absolute truths. One might argue that it does not matter how we get to know, what we know: All mathematical truths are absolute and universal, no matter how they are obtained. Such an absolutistic stance seems hard to defend. As I have demonstrated above, cognitive tools are not neutral. The cognitive tools we use are essentially anthropomorphic; our body and environment determine both which basic experiences that are available to us as source-domain for conceptual metaphors, and which physical artifacts it is possible for us to produce and operate. Human knowledge is in a non-trivial way embodied, and thus shaped by the nature and possibilities of our body and physical surroundings. And that goes for mathematical knowledge as well. Consequently mathematics cannot be absolute and universal—at least not in any strong sense of those terms.

# Bibliography

Campbell, J. I. and Epp, L. J. (2005). Architectures for arithmetic. In Campbell, J. I. D., editor, *Handbook of Mathematical Cognition*, pages 347–360. Psychology Press, New York NY.

Chrisomalis, S. (2004). A cognitive typology for numerical notation. *Cambridge Archaeological Journal*, 14(1):37–52.

De Cruz, H. (2005). Mathematical symbols as epistemic actions: An extended mind perspective. Unpublished on-line working paper.

Dehaene, S. (1997). *The Number Sense: How the Mind Creates Mathematics*. Oxford University Press, New York NY.

Dehaene, S., Dupoux, E., and Mehler, J. (1990). Is numerical comparison digital? Analogical and symbolic effects in two-digit number comparison. *Journal of Experimental Psychology: Human Perception and Performance*, 16(3):626–641.

Hilbert, D. (1926). Über das Unendliche. *Mathematische Annalen*, 95(1):161–190. Quoted from the English translation in (van Heijenoort, 1967, pp. 367–392).

Hittmair-Delazer, M., Sailer, U., and Benke, T. (1995). Impaired arithmetic facts but intact conceptual knowledge: A single-case study of dyscalculia. *Cortex*, 31(1):139–47.

Holland, J., Hutchins, E., and Kirsh, D. (2000). Distributed cognition: Toward a new foundation for human-computer interaction research. *ACM Transactions on Computer-Human Interaction*, 7(2):174–196.

Katz, V. (1998). *A History of Mathematics: An Introduction.* Addison Wesley Publishing Company, Reading MA, 2 edition.

Kirsh, D. and Maglio, P. (1994). On distinguishing epistemic from pragmatic action. *Cognitive Science*, 18(4):513–549.

Kline, M. (1990). *Mathematical Thought From Ancient to Modern Times.* Oxford University Press, Oxford.

Lakoff, G. and Johnson, M. (1980). *Metaphors We Live By.* University of Chicago Press, Chicago IL.

Lakoff, G. and Johnson, M. (1999). *Philosophy in the Flesh: The Embodied Mind and Its Challenge to Western Thought.* Basic Books, New York NY.

Lakoff, G. and Núñez, R. E. (2000). *Where Mathematics Comes From. How the Embodied Mind Brings Mathematics into Being.* Basic Books, New York NY.

Menninger, K. (1992). *Number Words and Number Symbols: A Cultural History of Numbers.* Courier Dover Publications, New York NY.

Nissen, H., Damerow, P., and Englund, R. (1993). *Archaic Bookkeeping: Early Writing and Techniques of Economic Administration in the Ancient Near East.* University Of Chicago Press, Chicago IL.

Núñez, R. (2004). Do real numbers really move? Language, thought, and gesture: The embodied cognitive foundations of mathematics. In Iida, F., Pfeifer, R., Steels, L., and Kuniyoshi, Y., editors, *Embodied Artificial Intelligence*, pages 54–73. Springer, Berlin.

Pickering, A. (1995). *The Mangle of Practice: Time, Agency, and Science.* University Of Chicago Press, Chicago IL.

Schlimm, D. and Neth, H. (2008). Modeling ancient and modern arithmetic practices: Addition and multiplication with Arabic and Roman numerals. In Love, B., McRae, K., and Sloutsky, V., editors, *Proceedings of the 30th Annual Meeting of the Cognitive Science Society*, pages 2097–2102, Austin TX. Cognitive Science Society.

Tymoczko, T. (1979). The four-color problem and its philosophical significance. *Journal of Philosophy*, 76(2):57–83.

van Heijenoort, J., editor (1967). *From Frege to Gödel. A Source Book in Mathematical Logic, 1879–1931.* Harvard University Press, Cambridge MA.

Zhang, J. (1997). The nature of external representations in problem solving. *Cognitive Science*, 21(2):179–217.

Zhang, J. and Norman, D. A. (1995). A representational analysis of numeration systems. *Cognition*, 57:271–295.

# Syntactic analogies and impossible extensions

## Brendan P. Larvor

Department of Philosophy, University of Hertfordshire, Hatfield, AL10 9AB, United Kingdom

E-mail: B.P.Larvor@herts.ac.uk

Mathematicians study shapes, structures and patterns. However, there are shapes, structures and patterns within the body and practice of mathematics that are not the direct objects of mathematical study. Rather, they are part of the explanation of how mathematical study is possible, and thus demand the attention of epistemologists and phenomenologists as well as mathematicians. Partial philosophical accounts of these enabling structures include heuristic in the senses of Polya and Lakatos; principles in the sense of Cassirer; ideas in the sense of Lautman and notions in the sense of Grattan-Guinness (Polya, 1954; Lakatos, 1976; Cassirer, 1956; Lautman, 2006; Grattan-Guinness, 2008). The study of these structures lies in the intersection of mathematics and philosophy because some of these shapes, structures and patterns may eventually submit to mathematical treatment, but others may have a 'Protean' quality that will always escape formal treatment.

The examples given here are heterogeneous in their origins and functions. Cassirer and Grattan-Guinness find their principles and notions (respectively) in applied mathematics and empirical science.[1] Lautman sought the same ideas (in his Platonic sense of 'idea') in mathematics and physics, though he looked longer and harder in pure mathematics than in physics (this doctrine, that mathematics and physics have a common root, is part of his Platonism). The heuristic patterns in Lakatos and Polya are more closely specific to pure mathematics, though this may be an artefact of their contingent mathematical interests. Regarding function, heuristic patterns are not necessarily the deepest of these shapes, precisely because they are evident before (or at least during) mathematical investigation, whereas Lautman's Platonic ideas typically come into view late in the day, when they are instantiated in several different mathematical theories. Cassirer's principles and Grattan-Guinness' notions seem to be intermediate, having both heuristic and ontological significance (reading 'ontological' here in something like Lautman's Platonic-Heideggerian sense).

The aim of this paper is not to undertake the large task of classifying and comparing these various kinds of enabling structures. Rather it is to

---

[1]Cassirer also gives an example of a principle in mathematics—the use of groups. I am grateful to David Corfield for this point, among others.

Benedikt Löwe, Thomas Müller (eds.). *PhiMSAMP. Philosophy of Mathematics: Sociological Aspects and Mathematical Practice.* College Publications, London, 2010. Texts in Philosophy 11; pp. 197–208.

*Received by the editors:* 30 August 2008; 25 September 2009; 30 November 2009.
*Accepted for publication:* 25 February 2010.

explore two kinds of pattern that are, I shall suggest, unique to mathematics: syntactic analogy and generalisation by extension. I hope thereby to illustrate some of the contrasts that such a classification would require.

# 1   Syntactic analogy

Mathematicians prior to Leibniz had already begun to reap the benefits of operating with mathematical expressions without waiting to find out what they mean. As early as 1545, Cardano calculated with complex numbers, though he could not use them in his proofs and had no way of understanding them. The development of symbolic algebra in the early decades of the seventeenth century introduced a distinction between the syntax of mathematical notation and the meanings of mathematical symbols. This allowed mathematicians to work with the syntax first and then, later, make sense of the new expressions it gave them. Of course, this gave rise to a debate among mathematicians about the value and reliability of results obtained this way. It was Leibniz who turned these *ad hoc* devices into something like a programme of research. He experimented with many new symbols in addition to those he introduced that are still in use today (see Cajori, 1925). When the mathematician Tschirnhausen objected to these typographical novelties, Leibniz reminded him that Arabic numerals had once been new, as had letters standing for numbers. Leibniz continued:

> In signs one observes an advantage in discovery which is greatest when they express the exact nature of a thing briefly and, as it were, picture it; then indeed the labour of thought is wonderfully diminished. (Gerhardt, 1899, vol I, p. 375); (Cajori, 1925, p. 416).

This advantage is not unique to mathematics; other discourses have notions that 'picture' the relationships among the elements of their objects. Chemical formulae picture the relations among the elements that constitute molecules. However, in chemistry, the relata matter. We would not suppose that $H_2O$ and $Ag_2F$ (silver subfluoride) must be, on some level, the same stuff just because their formulae picture the same relations between their elements. In contrast, mathematics abstracts from the relata, the better to study the relations. If two mathematical structures are isomorphic, then for many mathematical purposes they are identical. Therefore, if two mathematical formulae (in Leibniz's words) 'express briefly and as it were, picture' the 'exact natures' of their referents, and the formulae are isomorphic, then we should expect the referents to be, at some level of description, identical (or at least, systematically related).

A well-known example from Leibniz illustrates this point. In a letter to Johann Bernoulli of October 1695, Leibniz proposed his general formula for repeated differentiation of a product.[2] He started with Newton's binomial formula:

$$(x+y)^n = x^n y^0 + nx^{n-1}y + \frac{n(n-1)}{1\cdot 2}x^{n-2}y^2 + \ldots + x^0 y^n$$

Notice the inclusion of $x^0$ and $y^0$ to perfect the homogeneity, and thus the 'harmony', of the expression. Leibniz rewrote it as follows, with exponentiation treated as an operator, denoted by '$p$':

$$p^n(x+y) = p^n x p^0 y + np^{n-1}x p^1 y + \frac{n(n-1)}{1\cdot 2}p^{n-2}x p^2 y + \ldots + p^0 x p^n y$$

Now Leibniz made a pair of inspired substitutions. He replaced the sum on the left hand side with a product and replaced the '$p$'s (for 'power') by '$d$'s (for 'differential'). This gave him the general product formula for differentiation:

$$d^n(xy) = d^n x d^0 y + nd^{n-1}x d^1 y + \frac{n(n-1)}{1\cdot 2}d^{n-2}x d^2 y + \ldots + d^0 x d^n y$$

It is not clear what inspired Leibniz to make precisely these substitutions (note the insight necessary to replace sum by product on the left hand side only). Nor can we tell quite how much confidence he placed in this formula before he tested it. We do know that Leibniz must have considered this procedure respectable, because otherwise he would not have shared it with Bernoulli. What is clear is that the simple heuristic of reading $\frac{dy}{dx}$ as a quotient of infinitesimals does not, alone, lead to this result.

As he was unable to prove many of his most important results, Leibniz did have to rely on a combination of corroboration from successful applications and faith that similarities in syntax indicate a common structure. A modern mathematical eye may find irrationality in Leibniz's willingness to rely on syntactic similarities of this sort.[3] After all, the trick works only if the 'signs' do indeed express the 'nature of the thing' perfectly—but if in any given case we know that, then we probably do not need to appeal to features of the notational syntax at all. Leibniz's confidence in his procedure

---

[2](Gerhardt, 1863, vol. III, p. 221); (Coe, 1950, p. 459). Cf. also (Serfati, 2005, pp. 389–390).

[3] "The play of symbols can thus seem opposed, not only to usage, but also, in a certain sense, to rationality" ("Le jeu combinatoire peut ainsi sembler s'opposer, non seulement à l'usage, mais aussi, en un certain sens, à la rationalité.") (Serfati, 2005, p. 390). Strictly speaking, the status of 'knowledge' belongs to results that the mathematical community accepts into the common canon, so judgments of epistemic rationality attach properly to the practice of whole communities of mathematicians rather than of individuals.

arises from his rationalism, and specifically from the metaphysical principle that 'harmony' is an indication of reality (because maximal harmony is one of the merits of the best of all possible worlds). This is the mood, familiar from Pythagoras and Plato, in which rationalism becomes mystical. In any case, the mathematical community did not collectively share Leibniz's rationalism, knew that proofs and explanations were lacking and (eventually) supplied them.

Notice the difference between this case and the many structural analogies in empirical science. Take, for example, Rutherford's planetary model of the atom. This captures some gross features of atoms—the massy centre orbited by (relatively) tiny specks, and the emptiness of almost all of the space 'occupied' by an atom. But no-one would suppose that atoms and solar systems are instances of the same general phenomenon. The gross similarities sustain nothing more than a rough analogy. By contrast, the common binomial form of Newton and Leibniz's formulae, as 'pictured' in the notation, suggests that they are both instances of a single mathematical structure. Certainly, it is reasonable (even without sharing Leibniz's metaphysics) to seek a mathematical explanation of their common binomial form (for which cf. Coe, 1950). The reason is obvious: mathematics studies phenomena 'up to isomorphism'—including, especially, mathematical phenomena. If the notation in each of two cases 'expresses the exact nature of the thing briefly and, as it were, pictures it', and if there is an isomorphism between the *notations*, then we should expect to find a level of description where the two *phenomena* appear as instances of a common mathematical structure. In empirical science, a notational isomorphism would not demand an explanation of the same sort. Isomorphisms between formulae in (say) mechanics and economics might suggest a heuristically useful analogy, but they would not prompt a search for a common genus (unless the formulae were abstracted from their empirical origins and treated as purely mathematical objects). This intimacy between mathematical objects and mathematical notation is unique to mathematics and reflects its nature as the science of structures.

Leibniz was probably the first mathematician to work in this way, but by no means the last. In 1861, John Blissard introduced a general method of this sort. Blissard's technique was to take identities involving sequences of polynomials with powers $a^n$, and create new expressions by changing the polynomials to discrete values and replacing the $a^n$ with the falling factorial $(a)_n \equiv a(a-1)\ldots(a-n+1)$. Many of the resulting identities are both true and interesting. Sylvester named this method the 'umbral calculus' because the newly generated identities are 'shadows' of the generating identities. In the 1970s, Steven Roman and Gian-Carlo Rota developed Blissard's idea rigorously and offered a mathematical explanation of the successes it

brought to Blissard, Sylvester and others (Roman, 1984). Beyond pure
mathematics, there are mathematical practitioners who routinely exploit
syntactic (or 'formal') likenesses. Cartier observes that "For the physicists,
[the word 'formal'] is more or less synonymous with 'heuristic' as opposed
to 'rigorous'". Cartier goes on to give a survey of mathematics conducted in
this 'formal' spirit (including a detailed discussion of the Leibniz's formula;
Cartier, 2000, p. 4).

In terms of our wider enquiry, we have here a pattern with two features.
First, it offers a general heuristic rule: where there is a structural similarity between two mathematical expressions, seek a corresponding structural
similarity between the mathematical matters they express. Second, we can
articulate this pattern almost entirely in mathematical terms. Where Leibniz wrote of 'picturing', we can express our rule in terms of isomorphisms,
understood literally in the usual mathematical sense.

For the sake of contrast, I turn now to a pattern that lacks these two
features.

## 2   Impossible extensions

These are cases in which a function is defined for some limited range of
values in such a way that an extension beyond that range seems impossible.
To take one of the simplest and earliest examples, Descartes popularised the
current exponential notation $a^x$, to indicate $x$ multiples of $a$ (though he was
not the first to use this notation).[4] This notation abbreviated the previous
practice of writing $a^3$ (for example) as $aaa$. Thus defined, exponentiation
requires $x$ to be a natural number, but nothing in this syntax inhibits us from
replacing $x$ by a negative number, a fraction or a complex number.[5] In these
cases, it is relatively straightforward to extend the domain of a function.
For example, power series allowed mathematicians to calculate values of
$e^z$ and of trigonometric functions of $z$ where $z$ is a complex number. Of
course, this required mathematicians to revise their understanding of what
these functions mean. It was precisely the ease of calculating the new values
that demanded this semantic revision. One of the characteristic features of
this period of mathematics is that calculating practice ran ahead of the
semantics, so that there were (in Serfati's terms) 'meaningless forms', that
is, mathematical expressions in use that no-one could explicate, let alone
define.

More interesting are cases where it is not possible to extend a function
simply by calculating values for arguments outside the original domain. In
these cases, Serfati identifies the following pattern: shift attention from the

---

[4]Cf. (Stedall, 2007, p. 399).

[5]Serfati (2005, pp. 262–266) traces this development in some detail, recording the introduction of fractional (Newton), irrational (Leibniz) and imaginary (Euler) exponents.

definition of the function to one or more of its trivial consequences. This corollary may then serve as a definition of the new, expanded function, or at least as a 'bridge' from the original restricted function to its new, extended version. The difficult part is then to find a function on the new, expanded domain that (a) satisfies this corollary of the original definition, (b) coincides with the original function on the original domain, and (c) does some useful mathematical work. Serfati offers several examples,[6] the earliest and simplest of which is Euler's extension of the factorial function from natural numbers to positive real numbers (Serfati, 2005, pp. 366–368). As long as we concentrate on the definition of the factorial function, it seems impossible to extend it beyond the natural numbers. However, Euler found a function that satisfies conditions (a), (b) and (c). But how?

In a letter to Goldbach[7] of 1729, Euler observed that for a natural number $m$,

$$m! = \frac{1 \cdot 2^m}{1+m} \cdot \frac{2^{1-m} \cdot 3^m}{2+m} \cdot \frac{3^{1-m} \cdot 4^m}{3+m} \cdot \ldots \cdot \frac{n^{1-m} \cdot (n+1)^m}{n+m} \cdot \ldots$$

Euler's guiding heuristic here is the use of infinite products. From this equation, it is a short step to this:

$$x! = \lim_{n \to \infty} \left( \frac{1 \cdot 2 \cdot \ldots \cdot n}{(1+x)(2+x)\ldots(n+x)} (n+1)^x \right).$$

Euler explains in his letter that for natural numbers, this function coincides with the factorial function, but it is also well-defined for fractions. Having satisfied our condition (b), Euler then goes on to argue (c) (that this development is mathematically worthwhile). In fact, in a slightly different guise, this function becomes the Gamma function, so with hindsight we may agree that (c) is satisfied.

In terms of Serfati's pattern, the 'bridge' in this case between the original factorial function and the extended version is the trivial fact that factorial satisfies the functional equation $g(x) = x\,g(x-1)$. It is easy to see that Euler's new function satisfies this equation too. Call this function $f(x)$.

---

[6] Factorial of a positive real number; Exponentiation by a complex number or square matrix; Trigonometric functions of complex numbers; Matrix pseudo-inverses; Derivative of a non-differentiable function; Derivative of a function on normed vector-spaces; Union and intersection of $r$-partitions (Serfati, 2005, pp. 366–376). The second and third examples seem ill-chosen as there is no obvious 'bridging' identity.

[7] Eneström number 00715. Quoted from (Fuss, 1843, pp. 3–7).

Then:

$$f(x) = x \lim_{n\to\infty} \left( \frac{1 \cdot 2 \cdot \ldots \cdot n}{x(1+x)(2+x)\ldots(n+x)} (n+1)^x \right)$$

$$f(x) = x \lim_{n\to\infty} \left( \left( \frac{1 \cdot 2 \cdot \ldots \cdot n}{x(1+x)(2+x)\ldots(n+x-1)} (n+1)^{x-1} \right) \frac{n+1}{n+x} \right)$$

$$f(x) = xf(x-1) \lim_{n\to\infty} \left( \frac{n+1}{n+x} \right) = xf(x-1)$$

Despite Serfati's description of the case, Euler did not bother to make this argument. Euler was certainly trying to extend the domain of the factorial function, but he did not seem to be looking for a function that satisfies the functional equation $g(x) = x\,g(x-1)$. Rather, his procedure was to express the factorial function as an infinite product, and then observe that he could calculate this product for fractional values of $m$. Certainly, it took a moment of mathematical imagination to wonder what happens if $m$ is a fraction, and in this sense this case is like the extension of exponentiation to irrational and complex values. However, we do not have mathematical syntax operating ahead of the corresponding semantics as we had in the earlier cases. By Euler's time, addition and exponentiation were perfectly well understood for fractional values, which is all that Euler's infinite product requires. Moreover (as Serfati mentions) Euler did not use the ! sign, which Christian Kramp introduced in his work of 1808 (Serfati, 2005, p. 367n72). Nor, in this letter, did Euler use any other symbol in its place. Rather, he refers to this series, 1, 2, 6, 24, 120, etc. Serfati's chapter is entitled 'forms without meaning' but there is no meaningless form here. Moreover, the bridge Serfati identifies (the functional equation $g(x) = x\,g(x-1)$) does not seem to have played any role in Euler's thinking. The case certainly fits Serfati's 'bridging' pattern, but only in his *post hoc* analysis.

This case is almost three centuries old, and depends in its details on a style of mathematics that now seems antique. What about Serfati's more up-to-date cases? I shall consider his fourth example of an extension, namely, matrix pseudo-inverses (Serfati, 2005, pp. 370–372). Serfati considers two kinds of pseudo-inverse: Moore-Penrose pseudo-inverses and pseudo-inverses defined using a norm on the space of matrices.

Serfati introduces this example by asking "Can we give a meaning to the 'form' $A^{-1}$ if $A$ stands for any matrix, that is to say, possibly not square or if square, non-invertible?"[8] Here again, we see Serfati trying to fit the example

---

[8] "Peut-on fournir une signification à la 'forme' $A^{-1}$, où $A$ est le signe d'une matrice complexe quelconque, c'est-à-dire qui peut être non carrée, ou bien carrée non inversible?" (Serfati, 2005, p. 370).

into the seventeen-century mould of mathematicians using symbols without really knowing what they mean. In fact, Penrose defines a new 'form', $A^\dagger$, to stand for his new function, just as one would expect (Penrose, 1955, p. 407). By the twentieth century, Leibniz's attitude to mathematical symbols was commonplace and no-one would bother to complain as Tschirnhausen did about the introduction of a new symbol. Indeed, the separation of syntax from semantics was complete in theory as well as in practice by the end of the nineteenth century. Consequently, mathematicians seeking to introduce a new concept have no reason to do so surreptitiously by abusing the symbol for an already existing notion. The 'meaningless form' aspect of Serfati's analysis sits even less happily with the case of Moore-Penrose pseudo-inverses than it does with the Euler factorial.

On the other hand, Penrose's paper does illustrate the pattern that Serfati identifies in these extensions: using a trivial consequence of the original definition as a 'bridge' to extend the function to a wider domain. In this case, the bridge consists of four trivial identities. For any invertible matrix $A$, where $A^*$ is the conjugate transpose, or adjoint of $A$:

$$AA^{-1}A = A$$
$$A^{-1}AA^{-1} = A^{-1}$$
$$(AA^{-1})^* = AA^{-1}$$
$$(A^{-1}A)^* = A^{-1}A$$

These four identities follow immediately from the definition of the inverse, and from the fact that the identity matrix is its own adjoint. At the outset of his paper, Penrose proves by construction that for every matrix $A$ there is a unique matrix $X$ such that:

$$AXA = A$$
$$XAX = X$$
$$(AX)^* = AX$$
$$(XA)^* = XA$$

That is to say, for every matrix $A$ there is a unique matrix $X$ that plays the role of an inverse in these four identities. Having proved the existence and uniqueness of this $X$, Penrose labels it $A^\dagger$ and proceeds to prove corollaries and show applications. As this is a formal, published proof rather than a private letter, Penrose gives no account of how he chose just these four identities to serve as a bridge, nor of how he arrived at his construction. Examination of the proof does suggest some plausible guesses; Penrose derives the pseudo-inverse from the fact that $A^*A$, $(A^*A)^2$, $(A^*A)^3$, etc. cannot be linearly independent (rather than simply introducing it as a *deus ex machina* in the sense of Polya, 1954, volume II, *Patterns of Plausible Inference*, p. 148). These hints aside, the proof does not expose its heuristic background. In its proof at least, this case does follow Serfati's 'bridging' pattern. However, given the complexity of the construction it seems unlikely that Penrose first somehow selected the four trivial identities and then later went in search of something that would satisfy them in the general case.

Serfati further illustrates the 'bridging' pattern with matrix pseudo-inverses of another kind, which use norms defined on the space of matrices. Given such a norm, it follows from definitions that for any invertible matrix $A$, this trivial identity holds: $\|AA^{-1} - I\| = 0$ This serves as the bridge. However, in this case, we cannot simply replace $A^{-1}$ by $X$ (as in the Moore-Penrose case), because $\|AX - I\| = 0$ has solutions only if $A$ is invertible. However, for any non-zero complex matrix $A$ there is a unique matrix $X$ such that $\|AX - I\|$ is minimal. Thus by loosening the condition a little, we can define $X$ as the right-hand pseudo-inverse of $A$. As with the Moore-Penrose pseudo-inverse, it seems unlikely that the bridging condition came first. Indeed, whenever we find this pattern, we should expect the choice of bridging condition to be informed by some effective heuristic (the use of infinite products in the case of Euler; linear dependency of the $(A^*A)^n$ in the Moore-Penrose case and perhaps the presence of rounding errors in the case of the second kind of pseudo-inverse).[9]

Serfati's 'bridging' pattern is objectively present in the cases he describes, but (for the reason just given) it cannot function as a heuristic in the sense of Polya or Lakatos. 'Look for a bridge to a useful general function' is hardly a helpful hint. Rather, the pattern that Serfati identifies in these cases is more like a 'dialectical structure' in something like the sense of Lautman. That is to say, it is objectively present in the mathematics but discernible only *post hoc* (Lautman, 2006, especially pp. 228–9). In respect of heuristic usefulness, Serfati's pattern contrasts with the case of syntactic analogy. These two patterns also contrast in respect of mathematical tractability. While it may be possible to articulate the syntactic analogy pattern in mathematical terms, this seems less likely in the case Serfati's pattern.

## 3   Structures in practice

One of the obvious ways in which mathematics differs from other sciences is the freedom that mathematicians have to invent new mathematical objects. In mainstream philosophy of mathematics, this is often understood as a freedom to investigate whichever axiom-systems seem interesting. That

---

[9]I am grateful to Thomas Müller for this last suggestion.

is to say, insofar as philosophy of mathematics treats mathematics as a collection of axiomatic systems, the mathematician's freedom seems limited to the moment when the axioms are chosen. The choice of axioms fixes the theorems; thereafter the task is to identify them correctly. Finding and proving the theorems may take ingenuity, but the possibility of creating radically new mathematics has passed.

In practice, not all of mathematics is axiomatised and even where there are axioms, these do not prevent mathematicians from creating new mathematics (Rav, 1999, pp. 15-19).[10] Serfati's examples of functional extensions remind us that the creation of new mathematics is rarely a matter of choosing axioms. Rather, the growth of mathematics (in these cases) depended on shrewd choices of 'bridge'. These choices are constrained by the requirement that the new function should coincide with the old one on the original, restricted domain, and that it should lead to some insight or useful application. Nevertheless, as we saw in the case of the pseudo-inverses, these constraints do not always pick out a unique candidate. On such occasions, the mathematician is free to make a judgment. Freedom of invention distinguishes mathematics from most other sciences, but we get a better understanding of the scope and character of that freedom by attending to examples such as these than by reflecting on a foundationalist conception of mathematics as the investigation of 'interesting' axiom systems. 'Interesting' makes mathematical decisions sound like exercises of taste, when in fact they are more likely to be exercises of shrewdness (which is not to say that taste plays no role in mathematics). At the outset, I gathered heuristic in the senses of Polya and Lakatos; principles in the sense of Cassirer; ideas in the sense of Lautman and notions in the sense of Grattan-Guinness under the umbrella-term 'enabling structures'. What these structures—and patterns such as Serfati's functional extension—enable is mathematical growth that is free but not random. For philosophers, these patterns enable us to explicate for particular cases the meaning of terms like 'tasteful', 'shrewd' and 'interesting'.

Our first example was Leibniz's appeal to notational symmetry in his discovery of the formula for repeated differentiation of a product. Here too, the philosophy of mathematical practice can claim an advantage over philosophical approaches directed at 'foundational' questions. The thought that mathematics is a science of patterns (to take up the title of Resnik's 1997 book) or that mathematical objects are structures (as argued in Shapiro's work also of 1997) is attractive and plausible to anyone with experience of doing mathematics. However, this plausibility depends in part on cases such as this, in which structure or pattern plays a *heuristic* role. That

---

[10]Cf. also (Corfield, 1997) and (Corfield, 2003, p. 166). This is not to suggest that axiomatisation is never productive or heuristically valuable. Cf. (Schlimm, 2010).

is to say, the study of mathematical practice can account for some of the philosophical intuitions that guide philosophical enquiries that are remote from the practice-oriented approach.[11] Resnik and Shapiro both turn to mathematical practice to explain how a science of patterns is possible, but only as a supplement to their metaphysical arguments. Having argued for structuralism as a metaphysical thesis, they have to supply a corresponding epistemology. Their question is, how is it possible to make discoveries about structures? By contrast, the Leibniz case prompts us to ask, how do structures (in particular, syntactic structures) help us to make discoveries? Structures can play heuristic roles in empirical science, but in the Leibniz case (and in umbral calculus more generally), syntactic structure plays a role unique to mathematics.

## Bibliography

Cajori, F. (1925). Leibniz, the master-builder of mathematical notations. *Isis*, 7(3):412–429.

Cartier, P. (2000). Mathemagics. *Séminaire Lotharingien de Combinatoire*, 44. Article B44d.

Cassirer, E. (1956). *Determinism and Indeterminism in Modern Physics*. Yale University Press, New Haven CT. Translated by Otto Benfrey.

Coe, C. J. (1950). The generalized leibniz formula. *The American Mathematical Monthly*, 57(7):459–466.

Corfield, D. (1997). Assaying Lakatos's philosophy of mathematics. *Studies in History and Philosophy of Science*, 28(1):99–121.

Corfield, D. (2003). *Towards a Philosophy of Real Mathematics*. Cambridge University Press, Cambridge.

Fuss, P. H. (1843). *Correspondance mathématique et physique de quelques célèbres géomètres du XVIIIe siècle*. Académie impériale des Sciences de Saint Petersburg, Saint Petersburg. Quoted from the edition (Fuss, 1968).

Fuss, P. H. (1968). *Correspondance mathématique et physique de quelques célèbres géomètres du XVIIIe siècle*, volume 35 of *The Sources of Sciences*. Johnson Reprint Corporation, New York NY.

Gerhardt, K. I., editor (1863). *Leibnizens Mathematische Schriften*. A. Asher, Berlin.

---

[11] Colin McLarty (2008) argues a similar point with much greater detail and subtlety.

Gerhardt, K. I., editor (1899). *Der Briefwechsel von G. W. Leibniz mit Mathematikern.* Mayer & Muler, Berlin.

Grattan-Guinness, I. (2008). Solving Wigner's mystery: The reasonable (though perhaps limited) effectiveness of mathematics in the natural sciences. *The Mathematical Intelligencer*, 30(3):7–17.

Lakatos, I. (1976). *Proofs and Refutations: The Logic of Mathematical Discovery.* Cambridge University Press, Cambridge. Edited by John Worrall and Elie Zahar.

Lautman, J., editor (2006). *A. Lautman. Les Mathématiques, les idées et le réel physique.* Vrin, Paris.

McLarty, C. (2008). What structuralism achieves. In Mancosu, P., editor, *The Philosophy of Mathematical Practice*, pages 354–369. Oxford University Press, Oxford.

Penrose, R. (1955). A generalized inverse for matrices. *Proceedings of the Cambridge Philosophical Society*, 51:406–413.

Polya, G. (1954). *Mathematics and plausible reasoning.* Princeton University Press, Princeton NJ. Two Volumes.

Rav, Y. (1999). Why do we prove theorems? *Philosophia Mathematica*, 7(1):5–41.

Resnik, M. D. (1997). *Mathematics as a Science of Patterns.* Clarendon Press, Oxford.

Roman, S. (1984). *The Umbral Calculus.* Academic Press, New York NY.

Schlimm, D. (2010). On the creative role of axiomatics. The discovery of lattices by Schröder, Dedekind, Birkhoff, and others. *Synthese.* to appear.

Serfati, M. (2005). *La Révolution Symbolique: la constitution de l'écriture symbolique mathmatique.* Éditions Petra, Paris. With a preface by Jacques Bouverasse.

Shapiro, S. (1997). *Philosophy of Mathematics: Structure and Ontology.* Oxford University Press, Oxford.

Stedall, J. (2007). Symbolism, combinations, and visual imagery in the mathematics of Thomas Harriot. *Historia Mathematica*, 34(4).

# Perspectives on mathematical practice from an educational point of view

## Katja Lengnink[1] and Nikola Leufer[2,*]

[1] Fachbereich Mathematik, Universität Siegen, Emmy-Noether-Campus, Walter-Flex-Straße 3, 57068 Siegen, Germany

[2] Institut für Entwicklung und Erforschung des Mathematikunterrichts, Universität Dortmund, Vogelpothsweg 87, 44227 Dortmund, Germany

E-mail: lengnink@mathematik.uni-siegen.de; nleufer@math.uni-dortmund.de

## 1 Introduction

The mainstream research program in philosophy of mathematics is taking mathematical practice increasingly into consideration. Several publications during the last years serve as witnesses for this change (cf., e.g., Van Kerkhove, 2009; Mancosu, 2008; Heintz, 2000; Buldt et al., 2008)).

A practice-based philosophy of mathematics requires an understanding of what mathematical practice and mathematical knowledge actually are. This is not a trivial question with undisputed answers. Buldt et al. (2008) propose that discrepancies between philosophical interpretations should be approached by "supply[ing] more data from a more varied range of sources, including data established via accepted empirical methods" (p. 324).

We are encouraged by this proposal to provide some insight into the research field of mathematics education,[1] where studying mathematical practice—viz. the practice of elementary school and high school students students learning mathematics—is our everyday business. Mathematics education has established a variety of research techniques to pursue questions of educational interest. In this paper, we are providing an overview to philosophers of significant empirical data and our methods of interpretation.

However, mathematics education has more to offer to philosophers than just serving as a pool of raw material. The questions raised and results obtained by mathematics education are of relevance for philosophy of mathematics under the assumption that there is an analogy between scientific mathematical practice and the learning process of mathematics in school. One of the central claims of the introduction of (Bruner, 1977), the psychologist Bruner, a well-known representative of mathematics and mathematics education, is

---

*The authors acknowledge support of the *Wissenschaftliches Netzwerk* PhiMSAMP funded by the *Deutsche Forschungsgemeinschaft* (MU 1816/5-1).

[1] In the following, we use the expression "mathematics education" as a translation for the German term "Mathematikdidaktik".

Benedikt Löwe, Thomas Müller (*eds.*). *PhiMSAMP. Philosophy of Mathematics: Sociological Aspects and Mathematical Practice.* College Publications, London, 2010. Texts in Philosophy 11; pp. 209–233.

*Received by the editors:* 6 October 2009; 31 January 2010.
*Accepted for publication:* 14 February 2010.

that intellectual activity anywhere is the same, whether at the frontier
of knowledge or in a third-grade class-room. What a scientist does
at his desk or in his laboratory, what a literary critic does in reading
a poem, are of the same order as what anybody else does when he
is engaged in like activities—if he is to achieve understanding. The
difference is in degree, not in kind. (Bruner, 1977, p. 14).

This is certainly a strong assumption, but we would like to stress the
fact that mathematical research activity is a way to achieve understanding
which makes the analogy between learning mathematics and doing research
in mathematics very plausible. Results in mathematics education, derived
from the observation of learning processes, may therefore inspire and focus
research on mathematics as a research discipline.

In Section 2 we shall give a rather short—and necessarily not comprehensive—overview on the aims and methods of research in the field of mathematics education. In Section 3 we sketch our work as reseachers in mathematics education on the basis of two surveys and give some results from
our studies that have a direct connection with philosophical questions on
mathematics. We shall conclude with a short discussion.

## 2 A didactical view on processes of learning mathematics. Research questions and methods

### 2.1 Questions and interests in mathematics education

What are researchers in mathematics education interested in and how do
they approach their field of interest? A brief answer could be: Mathematics
education focusses on the relationship between mathematics, humans and
society (cf. Fischer and Malle, 1985). Thus, their research involves the background and the practice of teaching and learning mathematics throughout
all age groups and all school forms. We work to obtain descriptive, normative, and constructive results. The term "descriptive" in our context
refers to the description and analysis of learning and teaching processes.
With the term "normative", we refer to questions like "What do we want
to be taught and what do we want children to learn?" As we want our
work to have immediate consequences for the mathematics classroom, we
emphasize an additional constructive dimension; this means that results in
mathematics education are translated into ideas and concepts that can be
implemented in mathematics classrooms.

The German mathematical education society *Gesellschaft der Didaktik
der Mathematik* (GDM) states the covered fields of interest and enquiry as
follows (from their English-language webpage):

> Mathematical didactics focuses on the teaching and learning of math-
> ematics within all age groups. It seeks to answer questions such as
> the following:

- What could, or should, students learn in mathematics education?

- How could, or should, mathematical contents be taught, how should mathematical ability be developed?

- What would enable students to derive pleasure from mathematical activity?

Mathematical didactics tries to find answers to such questions by

- Critical investigation with regards to justifying the contents and specific teaching aims within the framework of the general goals of mathematical education,

- Conducting research on learning conditions and teaching and learning processes related to the development of suitable empirical methods and theoretical concepts,

- Investigation and preparation of mathematical content, with the aim of making it accessible to specific learning groups.

Mathematics education tries to approach these questions by employing a range of results and research methods, also taken from other disciplines such as pedagogy, sociology, psychology, the humanities and history. Fields of interest in mathematics education can be mathematical content and the individual learning process. Mathematical content is investigated in a complexity of contexts, such as individual learning, the transmission process, the curriculum and general aims of education. On the other hand, the individual learning process of mathematics, the transmission process and other learning conditions and contexts themselves become the object of our study.

## 2.2   Research methods in mathematics education

Research interests and methods in mathematics education range from subject specific considerations, interpretative studies monitoring school situations to quantitative studies investigating comparative tests and general questions on the relationship of human, mathematics and society. A listing of research questions and approaches cannot be complete; we state some of them to give the philosophical reader an idea of the variety of issues; in our list, we emphasize the proximity to philosophical issues. In our account, we focus on research activities in Germany. Details and an international overview can be found in the "Second International Handbook of Research in Mathematics Education" (Bishop et al., 2003). Since in mathematics education, our concern is to understand mathematical thinking and mathematical procedures, most of our fields of enquiry are strongly dedicated to researching practice. For a philosophical perspective on mathematical practice, our results may be of interest when they touch the process of understanding mathematics between social conditions and human construction. In the rest of this section, we shall give examples of the following approaches:

- *Stoffdidaktik* as an approach of connecting basic concepts of mathematics with intuitions of the learners; (Example 1)

- Empirical investigations (qualitative and quantitative studies);

- Semiotic approaches;

- Sociological approaches; (Example 2)

- Beliefs concerning mathematics and mathematics education.

One field of enquiry in mathematics education genuinely linked to philosophical questions is called *Stoffdidaktik*. The main intention of *Stoffdidaktik* was to simplify and adapt mathematical topics for children. Since an increasing amount of material was to be handled, it became necessary to select and analyse appropriate content during the last decades. Some basic ideas and intuitions of mathematical concepts recur throughout several fields and levels of mathematics. These basic ideas and intuitions are constitutive for learning mathematics as they give some orientation and structure in a complex and diversified field of knowledge. *Stoffdidaktik* also deals with linking mathematics to general human thinking by investigating copulative concepts that allow for the transition between the real and the mathematical world. These basic concepts, regarded as mental models or mental objects are called *Grundvorstellungen* (Vom Hofe, 1995). Under underlying research questions will be approached by means of reflecting mathematics at all levels and researching children's beliefs and conceptions to mathematical objects, e.g., "What are the basic ideas in a certain field of mathematics, to what extent do they constitute it and how do they relate to general human thinking?" "Which conceptions and beliefs do learners have of mathematical objects and how do they relate to the basic ideas of mathematical concepts that are mathematically intended?" "Which educational value can be given to the mathematical object?"

On the one hand, *Stoffdidaktik* is very pragmatic as it opens out in the design of curricula and the construction of special learning arrangements. On the other hand, by aiming at the relation between human thinking in general and mathematical thinking, this very basic field of inquiry implicitly makes use of philosophical ideas and directly translates into philosophical issues.

Another branch in mathematics education examines classroom processes in various ways. There are, e.g., video based studies, differently designed and interpreted in many different ways following different methods. The well-known TIMSS Video Study is one of them (cf., e.g., Stigler and Hiebert, 1997). TIMSS shows and describes different forms of teaching and investigates the impacts on mathematical abilities of children.

For an analysis of the classroom videos, interpretative techniques are applied (cf., e.g., Heinze et al., 2006). By this means, hidden structures and differences in framing that affect the classroom situations from behind the scenes are brought forward; as, e.g., the actors' tacit convictions, experiences and expectations. It is interesting that what the participants perceive as mathematics turns out to be very different in these situations. Hence, mathematics can be looked at as a concept that does vary subjectively and therefore cannot exist independently but is closely bound to humans. If one also takes into account historical studies of how mathematical contents developed, this may generate an interesting question for the philosophy of mathematics: "If mathematics is not looked at as an independent body of knowledge, how is mathematics established in human thinking?" There are numerous studies involving videos showing processes of teaching and learning mathematics used that can shed some light on the philosophical questions.

Fundamentally different from the mentioned approaches are quantitative studies of intervention which are implemented to investigate the impact of certain training types or the use of certain media in classroom (e.g., specific training in problem solving strategies, use of computers, proof, . . . ). Intervention studies typically try to provide evidence for a growth of the level of competence in the group of learners that have undergone the specific training. It is not possible to name all the projects here. To give one example, quantitative projects have emerged from the DFG-funded program BiQua (e.g., Bruder et al., 2004, for problem solving).

Research in education also tries to measure competencies of learners at a given point in time. Large-scale quantitative surveys such as IGLU, PISA, TIMSS, VERA are well-known. These surveys use indicators for the assessment of skills and capacities and generate significant information about the abilities of groups. It is particularly noteworthy to observe how these studies understand mathematical competencies and to analyse the underlying conception of mathematics (cf., e.g., the concept of mathematical literacy underlying PISA in (Baumert et al., 2001), and the critical remarks in (Jahnke and Meyerhöfer, 2007)). On the other hand, specific evaluation may be necessary for some competencies to be actually perceived as mathematical skills. Hence, testing makes use of conceptions of mathematics and may also affect them. Investigating competencies considered valuable in teaching and research can provide background for philosophical discussions.

In particular the use of the computer as a tool suggests the question to what extent thinking and practice depend on the instruments used; is the use of different tools likely to alter our concept of mathematics? These questions are central in mathematics education; e.g., the semiotic approach by Dörfler (2008) and Hoffmann (2006), drawing on Charles Sanders Peirce, or the analysis of the usage of software in class and the resulting changes in practice and mathematics (e.g., Hölzl, 1994, on Cabri-Gometr-Software). The general philosophical issue underlying these studies is "What instruments are allowed in the construction and verification of knowledge?"

As Fischer and Malle (1985) point out, mathematics education draws on research within the complex web of relations between human, mathematics and society. Over the last years more and more sociologically oriented approaches in mathematics education (cf., e.g., Jablonka et al., 2001; Gellert and Hümmer, 2008; Leufer and Sertl, 2009) relate to this. Their work accounts for the learning individual by taking into consideration and reflecting his living and learning conditions. Mathematics education transgresses the borders of the discipline with such approaches that consider the practice of mathematics education as part of the subject of mathematics education; these are typically questions raised by the sociology of education and researchers in this field often employ sociological methods.

The ethnomathematical approach (references can be found in François and Van Kerkhove, 2010, in this volume) raises the issue of the relationship between human, society and mathematics from an interdisciplinary perspective, drawing on ethnology. Ethnomathematics is concerned with the social conditions of mathematical development and the development of mathematical competencies. This field of enquiry suggests a relativist perception of mathematics, looking and accounting for specific and different characteristics of mathematics in different cultures, groups or situations. Understanding mathematics as being socio-culturally shaped and characterised by the conditions of practice is very instructive for research on learning processes. Obviously, the ethnomathematical approach emphasizes the question of contingency, i.e., the question to what extent socio-cultural structures and settings influence the construction of knowledge, and whether different conditions could have or have resulted in a different development of the discipline.

Another field of enquiry in mathematics education that can be seen as a practical application of philosophical theories is the research on beliefs of students, teachers and parents (cf., e.g., Felbrich et al., 2008). Beliefs can be defined as subjective knowledge and theories and are closely related to the perception of mathematics and the ascription of relevance in everyday life situations. This relates to questions of educational value including issues of general education, claims of relevance in everyday life and also the establishment of mathematical literacy.

# 3   Mathematics education from an inside perspective. Two examples.

We shall expect the style of approach of mathematics education to mathematical practice or the practice of learning mathematics via the following two case studies. The first example is from the field of *Stoffdidaktik* and investigates the learners' conceptions of negative numerals and the historical and mathematical background. The second example illustrates a sociologically oriented issue and demonstrates the impacts of situative contexts on learners' competencies when working on mathematical problems.

## Example 1: Basic concepts and student ideas—highlighting the relationship between humans and mathematics.

Looking at learning processes, researchers usually observe differences in the basic associations and concepts of learners and teachers. The intended mathematical understanding of theories and their concepts often is not self-evident. For example the calculation $2 - (-5) = 7$ is mostly calculated correctly, but appears mysterious to many children: The calculation rules for integers are used correctly, but from the viewpoint of the children the result does not make any sense. Subtraction in primary school is a process of reduction. Therefore, how can the result of a subtraction be more than the minuend? Such differences between the associations of the learners and the intended mathematical ideas and concepts are widespread and ingredients of every learning process. They give us pointers to the immanent obstacles of a mathematical subject, often the same that proved to be barriers in the historical development.

In the following example, we analyzed the ideas of 13 high-performing children in the fourth grade concerning negative numerals, focussing on the differences and semantic variances between intuitive and mathematical intended concepts. This investigation has been designed by Lengnink and Linnenbaum and implemented by Linnenbaum (2009). The students were asked to solve six different tasks and write down there solutions in a mathematical journal. The examples of the journal entries give an example of how the thinking processes of learners can contribute to clarifying the philosophical problems arising with the extension of the number range.

## The semantic field of integers

On the basis of the question "What do you associate with negative numerals (*Minus-Zahlen*)?" the children were supposed to unfold their intuitive knowledge and their previous experiences with this mathematical subject. The children were asked to write into a mathematical journal. The results are widely spread in respect to different interests and outlooks.

Some children are interested in specific mathematical topics: Where do the negative numerals come from? When did they occur first? How do people calculate with them and why? Even the representation of integers was raised as an issue. Consider the diary entry of Ben who draws a number line and explains: "Ein Zahlenstrahl ist nich an der Null zuende. Jeder Zahlenstrahl der Welt ist unendlich ($\infty$)."[2]

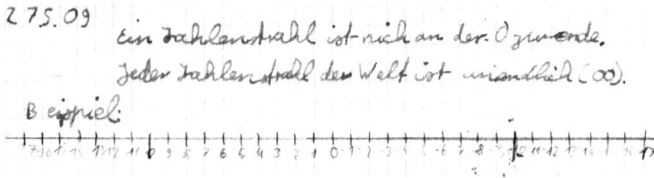

2 75. 09

*Ein Zahlenstrahl ist nich an der 0 zuende.*
*Jeder Zahlenstrahl der Welt ist unendlich ($\infty$).*

*Beispiel:*

Negative numerals are often seen as the result of a subtraction task. Fabian writes "$100 - 1000 = -900$", and then tries to calculate $500 - 634$. This process shows the difficulties and the uncertainty with the operation rules, as he writes: "$500 - 634 = -246$ glaub ich".[3]

For some students it is important to search for counterparts from everyday life. Negative numerals are discovered on the thermometer or as debts. Some children mentioned the CD-player and the alarm clock as an example for negative numerals, where you can go back by a push button to get the last title or a former time (we are not entirely sure what the children were referring to).

In the children's approach, a general phenomenon of development of number extensions arises. Whereas some learners accept the negative numerals as objects with certain calculation rules, others search for coherent contexts and sustainable associations for them in real life. They need such an analogue establishing a semantic foundation, which allows them to calculate appropriately.

The same phenomenon can be observed in the historical development. Whereas in China negative numerals occur as solutions of systems of linear equations (since 100 BCE), even Descartes did not want to accept them calling them "false solutions". It is not just a reflection of the individual intellectual development of children if they find cognitive access to negative numbers difficult: historically, the concept of negative numbers needed time to be accepted.[4]

The intuitive associations and the mathematically intended basic concepts of integers very often do not match; in *Stoffdidaktik*, the *basic concepts* of negative numeral are extracted from a mathematical perspective in order to bridge this gap.

---

[2] A number line does not end in 0; every number line of the world is infinite ($\infty$).
[3] $500 - 634 = -246$, I believe.
[4] Cf. (Seife, 2000; Alten et al., 2003).

Der folgenden Tabelle könnt ihr die Höhenlagen
weiterer Städte in den Niederlanden entnehmen:

**Niederlande**     N

0   25   50 km

| Alkmaar | 3,5 m unter NN |
|---------|---------------|
| Amsterdam | 0 m über NN |
| Apeldoorn | 8 m über NN |
| Arnhem (Arnheim) | 10 m über NN |
| Breda | 0,5 m über NN |
| Middelburg | 0,5 m unter NN |
| Rotterdam | 6,5 m unter NN |
| Sneek | 1 m unter NN |
| Utrecht | 1 m über NN |

A
B
C  Nordsee
D
E  W
F
G
H
I

Sneek
•-1

Alkmaar
•-3,5

**Amsterdam**
◉ 0

Utrecht
•1

•Apeldoorn
8

Arnheim
•10

-6,5•
Rotterdam

Middelburg     Breda
•-0,5        •0,5

O

DEUTSCH-
LAND

0,5 ≙ 0,5 m über NN
-3,5 ≙ 3,5 m unter NN

BELGIEN  S

## 2. Auftrag:

*Welcher Ort liegt am höchsten, welcher am Niedrigsten.*
*Stelle alle Orte in einer Reihenfolge übersichtlich dar.*
*Eventuell hilft dir eine Zeichnung dabei.*

FIGURE 1. English translation: "You can find the altitudes of more cities
in the Netherlands in the following table: Alkmaar 3.5 m below sea level;
Amsterdam 0 m above sea level; Apeldoorn 8 m above sea level; Arnhem 10
m above sea level; Breda 0.5 m above sea level; Middelburg 0.5 m below sea
level; Rotterdam 6.5 m below sea level; Sneek 1 m below sea level; Utrecht
1 m above sea level. Task 2: Which city is the highest; which is the lowest.
Present all cities clearly in an order. Maybe a picture helps."

### 3.1   Mathematical basic concepts and student ideas

What do children do when they are asked to solve mathematical tasks with
integers? The children were asked to work on the task listed in Figure 1.
The students came up with several different representations. For instance,
Ben drew a vertical section (Figure 2) and describes his figure as follows:

> ich habe mir die Aufgaben angeguck und sie nach dem Alphapet
> geordnett. So hatte ich es sehr leicht. Manche Striche gehen nach
> oben, manche gehen nach unten weil: es ein über und ein unter giebt[5]

[5]I looked at the tasks and ordered them [the cities] alphabetically. Therefore it was

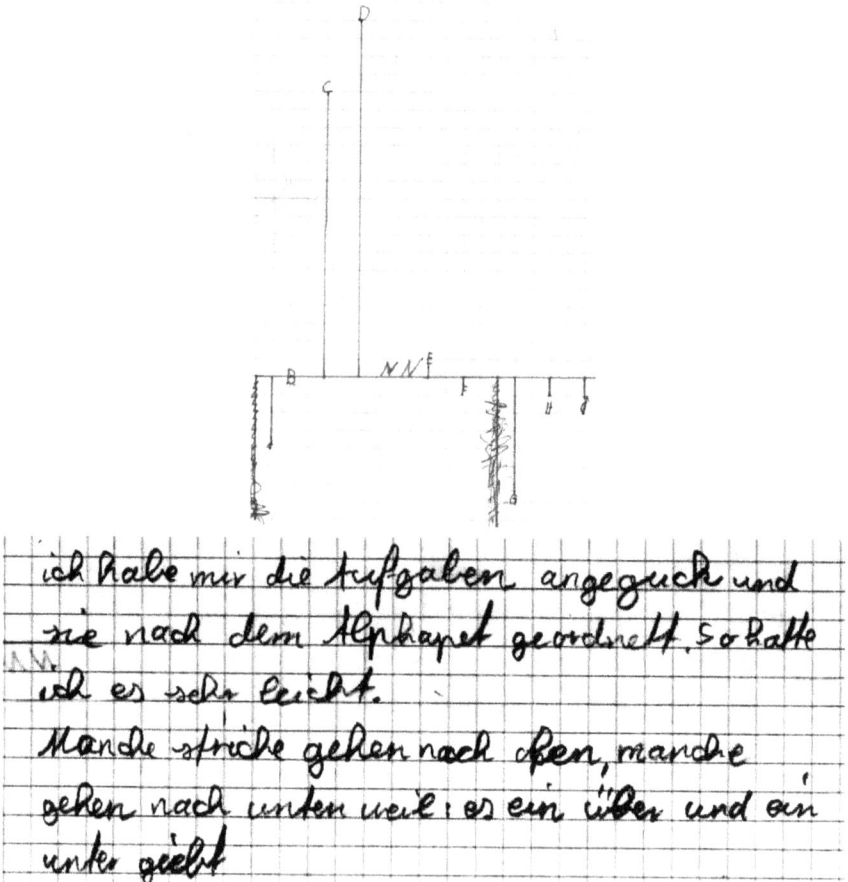

ich habe mir die Aufgaben angeguck und
sie nach dem Alphabet geordnet. So hatte
ich es sehr leicht.
Manche striche gehen nach oben, manche
gehen nach unten weil es ein über und ein
unter giebt

FIGURE 2.

With this representation he is able to solve all further tasks concerning altitude and temperature differences. Ben consistently interprets the subtraction as "up to".

This can be seen in Sven's solution as well who uses the analogy of height and temperature and represents the cities on a scale as given in Figure 3 (note that the zero is not written). Sven distinguishes between the sign of an integer and the arithmetic operator. The operator is marked in red and means "up to" and the sign is marked in blue and means "below sea level".

---

very easy for me. Some lines go up, some go down, because there is an 'above' and a 'below'.

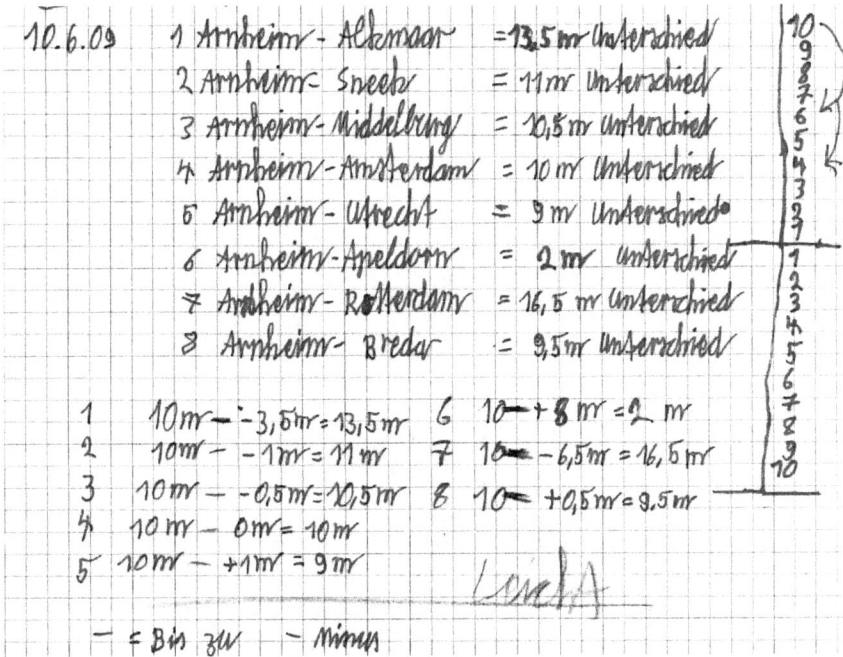

FIGURE 3.

This stresses two basic conceptions of integers which need to be developed in the learning process for an appropriate understanding: The description of a current state with integers (an actual temperature, height, ...) has to be distinguished from a change of state (variation of temperature, variation of account, variation of water-level, ...) which is also described by integers (cf. Postel, 2005). In the first interpretation − (minus) is a sign, in the second one we interpret − as an arithmetic operation. Sven seems to understand this difference intuitively.

It is an important step in the learning process to interpret negative numerals as natural numerals with a negative sign. This can also be detected as a step in the historical development of the integers (cf. Malle, 2007). As negative numerals occur in the process of solving systems of equations, it becomes necessary to find a notation and calculation rules for those numerals. This was a desideratum at one particular time in the historical development of mathematics and similarly, this is exactly what Fabians asks for: he wants calculation rules. Even though he calculates correctly, he comments (Figure 4: "ich weiss nicht wie mann über und unter zahlen

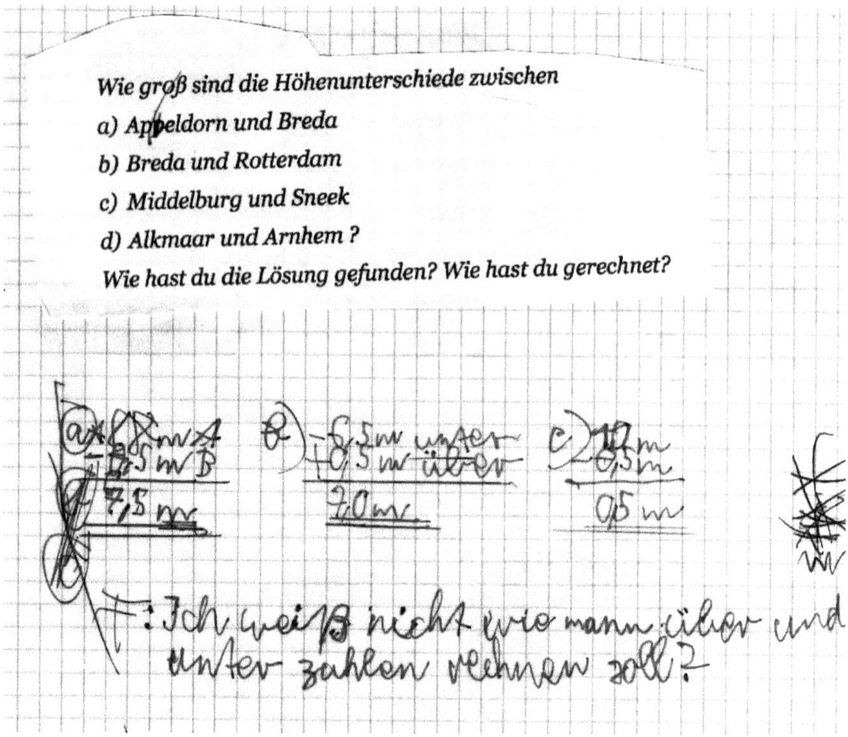

FIGURE 4.

rechnen soll".[6] He probably has problems dealing with the non-standard notation of his calculations. This problem is not shared by others. For example, Nadine confidently describes her handling of integers (Figure 5: "mann rechnet die umkeraufgabe und das ergebnist die − Zahl".[7] Nevertheless there are also children who have difficulties calculating with and interpreting integers. The order especially causes difficulties. For instance, Nils puts the cities above and below sea level in different categories (Figure 6) and comments: "Ich habe erstmal die Uberstehende Orte von 0 m–10 m, dann habe ich die unterstehende Orte von 1.5 m–6.5 m geordnet."[8] The numbers are interpreted as positive integers in both groups. Even after introducing the number line as a representation tool for integers, there is still room for interpretation. Being on the left-hand side is not automatically

---

[6] I do not know how to calculate with "above" and "below" numbers.

[7] One calculates the inverse task and the result is − numeral.

[8] I first ordered the "above" cities from 0 m to 10 m, and then the "below" cities from 1.5 m to 6.5 m.

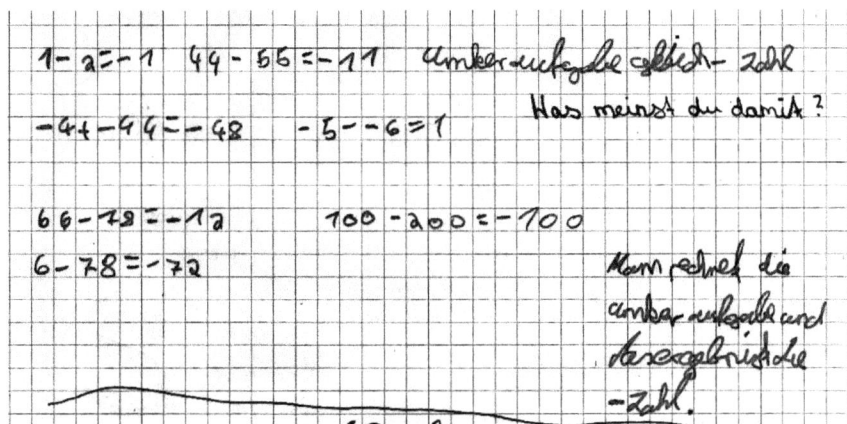

FIGURE 5.

interpreted as being smaller. For instance, Nils argues: "Für mich ist −5 größer als −3, denn es liegt weiter von der 0 entfernt."[9]

Considering the context of temperature and debts the mathematical order sometimes does not make any sense at all. The encountered tensions and frictions can be clearly seen in Jonas's journal (Figure 7). Even though his order of integers matches with the mathematical one, he is confused:

> Das stimmt nicht. Denn −2 C° ist höher als −11 C° Unter Null sind es mehr das stimmt aber ich finde trotzdem −2 C° ist höher. Das hatte ich ja auch schon in den Aufgaben dafor begründet. Wenn man schon Schulden hat und dann trotzdem noch mehr abhebt wird er größer. Wenn man beim Kontostand noch mer abhebt wird er kleiner weil er dann noch tiefer sinkt.[10]

Finally, the tension between everyday life and mathematics can be seen in the following comment as well (Figure 8): "Anja rechnet nach der Große der Zahlen. Bea rechnet mathematisch."[11] Postel (2005) points out that the usual interpretation of the order of integers has to be discussed in school lessons. The interpretation "is less than" has to be combined with the concept of the number line and the left-right-orientation.

---

[9]For me, −5 is bigger than −3 since it has greater distance from 0.

[10]This is not right since −2 C° is higher than −11 C°. It is more below zero, that's right, but I still think that −2 C° is higher. I had already argued for this in the last tasks. If you are in debt and withdraw even more money, then it becomes bigger. If you withdraw more from the account balance, it becomes smaller as it falls even more.

[11]Anja calculates with the size of numerals. Bea calculates mathematically.

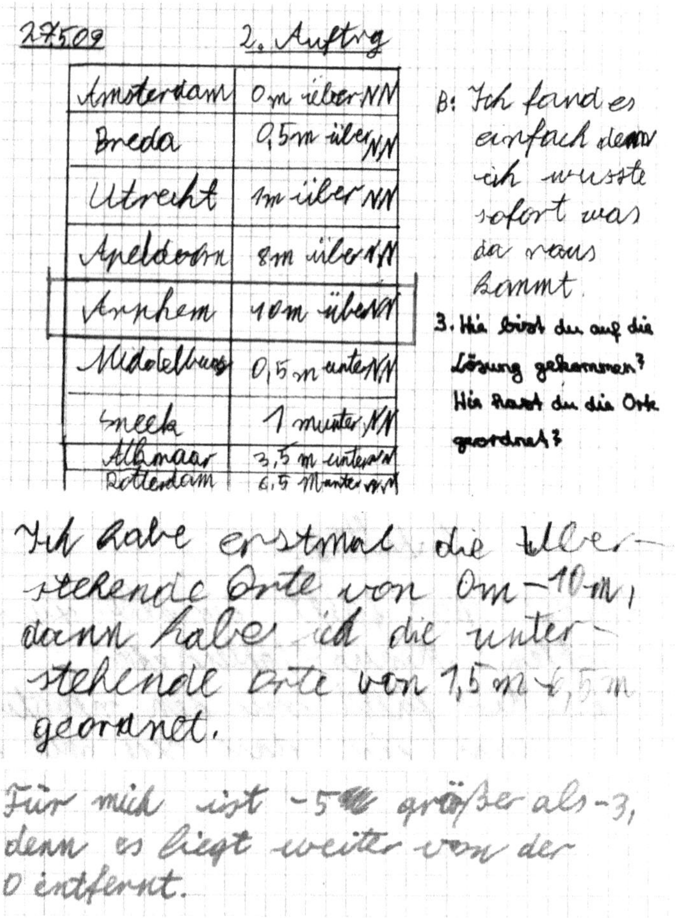

FIGURE 6.

## Conclusion Example 1

The problems of learners in the field of integers can be clarified by an analysis of the historical development and the intuitive student ideas. They are based on the fact that the basic concepts in this mathematical subject are very abstract. While it was possible for the number ranges introduced prior to the integers (natural and positive rational numerals) to illustrate the concept by way of concrete experiences and handling with concrete objects, this cannot be done here. Vom Hofe (1995) therefore distinguishes between primary and secondary basic concepts; the basic concepts occurring in the field of integers are secondary concepts. They have to be built

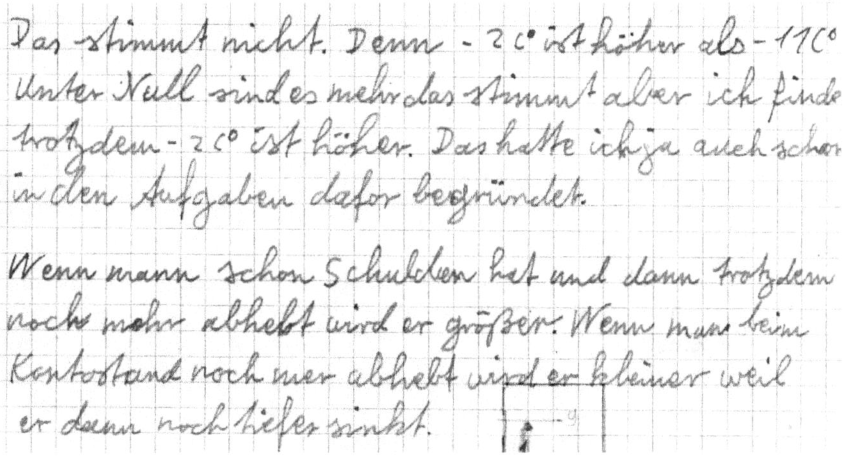
Das stimmt nicht. Denn -2C° ist höher als -11C°
Unter Null sind es mehr das stimmt aber ich finde
trotzdem -2C° ist höher. Das hatte ich ja auch schon
in den Aufgaben dafor begründet.

Wenn mann schon Schulden hat und dann trotzdem
noch mehr abhebt wird er größer. Wenn man beim
Kontostand noch mer abhebt wird er kleiner weil
er dann noch tiefer sinkt.

FIGURE 7.

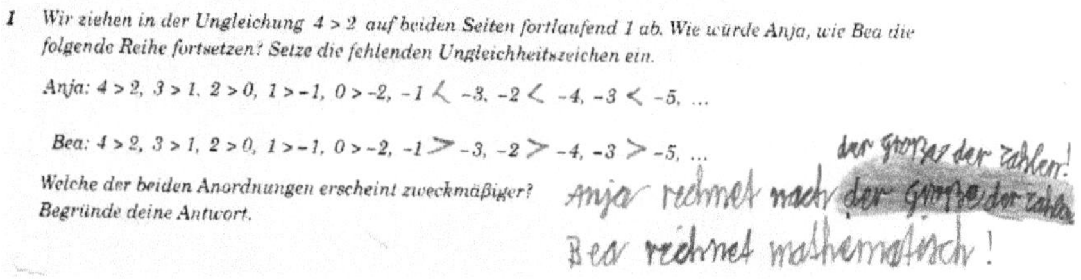
*1* *Wir ziehen in der Ungleichung 4 > 2 auf beiden Seiten fortlaufend 1 ab. Wie würde Anja, wie Bea die folgende Reihe fortsetzen? Setze die fehlenden Ungleichheitszeichen ein.*

*Anja:* 4 > 2, 3 > 1, 2 > 0, 1 > -1, 0 > -2, -1 < -3, -2 < -4, -3 < -5, ...

*Bea:* 4 > 2, 3 > 1, 2 > 0, 1 > -1, 0 > -2, -1 > -3, -2 > -4, -3 > -5, ...

*Welche der beiden Anordnungen erscheint zweckmäßiger? Begründe deine Antwort.*

der größe der zahlen!
Anja rechnet nach der größe der zahlen
Bea rechnet mathematisch!

FIGURE 8.

up by experiments with abstract representations such as the number lines, because a direct correspondent for integers does not exist in everyday life (cf. Postel, 2005, p. 196). The objects corresponding to negative numerals are not concrete and physical, they have a theoretical status. Taking this into consideration, the learning obstacles become obvious.

## Links to philosophy

In the learning process, natural links between general human thinking and mathematical thinking appear, although there are also differences and semantic displacements between the intended mathematical concepts and self-evident ideas. The presented analysis exemplifies how learner perspectives open up our view of the status of the objects in a certain fields of mathematics. Furthermore, the learners' confusions can clarify the discrepancy between intuitive thinking and mathematical formalization. They highlight the implicitness of semantic shifts in the understanding of numerals—

historically and in the personal development. In the example, this can be illustrated by the order of integers, which is counterintuitive and is in conflict with the order to cardinal extent. This issue was an obstacle in history as well and it needed some time to overcome it with an elegant conceptual design.

The semantic content of the integer symbols could be of philosophical interest as well. How is the meaning of integers established? Which semantic shifts occurred in history and how were they triggered? How did integers get accepted? Malle (2007) points out that usually in both the learning process and history there is often formal calculation without any interpretation, purely mathematical. Only later on do explanations and meaning emerge, as acceptance often comes along with familiarization and figural representation of mathematical objects. In the above example, the acceptance of the integers was reached together with the axiomatization of numerals: This progress may be ascribed to Ohm. His claim was not to define terms and numerals by their properties, but to look at their properties in operations and relationships. Emphasizing operational properties instead of some complex inherent nature has considerably eased, and hence promoted, the handling of integers in historical development (cf. Alten et al., 2003, p. 305). Today there are only few children who do not believe in the existence of negative numerals, but for many children they are still mysterious (cf. Malle, 2007, p. 53).

This sort of reflection about the relationship of humans and mathematics rests on the assumption that the growth of knowledge as well as the obstacles to it are analogous in personal and historical development (as indicated in the given example above). To find out which concepts that we have accepted by long-term exposure, it may be interesting to investigate learner perspectives on mathematical subjects, in order to get an exposure-free view of the mathematical concept itself. This methodology, the so-called didactical reconstruction, has stood the test in the didactics of natural sciences (cf., e.g., Duit et al., 2005). From this point of view, mathematics is not independent from the learners. Devlin points that out as well: music is not music if it is only written down on a sheet of paper. It has to be played and listened to, to become music; and the same holds for mathematics. Mathematical symbols aren't mathematics, they have to be thought about by living persons (Devlin, 2000).

## 3.2 Example 2: Mathematics and everyday knowledge: What's the right discourse?

For many reasons, including the mediocre German PISA results, there is increasing demand for a stronger orientation towards the application of mathematical concepts in the mathematics classroom. The OECD has coined

Aufgabe: „Schätzen" 2

Die Klasse 10a möchte einen Ausflug zum Düsseldorfer Stadttheater machen und dort eine Vorstellung besuchen. In der Klassenkasse sind noch 550 Euro. Reicht das Geld aus der Klassenkasse, um den Ausflug zu bezahlen?

FIGURE 9.

the popular notion of "mathematical literacy" as the individual's capacity to "[...] use and engage with mathematics in ways that meet the needs of that individual's life as a constructive, concerned and reflective citizen" (Baumert et al., 2001). Consequently, efforts have been made to fill mathematical syntax with content to emphasize a real-life relevance and prevent mathematics from degenerating to solely technical operations. The aim is to achieve a meaningful handling of mathematical symbols and concepts so that children will develop the competence to apply mathematics in a more confident and flexible way.

As so-called "realistic items" embed a mathematical problem in the context of a (more or less) authentic problem situation, this type of problem is of great importance in most of today's mathematics classrooms. Realistic items connect the real and the mathematical world as shown in the example[12] given in Figure 9. The task, entitled *Schätzen* (estimation) is about a tenth grade class that wants to undertake a field trip in a nearby city to go to the theatre. Ticket price and class size are not given. The question is: "There are 550 Euros left in the class treasury. Is this enough to pay for the trip?" This problem was part of a set of realistic problems given to tenth grade students of a secondary school. The whole interview session with each student took about one hour and took place in the school building. The study's initial interest was to analyse, from a mathematics didactical perspective, how students handled the realistic contexts and how fast and to what extent they would draw on their everyday knowledge to solve the problems. For the interpretation of the students' work, the theoretical model of pedagogic practice by the sociologist Basil Bernstein (e.g., Bernstein, 1996) was used.

In this paper, we shall use the example to focus on the relation between mathematics and real world argumentation and raise the issue of

---

[12]The item is taken from a workshop presentation, held by Andreas Büchter, Dortmund 2006.

legitimate reasoning. As becomes clear by adopting a sociocultural perspective throughout the example, the student's struggling with the mathematical problem turns out to be a—subjective—conflict of determining allowed methods. While in school this issue may be often settled by explicit teachers' expectations and a strong classroom culture that is—presumably—not so in mathematical research. The question of acceptable methods therefore remains an interesting challenge for studying mathematical practice.

## Mathematical capabilities and realistic items

Prija is a quiet immigrant girl with respectable marks in mathematics. This is how Prija worked on the problem:

| 1 | S | [reads the problem out loud] |
|---|---|---|
| 2 |   | But we do not know how much this costs [looks at interviewer] [7 sec silence] |
| 3 |   | It only says that it is ...er... there are 550 Euro in the class treasury, but how much the theatre is, per person, it does not say. |
| 4 | I | Hmm |
| 5 | S | [appears to read the title] Ok, maybe one is to estimate, how much the tickets are [8 sec pause] |
| 6 | I | Go ahead |
| 7 | S | Er ...but we do not know the number of people.... |
| 8 | S | [aborts working on the problems] |

What can we say about the student's mathematical competencies by studying the transcript? First, Prija most probably recognizes the underlying mathematical model of the problem. That is presumably why she needs the quantities "class size" and "ticket price", which must be multiplied to calculate the entire costs. Supposedly, Prija would then match the result with the money available and would solve the problem by this means. From this perspective, she probably does have the technical mathematical abilities to solve the problem, but fails to deal with the problem of the missing values.

Those missing values need to be estimated—as already specified in the title of the item. This means the student needs to come up autonomously with strong assumptions from her everyday experience. These assumptions are strong or rather crucial in the sense that it will follow from this data whether the answer to the problem is "yes" or "no".

Negotiating uncertainties, as required in this task, can be seen as official practice in today's mathematics classrooms. Handling crude data and estimation is, after all, a part and consequence of realistic problem solving and is therefore discussed, practised and also examined in the mathematics class. But, when seeing estimation as a mathematical competence or mathematical ability, then Prija does not have the competencies necessary to solve the problem.

## Mathematics in a social context

It is now interesting to take a closer look: What is it exactly that the student does not know or is not capable of? Prija definitely has problems with estimation. This becomes very clear when she reads the title of the problem "Estimation"—which in effect tells her exactly what she is supposed to do. But, although the interviewer confirms her idea, she does not put the hint into practice. What do these difficulties mean? After all, it is very improbable that a tenth grade student with respectable marks in mathematics cannot answer the question "Estimate how many students are there in a (your) tenth grade class?" It seems more likely that Prija does not understand what exactly she is supposed to do. And, because of her confusion, she stops working.

But, if the student is able to estimate values as required in the problem, but does not make use of this in this situation, then how can we describe the competencies that she lacks?

To do so, we have to look more thoroughly at what exactly is to do at that very point and what might be the cause that hinders the student to demonstrate her abilities. We shall now try to account for the whole setting: The student is used to a very traditional mathematics class and maybe therefore perceives it as strongly classified, which means strongly insulated from other disciplines. In this case study she finds herself in an exertive interview situation that she associates with mathematics class. And she has to work on problems that she may consider as strongly classified as well. From experience or by habit she expects this problem to be stated in an explicit way, requiring a precise and unique answer. To solve the problem however, she has to recognize that the item "Estimation" does not refer to this strong classification, but foils it. The linkage to an authentic realistic situation requires a new practice which, to the student, may not seem legitimate: She has to refer to her everyday life and bring in crucial information from her everyday knowledge. This means the item draws on the student's domestic context, her social background—that potentially differs so radically from the official school context that she is accustomed to ignore and repress it during her time at school. The required ability for the above item at this point is to successfully recognize the context and negotiate confidently between home and domestic or unofficial and official skills and contexts. As this does not match with what the student considers as legitimate reasoning, she stops working on the problem.

The analysis of the transcripts obviously ends up describing a competency that neither is part of the school curriculum nor is evenly spread among the population. Using the terminology of Basil Bernstein, we can reformulate the interpretation: Given the interview situation that Prija may perceive as strongly classified, the student runs the risk of not recognizing the weak classification of the problem. She does not select the "recognition rule" that allows her to recognize the situation and the required discourse. As Bernstein and others have found out, this is mainly a characteristic of children from the working class (e.g., Bernstein, 1996; Cooper and Dunne, 2000). Their findings include that differences in socialisation, particularly in language use (such as familial child communication) seem to have a strong impact on the distribution of recognition rules.

We can conclude from these considerations that Prija might have solved the problem under different conditions that would have helped her to recognize the weak classification of the problem. Different conditions could either mean a different interview framework or a slightly different wording of the task, pointing out the weak classification more explicitly. Other children from the same class but from different social backgrounds might have handled the problem more successfully.

## Conclusion Example 2

Videographing, transcribing and interpreting working processes using the transcript turns out to be a very reasonable method to obtain in-depth ideas and results from an interview situation. The consistent wish to achieve understanding of students' thoughts, workings and problems with this method can obtain interesting results and questions of educational interest. This may be a useful method to investigate in mathematical practice as well.

However, to explain, analyse or even predict Prija's problem, we have to abandon the level of the immediate situation and take more parameters into account: This is where Basil Bernstein's model of pedagogic processes can be of help: The model offers a language, which is capable of linking the microlevel of immediate interaction, the institutional level and the institutional macrolevel of society. The sociological perspective, drawing on Bernstein, systematically enables us to consider as many parameters.

When we stop focussing exclusively on the mathematical problem and instead take into account the active learner, his situative contexts and his societal involvements, boundaries between professional and non-professional, technical and non-technical, official and non-official competencies start to blur. Especially for researching realistic items, which is one example where school and domestic contexts and discourses explicitly and implicitly overlap, a powerful sociological perspective is beneficial (cf. also Gellert and Jablonka, 2009).

Adopting Bernstein's perspective and accounting for as many different levels in the above example, the legitimacy of mathematical knowledge not only becomes an issue, but an issue of social importance. This means, it is not only in question, to what extent it may be legitimate to employ real-life arguments in mathematics. But we can also ask: Who might use to what extent and why real life arguments—and who does not?

As a conclusion from the above considerations, we seem to get at least two results that may be of interdisciplinary interest: On the one hand, the example raises the question to what extent real-life argumentations are acceptable in mathematics classrooms. Generally spoken, this questions the disciplinary legitimacy of arguments and points to Fleck's notion of "thought-styles" in "thought-collectives" (cf. Fleck, 1980).

On the other hand, looking at mathematical practice thoroughly involves some uncertainty as to what we are actually observing and what contexts we are actually dealing with. The examples firmly suggest a situative perspective and an understanding of mathematics as always closely bound to persons—learners or scientists—and situations.

In mathematics education, such a perception of mathematics helps and requires to appreciate not only precise solutions but to account for the subjective constructions of our students. The mathematician and philosopher of mathematics Reuben Hersh (1997) offers a socio-constructive concept of mathematics that seems very suitable for our job: He consistently understands mathematics as human activity (for details, cf. also Prediger, 2004): "[...] mathematics must be understood as a human activity, a social phenomenon, part of human culture, historically evolved, and intelligible only in a social context: I call this viewpoint 'humanist'."

As shown by empirical research, mathematics turns out to be not clearcut at all, but subjective and of some social relativity. Understanding mathematics in a "humanist" way may be instructive and helpful not only for mathematics education. It will surely be an insightful prerequisite for studying mathematical practice as well.

## Summary

In this paper we have attempted to give a rough idea of the variety of research questions and approaches of (German) mathematics education. Our concern was to explain the specific interests of mathematics education to the non-specialist and to promote (some of) our approaches. We have outlined in two research examples how the interest in the process of handling mathematics and in the context in which mathematics is practiced leads to a deeper understanding of mathematical objects and mathematical practice itself.

The mentioned studies emphasize the proximity of didactical and philosophical questions and research in the field of mathematical practice. Maybe the philosophy of mathematics can use some methods and questions of math-

ematics education as a blueprint to investigate the practice of mathematicians. Assuming the analogy of learning and researching mathematics (cf.
Bruner, 1977) we expect helpful results for our field of understanding learning processes better.

# Bibliography

Alten, H.-W., Djafari-Nanini, A., Folkerts, M., Schlosser, H., Schlote, K.-
H., and Wußing, H. (2003). *4000 Jahre Algebra*. Springer, Heidelberg.

Baumert, J., Klieme, E., Neubrand, M., Prenzel, M., Schiefele, U., Schnei-
der, W., Stanat, P., Tillmann, K.-J., and Weiß, M., editors (2001).
*PISA 2000. Basiskompetenzen von Schülerinnen und Schülern im inter-
nationalen Vergleich*. Leske und Budrich, Opladen.

Bernstein, B. (1996). *Pedagogy, Symbolic Control and Identity. Theory,
research, Critique*. Routledge & Kegan Paul, London.

Bishop, A., Clements, K., Keitel, C., Kilpatrick, J., and Leung, F. K.
(2003). *Second International Handbook of Research in Mathematics Education*. Kluwer, Dordrecht.

Bruder, S., Perels, F., Schmitz, B., and Bruder, R. (2004). Die Förderung
selbstregulierten Lernens bei der Hausaufgabenbearbeitung – Evaluation
eines Schüler- und Elterntrainings auf der Basis von Prozessdaten. In Doll,
J. and Prenzel, M., editors, *Bildungsqualität von Schule. Lehrerprofession-
alisierung, Unterrichtsentwicklung und Schülerförderung als Strategien der
Qualitätsverbesserung*, pages 377–397. Waxmann, Münster.

Bruner, J. S. (1977). *Process of Education*. Harvard University Press,
Cambridge MA. 27th printing.

Buldt, B., Löwe, B., and Müller, T. (2008). Towards a new epistemology
of mathematics. *Erkenntnis*, 68(3):309–329.

Cooper, B. and Dunne, M. (2000). *Assessing Children's Mathematical
Knowledge: Social Class, Sex and Problem-Solving*. Open University Press,
Buckingham.

Devlin, K. (2000). *The Language of Mathematics: Making the Invisible
Visible*. Holt Paperbacks, New York NY.

Dörfler, W. (2008). Mathematical reasoning: Mental activity or practice
with diagrams. In Niss, M. and Emborg, E., editors, *Proceedings of the*

*10th International Congress on Mathematical Education, 4–11 July, 2004. Regular Lectures.* IMFUFA Roskilde University, Roskilde.

Duit, R., Gropengießer, H., and Kattmann, U. (2005). Towards science education that is relevant for improving practice: The model of educational reconstruction. In Fischer, H. E., editor, *Developing Standards in Research on Science Education*, pages 1–9. Taylor & Francis, London.

Felbrich, A., Müller, C., and Blömecke, S. (2008). Epistemological beliefs concerning the nature of mathematics among teacher educators and teacher education students in mathematics. *Zentralblatt für Didaktik der Mathematik*, 40:763–776.

Fischer, R. and Malle, G. (1985). *Mensch und Mathematik, Eine Einführung in didaktisches Denken und Handeln*. BI-Wissenschaftsverlag, Zürich.

Fleck, L. (1980). *Entstehung und Entwicklung einer wissenschaftlichen Tatsache: Einführung in die Lehre vom Denkstil und Denkkollektiv*. Suhrkamp, Frankfurt a.M.

François, K. and Van Kerkhove, B. (2010). Ethnomathematics and the philosophy of mathematics (education). In Löwe, B. and Müller, T., editors, *PhiMSAMP. Philosophy of Mathematics: Sociological Aspects and Mathematical Practice*, volume 11 of *Texts in Philosophy*, pages 121–154, London. College Publications.

Gellert, U. and Hümmer, M. (2008). Soziale Konstruktion von Leistung im Unterricht. *Zeitschrift für Erziehungswissenschaft*, 11(2):288–311.

Gellert, U. and Jablonka, E. (2009). "I am not talking about reality." Word problems and intricacies of producing legitimate text. In Verschaffel, L., Greer, B., Dooren, W. V., and Mukhopadhyay, S., editors, *Words and Worlds: Modelling verbal descriptions of situations*, pages 39–53. Sense Publications, Rotterdam.

Heintz, B. (2000). *Die Innenwelt der Mathematik. Zur Kultur und Praxis einer beweisenden Disziplin*. Springer, Vienna.

Heinze, A., Lipowsky, F., and Clarke, D., editors (2006). *Video-based Research in Mathematics Education*. Special issue of the journal *Zentralblatt für die Didaktik der Mathematik*; Volume 38, Issue 5.

Hersh, R. (1997). *What is mathematics, really?* Oxford University Press, Oxford.

Hoffmann, M. (2006). What is a "semiotic perspective," and what could it be? Some comments on the contributions to this special issue. *Educational Studies in Mathematics*, 61(1/2):279–291.

Hölzl, R. (1994). Eine empirische Untersuchung zum Schülerhandeln mit Cabri-géomètre. *Journal für Mathematikdidaktik*, 16(1/2):79–113.

Jablonka, E., Gellert, U., and Keitel, C. (2001). Mathematical literacy and common sense in mathematics education. In Atweh, B. and Forgasz, H., editors, *Sociocultural Research on Mathematics Education: An international perspective*, pages 57–73. Lawrence Erlbaum Associates, Mahwah NJ.

Jahnke, T. and Meyerhöfer, W., editors (2007). *Pisa & Co. Kritik eines Programms*. Franzbecker, Hildesheim.

Leufer, N. and Sertl, M. (2009). Kontextwechsel in realitätsbezogenen Mathematikaufgaben. Zur Problematik der alltagsweltlichen öffnung fachunterrichtlicher Kontexte. In Bremer, H. and Brake, A., editors, *Alltagswelt Schule. Bildungssoziologische Beiträge*, pages 111–134. Juventa, Weinheim.

Linnenbaum, K. (2009). *Vorstellungen zu negativen Zahlen in der Grundschule – Eine Untersuchung mithilfe von Reisctagebüchern*. Universität Siegen, Staatsarbeit.

Malle, G. (2007). Zahlen fallen nicht vom Himmel—Ein Blick in die Geschichte der Mathematik. *Mathematik Lehren*, 142:4–11.

Mancosu, P., editor (2008). *The Philosophy of Mathematical Practice*. Oxford University Press, Oxford.

Postel, H. (2005). Grundvorstellungen bei ganzen Zahlen. In Henn, H. W. and Kaiser, G., editors, *Mathematikunterricht im Spannungsfeld von Evolution und Evaluation*, pages 195–201. Franzbecker, Hildesheim.

Prediger, S. (2004). *Mathematiklernen in interkultureller Perspektive. Mathematikphilosophische, deskriptive und präskriptive Betrachtungen*, volume 6 of *Klagenfurter Beiträge zur Didaktik der Mathematik*. Profil, München.

Seife, C. (2000). *Zero. The Biography of a Dangerous Idea*. Viking, New York NY.

Stigler, J. and Hiebert, J. (1997). Understanding and improving classroom mathematics instruction. An overview of the TIMSS video study. *Phi-Delta-Kappan*, 79:14–21.

Van Kerkhove, B., editor (2009). *New Perspectives on Mathematical Practices: Essays in Philosophy and History of Mathematics.* World Scientific Publishing Co., London.

Vom Hofe, R. (1995). *Grundvorstellungen mathematischer Inhalte.* Spektrum Akademischer Verlag, Heidelberg.

# Learning and understanding numeral systems: Semantic aspects of number representations from an educational perspective

Katja Lengnink[1] and Dirk Schlimm[2,*]

[1] Fachbereich Mathematik, Universität Siegen, Emmy-Noether-Campus, Walter-Flex-Straße 3, 57068 Siegen, Germany

[2] Department of Philosophy, McGill University, 855 Sherbrooke Street West, Montreal QC, H3A 2T7, Canada

E-mail: lengnink@mathematik.uni-siegen.de; dirk.schlimm@mcgill.ca

## 1 Introduction

### 1.1 Mathematical concepts and notation

In recent years philosophers of mathematics have begun to show greater interest in the activities involved in doing mathematics.[1] This turn to mathematical practice is motivated in part by the belief that an understanding of what mathematicians *do* will lead to a better understanding of what mathematics *is*. One obvious activity that mathematicians engage in is that of writing and manipulating meaningful symbols like numerals, formulas, and diagrams. These notational systems are used to represent abstract concepts and objects, and by operating with their symbolic representations we can learn about the their properties. Since such notational systems are crucial ingredients of mathematical practice, a better understanding of such systems and the way we handle them also contributes to a more encompassing understanding of mathematics.[2]

Natural numbers are among the most fundamental mathematical objects. In the history of mankind different linguistic systems for their representation have been invented, used, and forgotten. Most readers will have some familiarity with the system of Roman numerals, which was widely used throughout the Roman empire, but was replaced in the period between 1200 and 1500 CE by a decimal place-value system using what have come to be known as Hindu-Arabic numerals.[3] The exact reasons for this transition are still largely in the dark, although popular accounts of this development speculate frequently about certain deficiencies of the Roman numeral

---

*The authors would like to thank Karen François, Thomas Müller, and Rachel Rudolph for helpful comments on an earlier version of this paper.

[1] See, e.g., the collections (Ferreirós and Gray, 2006) and (Van Kerkhove, 2009).

[2] The influence of numeral systems on the performance of mental numerical tasks has been studied in (Zhang and Norman, 1995); see also (Campbell and Epp, 2005, p. 350).

[3] See (De Cruz et al., 2010, fn. 14) for a brief discussion of this nomenclature.

Benedikt Löwe, Thomas Müller (*eds.*). *PhiMSAMP. Philosophy of Mathematics: Sociological Aspects and Mathematical Practice.* College Publications, London, 2010. Texts in Philosophy 11; pp. 235–264.

*Received by the editors:* 8 October 2009; 3 February 2010.
*Accepted for publication:* 25 February 2010.

system.[4] However, as Schlimm and Neth (2008) have shown, the Roman system has no limitations in principle with regard to the basic arithmetic operations. In the present paper we discuss a different aspect of numeral systems other than that of their aptness for algorithmic computations, namely how different systems of numerals embody semantic information about the represented concepts.

In contrast to natural languages, notational systems are introduced for particular, often very specific purposes. Numerals, for example, are used to represent arbitrary numerosities (collections of one or more objects) and to compute efficiently. In general, being able to use a notational system effectively involves (a) understanding the relation between the symbolism and the represented concepts, and (b) knowing the correct rules for manipulating the notation. For an experienced user of a notational system these two go hand in hand, but for somebody who is starting to get acquainted with such a system, the situation can be very different. In the process of mastering a notation, being able to manipulate the notation correctly can lead to a better grasp of the represented concepts, while at the same time, a good understanding of the concepts can help to differentiate between correct and incorrect usage of the notation. Conversely, persistent incorrect usage of the notation reveals a fundamental lack of understanding of the relationship between the symbols and their meanings.

This latter discrepancy forms the basis of our analysis of two different numeral systems. In particular, we use systematic errors, as opposed to random errors due to carelessness or inattentiveness,[5] in computations by children who learn the decimal place-value system as indicators of a lack of full understanding of the relation between numerals and numbers. Such faulty procedures are also referred to as 'bugs' by Brown and Burton (1978). Here, the operations are carried out mainly as purely syntactic manipulations, without the student having a proper understanding of the semantics of the notation. In his developmental studies, Hughes also found 'serious limitations in children's understanding of arithmetical symbols' (Hughes, 1986, p. 111). By analyzing the kinds of mistakes that are made most frequently we show what semantic information is often not being taken into consideration by the student, and we conclude that the notational system does not straightforwardly convey this particular kind of information.

## 1.2 Semantic content

A few words on the *semantic content* of systems of numerals are in order at this point. The Arabic digit '5', the Roman numeral 'V', and the English

---

[4]See, e.g., (Menninger, 1992, p. 294) and (Ifrah, 1985, p. 431).

[5]We also count wrongly memorized basic addition and multiplication facts as such 'random errors.' Although they might be systematic for the individual they are not related to the structural features of the numeral systems.

word 'five' denote a particular natural number, namely 5. The first two symbols are constituents of a numerical notation system, i.e., a structured system of representation for numbers.[6] If these symbols are considered in isolation, the connection to their referents, the number 5, is by no means obvious. These symbols are *ciphers*: more or less arbitrarily chosen marks that are intended to represent a particular numerosity.[7] Numerals, however, do not usually come in isolation, but as parts of a system. The structure of such a system of numerals allows for the determination of the value of a numeral, i.e., the number it denotes, from knowledge about the basic symbols and the structural principles that are used to connect them. For example, in our decimal place-value system, the first occurrence of the symbol '5' in the numeral '505' denotes 500, while the second occurrence represents 5, and the number represented by '505' is obtained by adding 500 and 5. The situation is different, however, with the Roman numeral 'XX'.[8] Here, both occurrences of 'X' stand for the value 10, and the referent of the entire numeral is obtained by adding 10 and 10, resulting in 20.

In general, the symbols in a place-value system have a base value, but the value of a symbol within a specific numeral also depends on its position. We shall refer to the base value as the *explicit meaning* of the symbol, and to the value that it represents within a numeral as its *implicit meaning*.[9] Grasping the implicit meanings of the symbols is a difficult task for children, and educators have devised many *semantic tools* to make these meanings more explicit, and thus easier to understand. In an additive numeral system, like the Roman one, each symbol has a fixed meaning, regardless of where the symbol occurs in a numeral. In such a system less information is encoded implicitly, making it more concrete than our decimal place-value one. As we shall argue below, this difference has a great impact on the ease with which the place-value and the Roman numeral systems are learned.

## 1.3  Learning arithmetic in the decimal place-value system

In the development of children's arithmetic competency several phases can be distinguished, which have been studied extensively by developmental psychologists and cognitive scientists.[10] Usually at the age of five years[11] children learn a sequence of number words that are associated with small numerosities. While they might be able to recite the number words and

---

[6]This terminology is based on (Chrisomalis, 2004, p. 38).

[7]On the importance of cipherization in numeral systems, see (Boyer, 1944, p. 154).

[8]For more information about place-value systems and additive systems, like that of the Romans, see (De Cruz et al., 2010), this volume. For a brief introduction to the Roman numeral system, see Section 2.3, below.

[9]This implicit meaning can be understood as hidden information about the syntactic notation which enables us to interpret it correctly.

[10]See, e.g., (De Cruz et al., 2010), this volume, for an overview.

[11]See (Hasemann, 2007, p. 9).

write the corresponding numerals, they are initially unaware of the deeper recursive structure of the natural numbers and of their place-value representations. Instead, they develop different conceptions of numbers, which are connected with the symbolic representations of the numerals. At this stage, '21' is read as if it were a single symbol that represents a collection of 21 elements, just as '2' is a symbol for collections of two elements, but no structural relation between the occurrences of the '2' is perceived. An understanding of the internal structure of the place-value representation is achieved gradually and goes hand in hand with the learning of strategies for computing with the basic arithmetic operations.

To illustrate the difficulties that a student encounters when learning the place-value system, let us briefly look at Padberg's suggestion to teach the representation of two-digit numerals in five different ways (Padberg, 2005, p. 65). First, numbers like 23, are to be described as '2 tens and 3 units'. Second, using a table the place-values should be visualized as being distinct; e.g., $\dfrac{\text{T} \mid \text{U}}{2 \mid 3}$. Third, the value of the tens is to be determined and the number written as a sum: $20 + 3$. Fourth, writing the corresponding number words ('twenty-three') is practiced. Finally, fifth, the numeral '23' is introduced. Indeed, Padberg emphasizes that all of these representations are necessary for a successful introduction of the place-value notation and a full understanding of its workings. They are intended to separate the numerals, which are first conceived as single entities, into their constitutive parts. The steps also provide a careful explanation and visualization of how these parts contribute to the value of the numeral.

Also as part of the process, the number words are decomposed and the structural similarities between the number words and the numerals are highlighted. The more regular the number words are in a language, the easier the student can understand this relationship. Chinese and Japanese number words correspond exactly to the symbols in the place-value system (e.g., eleven and twelve being literally translated as 'ten-one' and 'ten-two'), and research has shown that they are learned faster than the number words in English (Miller et al., 1995). German and Dutch number words introduce an additional level of difficulty, since they inverse the positions of the tens and units; e.g., the German word for 23 is '*dreiundzwanzig*' ('*drieëntwintig*' in Dutch, literally: three-and-twenty). As a consequence, by following the order in the number word, German and Dutch students are initially misled into writing 23 as '32', which is also the source of frequent errors in written calculations.[12]

---

[12]This source of error is so common, that there even is a term in German for getting two numbers the wrong way round ('*Zahlendreher*').

As we shall see, other common sources of mistakes made by learners concern the proper handling of zeros, empty places as well as places in general, in particular when operations like 'carrying' and 'borrowing' require the movement of digits from one place to another. Students often take refuge in totally syntactical execution of algorithms often resulting in notational problems and further mistakes.

## 1.4 Overview

In the next three sections we look at the basic arithmetic operations of addition, subtraction, and multiplication. For each of these we first describe the main difficulties that students reveal by making systematic errors in their computations in the decimal place-value system. Then, we discuss semantic tools that have been proposed in the didactical literature and are used in schoolbooks to develop a better semantic understanding of the basic written algorithms for these operations. Finally, we present how the computations could be carried out with Roman numerals, which is a purely additive system of numerals.

Section 2, on addition, will contain the most detailed presentation and discussion, while the other two will deal mainly with matters that haven't been covered before. For this reason, we also omit a discussion of division, since the description of even the most basic common algorithms would be quite cumbersome without bringing about any essentially new features.

# 2  Addition

## 2.1  Examples of difficulties with addition in the decimal place-value system

In the following we present eleven examples of common systematic errors that are made by students learning written addition with the decimal place-value system. Ten are taken from the book *Error patterns in computation: Using error patterns to improve instruction* by Robert Ashlock (1998),[13] and one example (A5) is from (Padberg, 2005, p. 99). These mistakes are the result of strategies that elementary school children have developed, often by abstracting from a small set of examples and over-generalizing. In other cases the students stumbled upon a situation in which they did not know how to continue and fixed the problem in an ad hoc manner, for example by reusing a symbol from the ones column when the tens column was empty. Thus, they have adopted a 'repair' strategy (VanLehn, 1983). Note that these strategies do not always lead to incorrect results, which explains how it is possible that the students have actually learned them: in those cases in

---

[13]We have used different labels for the exercises than Ashlock. We list here how our labels correspond to those of Ashlock: A8: AW1; A3: AW2; A6: AW3; A7: AW4; A1: P4; A4: P5; A10: P6; A2: P7; A9: P8; A11: P9.

which an incorrect strategy led to a correct result the student was reinforced in believing that the strategy was indeed correct.

We now analyze the types of frequent systematic errors made when performing addition and discuss some suggestions that have been put forward in the literature to help the students correct their incorrect grasp of the semantics of the operation (see Figures 1 and 2).

| A1: | A2: | A3: | A4: | A5: |
|---|---|---|---|---|
|  | 9 | 32 |  |  |
| 43 | 8 | 618 | 48 | 27 |
| +26 | + 7 | +782 | +37 | +11 |
| 15 | 105 | 1112 | 75 | 36 |

FIGURE 1. Examples of common addition errors: Columns, direction, operation.

### 2.1.1 Difficulties with columns in general (A1, A2)

In example A1 the student determines the sum of all digits occurring in the problem, regardless of their place-value. The strategy pursued in A2 is more difficult to detect. One possibility is that the top two numbers are perceived as a single two-digit number to which the third number is added (i.e., $98 + 7$ is computed). An alternative strategy to obtain the same result would be first to add the lower two numbers and record the units digit in the result, and then add the carry to the top number.

Students who use the strategy A1 have clearly not understood the essence of the place-value notation. The single digits in a numeral like '43' are considered to be on par with each other. As a consequence, to add two numerals amounts to adding the values of each of their digits. We see clearly in this example that it is possible repeatedly to apply the single digit addition facts (possibly, by just 'counting on'), such that $4 + 3 + 2 + 6$ yields 15, and to know that 15 is represented by '15', without being aware of the internal structure of the place-value notation for numbers. The situation in A2 is similar, but almost in the opposite direction. Here two separate one-digit numerals are read as a single two-digit numeral, despite the fact that the digits are written one underneath the other, instead of one next to the other.

### 2.1.2 Difficulties with direction (A3, A4)

The next two examples illustrate difficulties with the direction in which the columns are dealt with. In A3 the digits are added column-wise, but from left to right, the tens are recorded, while the units are 'carried' to the next column on the right. Example A4 is similar, but the student simply ignores the tens digit.

Learning and understanding numeral systems 241

Both strategies A3 and A4 indicate a good grasp of single-digit addition as well as some understanding of the importance of column-wise addition, but also betray a lack of understanding of the semantics of the numerals. The algorithms used are purely syntactic manipulations.

A related source of errors when two small numbers are added, according to Padberg (2005), is the inversion of two-digit numbers. For example, if the result of $54 + 4$ is given as '85'. In this case, the student has calculated the result of 58 correctly, but recorded the solution in the wrong way. This is an example of a *'Zahlendreher'* (see footnote 12), which is a quite frequent mistake made by German and Dutch students.

### 2.1.3 Difficulties with operation (A5)

In addition to the mistake just mentioned, a second typical mistake that Padberg points out reveals general difficulties with the operation of addition for multi-digit numerals. In A5 the tens are added correctly, but then the smaller unit is subtracted from, instead of added to, the larger unit.

| A6: | A7: | A8: | A9: | A10: | A11: |
|---|---|---|---|---|---|
| | | | | 11 | |
| | 1 | | | 457 | // |
| 26 | 98 | 8 8 | 3 5 9 | 368 | 775 |
| + 3 | + 3 | + 3 9 | + 5 6 | +192 | + 483 |
| 11 | 131 | 1117 | 81115 | 927 | 2158 |

FIGURE 2. Examples of common addition errors: Empty places and carries.

### 2.1.4 Difficulties with empty places in columns (A6, A7)

The error shown in A6 is made by a student who can add two two-digit numbers correctly, but is baffled if one place in a column is empty. In this situation, all digits occurring in the problem are simply added (like in A1). A different way of coping when confronted with an empty place is to look for the 'next best' number and add that one. A calculation that results from this strategy is shown in A7, where the units are added correctly, but when the tens are added, the unit (3) — with an empty place in the tens column — is added again (together with a correct carry).

These two examples illustrate how empty places can be confusing; the students have learned how to manipulate the symbols correctly, but are puzzled about what to do if there is no digit that they can operate on. To resolve the impasse, they change the problem task or take the 'next best' digit, i.e., the one that is in a position close to the empty place. Although the students who employ these strategies would not produce errors in calculations that do not contain empty places, their behavior reveals that they have not yet fully understood the semantics of the numerals.

### 2.1.5 Difficulties with carries (A8, A9, A10, A11)

The most frequent errors deal with the handling of carries. According to Ashlock (1998, p. 101), these mistakes amount to 67% of all errors.[14]

In A8 the digits are added column-wise, but the carries are ignored and written in the row for the solution instead. In other words, the intermediate sums are written out directly in full. A similar behavior is the cause of the erroneous calculation in A9. Here, however, the carry of the units column is used for the tens column, but the empty place in the hundreds column causes the student to reuse the second value of the tens column (like in A7) and to disregard the carry. In the next two examples the carries are used at a later stage of the calculation, but incorrectly. In A10 the student always records the greater digit as the result and uses the lesser digit as the carry to the next column (except for the final, left-most column). In example A11 all carries are collected above the left-most column and are then recorded in the thousands place.

### 2.2 Semantic tools for addition algorithms

The mistakes discussed in the previous section, made by students who learn to compute with the decimal place-value system, are caused mainly by a lack of understanding of the semantics of numerals, and in particular by a lack of understanding of the implicit meanings of the digits. We now present some semantic tools that have been devised to help students to overcome these difficulties. These tools are intended to link the written, formal algorithms with a conceptual understanding of the numerals. In other words, they are intended to build bridges between the learners' conceptions of numbers and the mathematical content of numerical notation. We discuss how various schoolbooks and the relevant didactical literature propose to cope with this semantic gap. In particular we look at two schoolbooks for elementary schools (Becherer and Schulz, 2007; Böttinger et al., 2008) as well as the illustrations from the popular books on mathematics education by Padberg (2005) and Wittmann and Müller (2005).

The aim of the semantic tools discussed below is to make the implicit meanings of the symbols explicit. This is done by offering a different format for representing the values of the digits that occur in the numerals. In particular, this new format is chosen in such a way that it shows the grouping structure of the notation, i.e., the bundling of units into tens, of tens into hundreds, etc., in a visual and compelling fashion.[15]

---

[14]See also (Cox, 1975).

[15]See also the pictographic and iconic examples in (Hughes, 1986, p. 123).

## 2.2.1 Graphically representing the grouping of numbers

Figure 3 demonstrates various ways of displaying, in decreasing levels of concreteness, the semantic content of numerals in simple additions. In the left-hand diagram the values of the digits are represented by *number portraits*, which use boxes of different sizes to show their magnitudes. Here, the repeated occurrences of the basic elements '■', '▌', and '.' function as icons (Pierce, in Hartshorne and Weiss, 1932, p. 247) In addition, analogous three-dimensional representations have been developed to be used in the classroom. These 'Dienes Multi-base Arithmetic Blocks,' which mirror the decimal grouping structure of our numeral system, include small cubes, ten of which are grouped into sticks, and 100 of which fit into a box.[16]

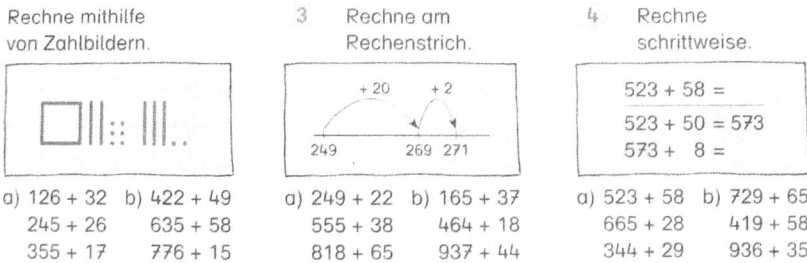

| Rechne mithilfe von Zahlbildern. | 3  Rechne am Rechenstrich. | 4  Rechne schrittweise. |
|---|---|---|
| □▌::: ▌▌▌.. | +20     +2 <br> 249        269 271 | 523 + 58 = <br> 523 + 50 = 573 <br> 573 +   8 = |
| a) 126 + 32  b) 422 + 49 <br>   245 + 26      635 + 58 <br>   355 + 17      776 + 15 | a) 249 + 22  b) 165 + 37 <br>   555 + 38      464 + 18 <br>   818 + 65      937 + 44 | a) 523 + 58  b) 729 + 65 <br>   665 + 28      419 + 58 <br>   344 + 29      936 + 35 |

FIGURE 3. Number portraits to support understanding. The headings above the pictures mean from left to right: 'Calculate with number portraits', 'Calculate with the number line', and 'Calculate stepwise'. Source: (Becherer and Schulz, 2007, p. 34).

Calculations using number portraits like in Figure 3 (left-hand side) are exactly the same procedures as in the Roman system, the only difference is the bundling. Roman numbers have a five and two bundling, our decimal place-value system has only a ten bundling. The middle diagram in Figure 3 appeals to an ordinal understanding of numbers represented as points on the number line. Finally, in the diagram on the right-hand side no additional semantic information is provided, but the values of the single digits are represented directly. The sequence of these three diagrams, from left to right, also shows how semantic information is reduced in the process of formalization.

## 2.2.2 Splitting up numerals

The splitting up of the numerals is also shown in the place-value table in Figure 4, where the place-values are indexed by their grouping (H =

---

[16]See (Davey, 1975).

'Hunderter' = hundreds, Z = 'Zehner' = tens, E = 'Einer' = ones). In addition, the student can also see the relation to the iconic number portraits. Again, this explication of the implicit meanings of the numerals is intended to help students understand the formal addition algorithm.

FIGURE 4. Visualization of the place values as a semantic tool. Source: (Padberg, 2005, p. 210).

### 2.2.3   Providing meaningful context

An alternative kind of semantic tool embeds the computations into a context with familiar objects that have a natural grouping structure, like money in different denominations or fruit that comes in units, crates, or containers. Figure 5 shows the combination of a place-value table with money-substitutes intended to provide a bridge between the student's everyday reasoning in a meaningful contexts and formal operations with numerals.

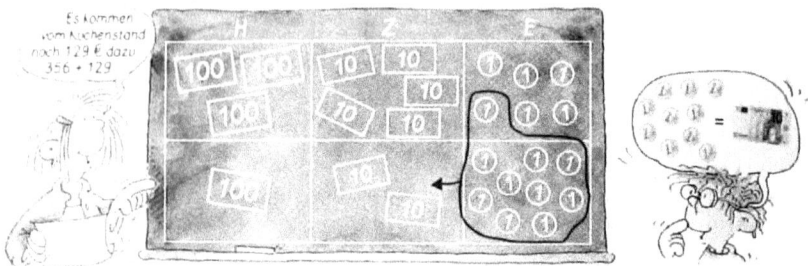

FIGURE 5. Money — a meaningful context for addition. Source: (Becherer and Schulz, 2007, p. 34).

## 2.3   Addition with Roman numerals

We now turn to an alternative notational system for numbers, namely the
Roman numeral system, which is purely additive. Similar systems have
been used in Egypt, Greece, and many other parts of the world.[17] Its basic
constituents are the symbols I, V, X, L, C, D, and M, which represent the
values 1, 5, 10, 50, 100, 500, and 1000, respectively. A numeral is written
as a sequence of basic symbols and its value is obtained by adding the
individual values of the symbols. For example, the numeral 'MDCCLXXI'
stands for $1000 + 500 + 100 + 100 + 50 + 10 + 10 + 1 = 1771$.

The reader might be familiar with the contemporary conventions of writing the Roman numerals IV for 4, IX for 9, XL for 40, XC for 90, etc., but
these were in fact only introduced during the Middle Ages, mainly to ab-
breviate the numerals on inscriptions. Since these subtractive conventions
violate the semantics of a purely additive system, and since they were apparently not used by the Romans themselves, we do not consider them here
to be part of the Roman numeral system under discussion.[18]

In an additive system, addition can be accomplished by simply writing all
of the symbols together and then performing simplifications, i.e., replacing
a group of symbols by a single symbol. For example, five 'I's are replaced
by a 'V', two 'V's by an 'X', etc. Let us consider the problem A8 in Roman
numerals, using a very crude strategy according to which all operations are
executed explicitly:

$$
\begin{array}{r}
\text{LXXXVIII} \\
+ \quad \text{XXXVIIII} \\
\hline
\underbrace{\text{LXXXXXX}}\,\underbrace{\text{VV}}\,\underbrace{\text{IIIIIII}} \\
\end{array}
$$

$$
\underbrace{\quad\quad}_{\text{C}}\ \ \text{L} \quad\ \text{X}\quad\ \text{V}
$$

(copy addends)
(simplifications)

$$
\hline
\text{CXXVII}
$$

While the simplification process may look messy if intermediate steps are
written out, it is accomplished by very simple rules of the form 'if you have
five I's, replace them by a V,' etc., which reflect the grouping structure of the
Roman numerals. As we have seen in Section 2.2, making such groupings
explicit is in fact used in semantic tools to help students understand the
relation between numerals and their values. Thus, they are regarded as
somewhat 'intuitive' in the pedagogical literature, and we can assume that
they are learned easily.

The addition algorithm for Roman numerals can be considerably simplified if the numerals are not written in a linear fashion, but in a more

---

[17]See (Ifrah, 1985).

[18]See (Cajori, 1928, pp. 30–37) for a discussion of the history of Roman numerals and
the subtractive conventions.

structured way, as they would be organized on an abacus. The above problem would then be represented as:

$$
\begin{array}{rlll}
 & \text{L} & \text{XXX} & \text{V} & \text{III} \\
+ & & \text{XXX} & \text{V} & \text{IIII} \\
\hline
 & \text{L} & \text{X} & \text{V} & & \text{('carries')} \\
\hline
\text{C} & & \text{XX} & \text{V} & \text{II} & \text{(simplification)}
\end{array}
$$

Here the simplifications are done while the single occurrences of a symbol are processed and the intermediate results have been written down as 'carries.' Notice how the structure of an abacus, where symbols of the same numeric value are written in a column, resembles the place-value table shown in Figure 4, which is used as a tool for conveying the semantics of the Arabic numerals. We see here that the Roman numeral system reflects more closely the operations on an abacus than our decimal place-value system and thus seems to be more apt to convey the semantics of the numerals.

In our analysis of the common systematic mistakes that are made by students who learn the decimal place-value system (Section 2.1) we have identified five classes of difficulties: with the use of columns in general, with the direction of the algorithm, with the operation, with empty places in a column, and with carries. How would these kinds of difficulties, which can lead to mistakes in calculations in a place-value system, fare with an additive system? Since the Roman numeral system is purely additive, there is no need to keep track of columns; all numerals of the same shape have to be treated together, regardless of where they are positioned. A1's strategy to simply add the explicit values of all symbols occurring in the addends would work perfectly well with the Roman system.

Also the direction in which the single-digit additions are performed plays only a minor role in an additive system. Adding from left to right could possibly lead to a few forgotten simplifications, but they could be detected by inspecting the end result. Transferring carries from left to right, as done by A3, would also be an erroneous strategy in the Roman system. However, it seems unlikely that it would occur to a student who has learned the rule 'two Vs yield one X' to replace two or five 'X's by a 'V.' (This mistake would correspond to the use of a wrong addition fact.) In a place-value system the digits in each column and the carries have the same shape. Thus, one must understand the semantics of the numerals to judge whether a digit is being dealt with correctly. In an additive system, however, the shape alone of a symbol conveys its numerical value so that the semantic content of a symbol does not have to be extrapolated from the position of the symbol within a numeral.

In additive systems there is no symbol to mark an empty place. Thus difficulties that arise in the decimal place-value system in connection with the zero have no analog in the Roman system. A related difficulty discussed above results from an inability correctly to handle empty positions. This difficulty seems to arise because the students have learned to process each digit in a column and they are thrown off course if there is no digit in a certain row of the column they are working on. To solve this problem, they use a digit from a neighboring column. In the Roman system, even if the symbols are arranged in columns as in the previous algorithm, each column contains only symbols of the same shape. The following calculation is analogous to the one performed by A6:

$$
\begin{array}{r}
\text{XX} \quad \text{V} \quad \text{I} \\
+ \quad\quad\quad\quad\quad \text{III} \\
\hline
\text{XX} \quad \text{V} \quad \text{IIII}
\end{array}
$$

In other words, which symbols are to be processed together is determined not only by the columns in which they appear, but also by the shapes of the symbols. In fact, the shape is the only relevant indicator; the columns are merely additional but dispensable aids. To 'borrow' a numeral from a different column would thus violate the rule that symbols of the same shape have to be processed together, and thus is an unlikely strategy for a student to adopt.

Failure to use the carries in the place-value addition algorithm would correspond in the Roman system to either (a) not applying the simplification rules, (b) applying them, but without recording the results of the simplification, or (c) applying them and using the results in an incorrect way. Since the simplification rules are most basic in dealing with Roman numerals, they are akin to the single-digit additions in the decimal place-value system. But the students who have learned the incorrect algorithms discussed above have managed to learn the single-digit additions, so we might assume that they would also have been able to learn the simplification rules, especially since there is only one rule to learn for each different symbol, while there are 10 single-digit addition facts to be mastered for each of the 10 digits in the decimal place-value system. Applying the simplification rule without recording the outcome (case b) can also result in an incorrect strategy for an additive system. However, since all symbols of the same shape are written in the same column, there is little room for ambiguities regarding where to write the result, i.e., even if the operations are understood purely syntactically, the additive system promises to be easier to learn and to apply.

## 2.4   Intermediate conclusion

Let us briefly summarize our findings so far. We have presented various kinds of systematic errors that students make when learning written additions in the decimal place-value system that reveal a lack of proper understanding of the semantic content of the numerals. As a consequence, the additions are performed as purely syntactic operations, which can easily lead to undetected mistakes. To overcome these problems, mathematics educators have developed semantic tools designed to convey the implicit meanings of the numerals. Our investigation of addition with Roman numerals has led to two observations. First, the Roman numeral system embodies already some of the principles that are used in the semantic tools discussed above, since the grouping structure is explicit in the symbolism (compare: ten 'I's yield an 'X', *versus* ten '1's yield a '1' in the column to the left, followed by a '0' in the current column). Because of this, the Roman numerals are often considered to be less abstract and thus easier to learn than our place-value numerals. Second, we have seen that the erroneous strategies that are mistakenly adopted by students using the decimal place-value system would either yield correct results when used with an additive numeral system, or would be less likely to be adopted.[19]

## 3   Subtraction

### 3.1   Examples of difficulties with subtraction in the decimal place-value system

The following subtraction example, given by Spiegel and Selter (2003, p. 24), shows how little semantic understanding can go into formal computations performed by children. When Malte is asked to calculate $701 - 698$ he computes the result via the formal subtraction algorithm, which yields:

$$
\begin{array}{r}
701 \\
-\phantom{0}698 \\
\hline
197
\end{array}
$$

The interviewer asks: 'Do you know another possibility of computing?' Malte answers: 'From 698 to 700 are 2 and from 701 to 700 it is 1, therefore it's 3.' The interviewer inquires about the correct answer, and Malte decides to accept the result of the written computation, explaining: '197. This I have calculated and 3 was only hopp di hopp generated by thinking.'

This little episode shows how fundamental the lack of semantic understanding can be. The dissociation of written computations from an otherwise present understanding of numbers is described as 'a large gap between the children's concrete numerical understanding and their use of formal written symbolism' (Hughes, 1986, p. 95).[20] This is also supported by more

---

[19] It would be very interesting to have empirical studies on the use of additive numeral systems. Schlimm and Neth (2008) have shown differences in the complexity of the arithmetic operations and we imagine that these differences would also have some effect on the occurrences of computational errors. Unfortunately, however, we are not aware of any empirical studies of these issues.

[20] See in particular the discussion in (Hughes, 1986, pp. 95–133).

recent investigations concerning 'street mathematics,' which show that children are able to calculate correctly when the context is outside of school mathematics (Nunes et al., 1993).

As in the discussion of addition, let us now review briefly the systematic mistakes that are made frequently by elementary-school students, according to Ashlock (1998) (see Figures 6 and 7).[21]

$$
\begin{array}{cccc}
\text{S1:} & \text{S2:} & \text{S3:} & \text{S4:} \\[2mm]
241 & 52 & 47 & 446 \\
-\ 96 & -27 & -\ 3 & -302 \\
\hline
255 & 30 & 14 & 104
\end{array}
$$

FIGURE 6. Examples of common subtraction errors: Subtraction of smaller from larger, empty places, zero.

### 3.1.1 Difficulties with what to subtract from what (S1, S2)

In contrast to addition, which is commutative, the order in which subtractions are performed is crucial. In other words, switching the minuend and the subtrahend yields a different result, which is not the case when addends are interchanged. This peculiarity of subtraction comes with the fact that it is much easier to subtract a smaller number from a larger one than the other way around. After all, in the realm of concrete objects, we can take away two apples from seven apples, but not seven from two. This can lead students to adopt the rule 'always subtract the smaller digit from the larger one,' if they do column-wise written subtractions.[22] The boy in the initial example did exactly this in his calculation, and S1 presents another example.[23]

An alternative solution to the difficulty of subtracting a larger number from a smaller one is to adopt the result from the concrete example. Given two apples, taking away any number greater or equal than two leaves no apples at all, i.e., zero apples. This way of thinking seems to lie behind the reasoning exhibited in S2, which appears to follow the rule 'If a larger digit is subtracted from a smaller one, the result is zero.'

---

[21]We have renamed the labels that are used by Ashlock (1998). Here are the correspondences: S1: SW1; S6: SW2; S4: SW3; S7: SW4; S3: P10; S9: P11; S2: P12; S10: P13; S8: P14; S5: P15.

[22]See also (Hughes, 1986, p. 121).

[23]This is a mistake that could also occur within the Roman numeral system. But, it seems unlikely, since in this case it would be much more natural to adopt the rule 'subtract only equal symbols.'

### 3.1.2  Difficulties with empty places in columns (S3)

Like in the case of addition, the occurrence of an empty place can put the student into a situation where the most rudimentary algorithm fails and a repair strategy has to be invoked (see A7). In S3, when the place of the minuend in the tens column is empty, the student's way out is to reuse the minuend from the ones and subtract it again.

### 3.1.3  Difficulties with zero (S2, S4, S8)

The idea of adding zero appears to be more easily understood than subtracting zero.[24] In the case of S4, if a zero appears in the subtrahend then the result is also written as zero. Thus, it seems that the zero is not understood as representing the null quantity, or 'nothing,' in a purely formal fashion. Otherwise, taking away 'nothing' should not yield a change in the numerosity being denoted by the digit in the minuend.

| S5: | S6: | S7: | S8: | S9: | S10: |
|---|---|---|---|---|---|
| $^{4\,5}$ | $^{7}$ | $^{4}$ |  |  | $^{2}$ |
| $3\!\!\!/63$ | $2\,8^{1}5$ | $6^{1}2^{1}5$ | $602$ | $437$ | $43^{1}6$ |
| $-341$ | $-\ 63$ | $-3\ 4\ 8$ | $-238$ | $-\ 84$ | $-21\ 8$ |
| $\overline{112}$ | $\overline{2112}$ | $\overline{1\ 8\ 7}$ | $\overline{274}$ | $\overline{453}$ | $\overline{24\ 8}$ |

FIGURE 7. Examples of common subtraction errors: Direction, borrows.

### 3.1.4  Difficulties with direction (S5)

Also in written subtraction, the direction in which the columns are to be processed can be mistakenly reversed (see A3, A4). In S5 for example, the student proceeds from left to right. Moreover, whenever the same digit appears in both minuend and subtrahend, the student borrows a unit from the next column to be able to subtract without getting a result of zero. So, while the formal operation of borrowing is performed correctly (although in a case when it shouldn't be applied and in the wrong direction), the semantics of the numerals is clearly not understood at all.

### 3.1.5  Difficulties with borrows (S5, S6, S7, S8, S9, S10)

By far the greatest difficulties in written subtraction problems are caused by 'borrowing,' i.e., when the operation cannot be carried out within a single column. This situation is similar to that of addition, where the 'carries' present the greatest problems. In the example S6, the student has learned to take borrows, but has over-generalized this strategy, and now applies it when subtracting the units whether it is needed or not. The borrows are

---

[24]General difficulties with learning the digit zero are presented in (Wellman and Miller, 1986).

always taken from the left-most digit in S7. This kind of mistake might also be motivated by the presence of a zero. For example, in S8 the student regroups directly from the hundreds to the ones, i.e., all borrows are taken from the hundreds, since there is a zero in the position of the tens. The borrows are taken correctly in S9, but the student does not remember to subtract one ten (or one hundred) when she regroups. Finally, in S10 the student borrows one from the '3' in the tens and writes the result as a crutch on top. But, then he adds the '2' (i.e., the crutch) and the '3' before subtracting.

## 3.2 Semantic tools for subtraction algorithms

The problems described above show that applying the formal algorithms and understanding the semantics behind them do not always go together. This becomes very obvious if the algorithms involve steps that prima facie go against previously acquired knowledge. For example, from the first grades in elementary school it is learned that there is nothing less than zero, so subtracting a larger number from a smaller one does not make much sense. The strategies applied in S1 and S2 result from the students attempting to integrate the formal algorithm with their existing knowledge of numbers. Here the individual columns are seen in isolation and dealt with independently from each other, which reveals that the semantics of the place-value system hasn't been grasped yet.

The strategies found in the didactics literature for coping with these difficulties are often based on ideas similar to those discussed above in the context of addition. Figure 8 shows a detail from the schoolbook *Duden* in which children are encouraged to discuss their calculating strategies. Again the iconic representation of numerical quantities in terms of dots, sticks, and areas is used, as well as the pictographic representation on the number line.

Additional support for the semantical understanding of the number portrait, in particular of the debundling of the tens into the ones, could be provided by annotations, which are missing in the figure.

In this particular example, students are instructed first to subtract the tens and then add the units. For students who are still struggling with understanding the workings of the place-value system, the direction of the calculation, which starts from the left, could lead to further misunderstandings. The changes of number representations are sometimes also illustrated with an abacus, as can be seen in Figure 9.[25] Here the grouping and ungrouping is made explicit by the number of pebbles on the different lines: two pebbles on the 50-line can be replaced by one pebble on the 100-line,

---

[25]See also (Hughes, 1986), who advocates the introduction of different number representations to gain a better, semantically richer, understanding of the place-value system.

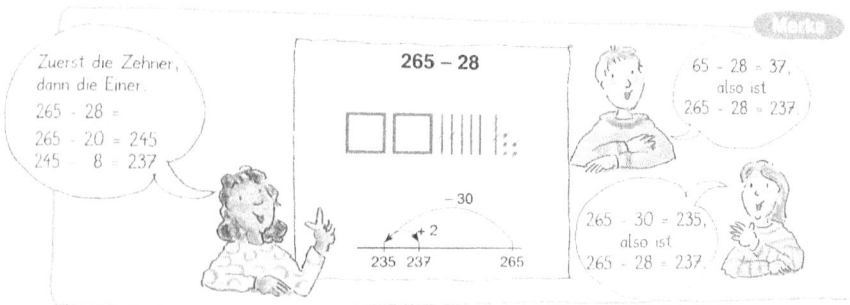

FIGURE 8. Subtraction strategies. Source: (Becherer and Schulz, 2007, p. 37).

and vice versa, without changing the numerical quantity that is being represented. A related approach is shown in (Lengnink, 2006), where students are encouraged to compare different number representations and discuss their benefits.

## 3.3 Subtraction with Roman numerals

As might be guessed from Figure 9, the grouping and ungrouping of numerals is more explicit in the additive numeral system of the Romans. Subtracting one numeral from another in such a system amounts to deleting all symbols that occur in the subtrahend from the minuend. For example, $38 - 12 = 26$ in the Roman system is computed as follows:

$$
\begin{array}{r}
\text{XXX} \quad \text{V} \quad \text{III} \\
- \quad \text{X} \qquad\quad \text{II} \\
\hline
\text{XX} \quad \text{V} \quad\ \text{I}
\end{array}
$$

Here it is not essential to the algorithm that the symbols of the same value be written in the same column. In the decimal place-value system, if a digit of the subtrahend is greater than that of the minuend of the same magnitude, one has to 'borrow' a unit from the next magnitude. This process, as we have seen in Section 3.1, is the origin of many of the erroneous strategies developed by students. In the Roman system, if there are not as many different instances of a number symbol in the minuend as there are in the subtrahend, the simplification rules have to be applied in reverse. For example, to subtract II from V, the 'V' of the minuend has to be rewritten as 'IIII'. Then, two of the 'I's can be deleted, leaving III. A slightly more complicated situation is shown in the following example, in which $16 - 7 = 9$ is calculated.

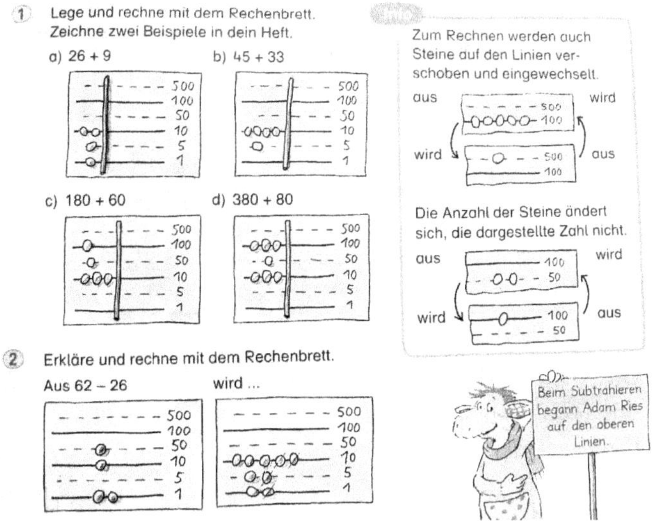

FIGURE 9. Historical exercise. The headings above the pictures mean: 'Place and calculate with the abacus. Draw two examples in your exercise book'. In the info box: 'To calculate, pebbles are also put on the lines and shifted: from ... you get ....  The number of pebbles is changed, but not the represented number.' The sign in the bottom right corner reads: 'When subtracting, Adam Ries always began with the top-most lines.' Source: (Becherer and Schulz, 2007, p. 119).

$$
\begin{array}{ccc}
\begin{array}{rcl}
\text{X} & \text{V} & \text{I} \\
- & \text{V} & \text{II} \\
\hline
\end{array}
\ \Longrightarrow\
\begin{array}{rcl}
\text{X} & & \text{IIIII} \\
- & \text{V} & \text{II} \\
\hline
\end{array}
\ \Longrightarrow\
\begin{array}{rcl}
\text{VV} & & \text{IIIII} \\
- & \text{V} & \text{II} \\
\hline
& \text{V} & \text{IIII} \\
\end{array}
\end{array}
$$

Thus, the 'borrowing' in the additive system simply amounts to an application of the simplification rules in reverse. If a student has understood that an X stands for two V's, this should not create as many confusions as the borrowing in the place-value system does. However, the potential danger of subtracting the smaller value from the larger one within a column (i.e., subtracting 'I' from 'II' in the example above) remains.

Figure 10 shows a sample subtraction with Roman numerals that involves regrouping in which the single steps are presented explicitly. To avoid the impression that a representation of the numerals in columns is necessary, the numerals are written linearly. The single steps consist in either deleting the same amount of occurrences of the same symbol from the minuend and subtrahend, or in unbundling a symbol into symbols representing a smaller value. It should be noted that the order in which the symbols are being

|   |   241      |   −   |      96     |                                                    |
|---|------------|-------|-------------|----------------------------------------------------|
| 1 | CCXXXX$\underline{\text{I}}$ | − | LXXXXV$\underline{\text{I}}$ | Problem |
| 2 | CCXXXX     | − | LXXXXV      | Crossed out one I                                  |
| 3 | CCXXX$\overline{\text{VV}}$ | − | LXXXX$\underline{\text{V}}$ | Unbundle one X into two Vs |
| 4 | CC$\underline{\text{XXX}}$V | − | L$\underline{\text{XXXX}}$ | Crossed out one V |
| 5 | CCV        | − | LX          | Crossed out three Xs, one still to be done         |
| 6 | C$\overline{\underline{\text{LL}}}$V | − | LX | Unbundle one C into two Ls |
| 7 | CLV        | − | X           | Crossed out one L                                  |
| 8 | C$\overline{\text{XXXX}\underline{\text{X}}}$V | − | $\underline{\text{X}}$ | Unbundle one L into five Xs |
| 9 | CXXXXV     |   |             | Crossed out one X, nothing left to subtract        |
|   |   145      |   |             |                                                    |

FIGURE 10. Example of subtraction with Roman numerals (S1). The symbols with a line above (e.g., $\overline{\text{VV}}$) are the result of unbundling; symbols that are underlined (e.g., $\underline{\text{L}}$) are to be crossed out in the next step.

crossed out does not matter at all in this algorithm. This can be seen in the example, where an L-symbol (standing for 50) has been crossed out in line 7 before the final X-symbol (with value 10) is deleted. Alternatively, the X could have been deleted before the L, after the unbundling operation (line 8). In this particular problem, it might have been more efficient to delete four Xs in the first step, to obtain numerals that consisted of fewer letters.

The algorithm shown in Figure 10 is much closer to the intuitive understanding of subtraction as 'taking away.' The rules to be followed are fairly simple: (a) always delete the same symbols from the minuend and subtrahend, (b) if this is not possible, unbundle a symbol in the minuend to obtain the symbols that need to be deleted, according to the grouping rules for each symbol, which are just the reverse rules of those employed in addition (e.g., X→ VV and V→ IIIII). Since here no single-letter digits are subtracted from others, there is no danger that a student might adopt a rule 'always subtract the smaller from the larger number,' as was done in S1 and S2. Also, we have seen that the particular order in which the operations are carried out in the Roman subtraction algorithm does not alter the final result, so that no difficulties with direction, as those exhibited in S5, can arise. As a consequence of the absence of any symbol for zero or of empty places in columns, problems that lead to the erroneous strategies shown in S2, S3, S4, and S8 do not arise. Finally, the handling of 'borrows,' the main source of difficulties when learning subtraction in the decimal place-value system, is reduced to simple applications of the grouping rules.

It is difficult to know whether students who learn subtraction with Roman numerals would make different kinds of mistakes or would also show behavior that reveals gaps between the syntactic manipulation of symbols and the semantic understanding of the numerals. However, our discussion has shown that the common systematic mistakes that students make when learning subtraction with the decimal place-value system would most likely not be made with a purely additive numeral system, and, moreover, that the strategies that are used in the literature to help students understand the place-value system are already embodied in the Roman system of numerals. Thus, from a purely educational perspective, it seems very plausible that the Roman numeral system would be easier to learn and that it would not generate such a huge gap between formal, written computations and the numerical meanings of the notation.

# 4 Multiplication

## 4.1 Examples of difficulties with multiplication in the decimal place-value system

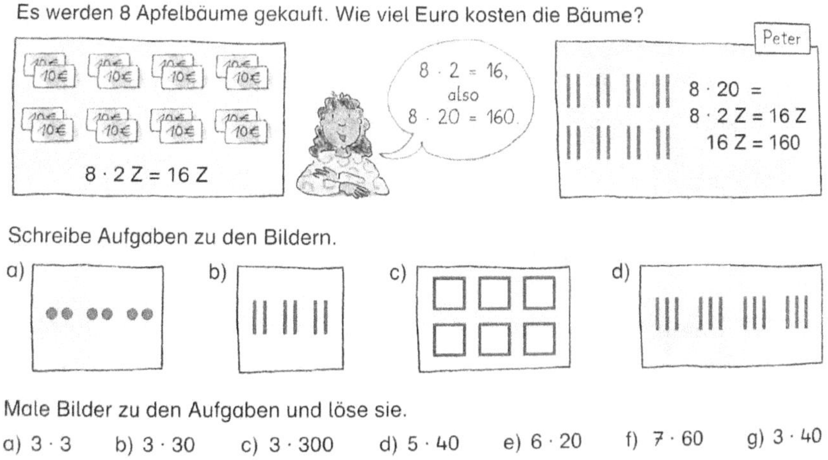
Es werden 8 Apfelbäume gekauft. Wie viel Euro kosten die Bäume?
8 · 2 Z = 16 Z
8 · 2 = 16,
also
8 · 20 = 160.
Peter
8 · 20 =
8 · 2 Z = 16 Z
16 Z = 160
Schreibe Aufgaben zu den Bildern.
a)
b)
c)
d)
Male Bilder zu den Aufgaben und löse sie.
a) 3 · 3 b) 3 · 30 c) 3 · 300 d) 5 · 40 e) 6 · 20 f) 7 · 60 g) 3 · 40

FIGURE 11. Pragmatic and semantic background of multiplication. The headings above the pictures mean: 'Eight apple-trees are bought. How many Euros do they cost?' and below 'Write down problems for the pictures' and 'Draw pictures for the following problems and solve them.' Source: (Becherer and Schulz, 2007, p. 56).

Different versions of the written multiplication algorithm are taught in different countries, sometimes even in different schools within a country, and the algorithms are also motivated in different ways. As an example, Figure 11 shows how multiplication in the special case with multiples of 10 is introduced in a German textbook for the third grade. The pictographic and iconic representations are intended to provide a bridge between the formal algorithm and reasoning with more concrete objects. However, such descriptions are often forgotten, even in the process of learning the algorithms. Consequently, students frequently learn to manipulate the numerals without gaining a deeper understanding of what they are doing. This leads to the typical mistakes discussed in (Ashlock, 1998), which are presented next (see Figures 12 and 13).[26]

### 4.1.1  Only column-wise operations (like for addition) (M1, M2)

One of the systematic mistakes that some students exhibit can be traced back to a generalization from the algorithms for addition and subtraction, which proceed by processing each column separately (with the occasional violations of this principle in the form of carries and borrows). In M1, for example, the student approaches each column as a single multiplication problem, with carries as commonly used in addition. In this particular case, a further complication arises because of the empty place in a column (see also A7), which is repaired by continuing to use the left-most digit of the second factor. Column-wise multiplication is also performed in M2, but here, if the resulting product has two digits, only the tens figure is recorded.

### 4.1.2  Difficulties with placement of intermediate results (M3)

In the usual multiplication algorithms the partial products that arise as intermediate results must be written in a fixed format to guarantee the correct result. If the student does not understand why this format is necessary, it will be easy for him to violate it. For example, in M3, the intermediate results are simply placed one underneath the other.

### 4.1.3  Difficulties with zero (M4, M5)

We have seen earlier that the digit that creates the most difficulties is zero. In M4, if a zero occurs in one of the factors it is inserted into the result, before regrouping is performed. Zero also creates problems if it occurs in an intermediate result. For example, in M5, the student correctly moves over one place, but incorrectly also writes down a zero.

### 4.1.4  Difficulties with operating with the crutch figure or carries (M9, M8, M10)

As we have seen in the discussion of addition and subtraction, if an operation requires the transgression of a column-boundary this can lead to major confusions for students who have not yet grasped the inner workings of the

---

[26]We have renamed the labels that are used by Ashlock in the following way: M9: MW1; M8: MW2; M1: MW3; M3: P16; M6: P17; M10: P18; M7: P19; M2: P20; M4: P21; M5: P22.

|        | M1: | M2: | M3: | M4: | M5: |
|--------|-----|-----|-----|-----|-----|

$$
\begin{array}{ccccc}
\overset{1}{524} & 837 & 56 & 5402 & 57 \\
\times\ \ 34 & \times\ 294 & \times\ 32 & \times\ \ \ 6 & \times\ \ \ 34 \\
\hline
1576 & 122 & 112 & 32502 & 228 \\
     &     & 168 &       & 1710 \\
\cline{3-3}\cline{5-5}
     &     & 280 &       & 17328 \\
\end{array}
$$

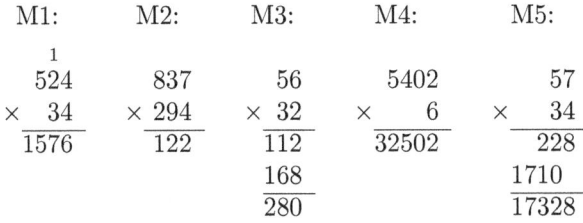

FIGURE 12. Examples of common multiplication errors: Column-wise operations, placement of intermediate results, zero.

place-value system. A few bugs that arise with multiplication out of such a situation are presented next.

The student in M6 has over-generalized the fact that often, and in particular in introductory examples, only the digit one has to be carried into the next column, such that she always adds one, regardless of the correct amount. Thus, after computing $6 \times 8 = 48$, the '8' is recorded for the units, but a '1' is carried into the tens, instead of a '4,' and similarly for the hundreds. M7 also shows difficulties with keeping track of the numbers to be carried. The sequence of computations in this example is as follows: $6 \times 2 = 12$, write '2' in the ones column and remember the 1; $6 \times 3 = 18$, plus the remembered 1 is 19, write '1' and remember the 9; finally, $6 \times 1 = 6$, plus the remembered 9 is 15, which is written down, yielding the final result of '1512.'[27] The next students use 'crutches', or recorded intermediate results, as a memory-aid, but when performing the computations they use these crutches incorrectly. For example, M8 adds the number recorded as crutch before multiplying the tens figure, instead of adding it to the result after the multiplication. The next student shows how a symbol written down during the computation can be misleading at later stages. M9 correctly calculates the result of the multiplication with the ones, but then uses the crutch recorded when multiplying the ones again when multiplying the tens. A similar mistake is made by M10, who also computes the multiplication with the ones without problems, but uses both crutch figures when multiplying with the tens digit (i.e., she adds $1+3$ to the result of $3 \times 6$, though only the 1 should be added, the '3' belonging to the multiplication with the ones).

## 4.2  Semantic tools for multiplication algorithms

As mentioned above, the different algorithms employed for multiplication may differ in different locations where the decimal place-value system is

---

[27]This is not a one-time miscalculation, but a systematic error that is made by this student.

```
 M6:        M7:        M8:        M9:        M10:
                         2          2
  368        132        27         46        ¹³65
×   6      ×   6      ×   4      ×  24       ×  37
─────      ─────      ─────      ─────       ─────
 1978       1512        168        184         455
                                   102         225
                                  ─────       ─────
                                   1204        2705
```

FIGURE 13. Examples of common multiplication errors: Crutch figures, carries.

taught. However, the difficulties noted by Ashlock are fairly general. Consult, e.g., (Padberg, 2005) for a discussion of similar systematic errors made by German students. The common multiplication procedures are quite elaborate and depend crucially on writing the intermediate results in the correct positions. A student who does not know why the particular positions are used, i.e., who has not fully understood their meanings, can easily adopt erroneous strategies without being aware of it. Particular difficulties are posed by the zero, which is often not properly recognized as a place value or which is erroneously interpreted as one, yielding $a \times 0 = a$.

An alternative algorithm, developed with the intention of being closer to the semantics of the symbols but without sacrificing any computational efficiencies, is discussed in (Wittmann and Müller, 2005, Vol. 2, p. 135); empirical studies have shown that this algorithm helps to lower the error rate.[28] Their representation, shown in Figure 14, can also be supplemented by additional semantic information (the 'H,' 'Z,' and 'E', standing for hundreds, tens, and ones) during early learning stages, which can be discarded later. From the notation alone, the student can see how multiplying a digit in the tens place with one in the ones place gives a numeral that occupies the tens (and possibly hundreds) place in the result.[29] Moreover, the carries, or crutches, are recorded in determinate positions, so that misuses as those exhibited in M9, M8, and M10 are rendered impossible. Due to the rigid format, it seems that the notorious difficulties with empty places and zeros are also reduced.

## 4.3   Multiplication with Roman numerals

Although multiplication with Roman numerals has often been characterized as exceptionally difficult or even impossible,[30] an algorithm for multipli-

---

[28]See also (Padberg and Thiemann, 2002).

[29]Compare this representation with the examples in Figure 11.

[30]See, e.g., the very popular (Menninger, 1992, p. 294) and (Ifrah, 1985, p. 431).

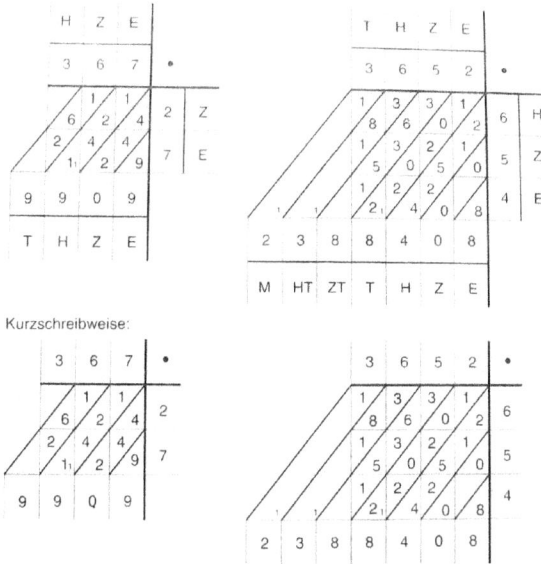

FIGURE 14. Linking formal multiplication algorithms with semantic information. Source: (Wittmann and Müller, 2005, p. 135).

cation in the Roman numeral system that is very similar to the familiar
one used for the decimal place-value system is presented and discussed in
(Schlimm and Neth, 2008). The general method is to multiply each digit
of one numeral with each digit of the other and then to add up the intermediate results. This is exactly how the common multiplication algorithms
for place-value systems proceed, too.[31] However, in the case of the place-
value system one has to make sure to assign the correct magnitude to these
intermediate results, i.e., to write them exactly in the right place. Indeed,
the elaborate positioning-schemes of common algorithms as well as that of
(Wittmann and Müller, 2005) are designed to achieve exactly this. As we
have seen above, the resulting formal complexities can be perceived as being
dissociated from any underlying semantics, and this is the origin of many
systematic mistakes. With the Roman system, however, the magnitude of
the value is inherent in the symbols (e.g., X×V = L and C×V = D), so that
no special attention has to be devoted to this matter. Also single-letter operations that result in numerals ranging over more than one magnitude are
handled in this way (e.g., V×V = XXV). In other words, the intermediate
results can be gathered in any way and then simply added together. The

---

[31]See the examples in Figures 12, 13, and 14.

student only has to learn the multiplication table to be able to multiply.[32]

Here is an example of the computation of $106 \times 18 = 1908$ in Roman numerals in a format similar to that discussed in (Wittmann and Müller, 2005):

1. Compute intermediate results by single-letter multiplications:

| $\times$ | X | V | I | I | I |
|---|---|---|---|---|---|
| C | M | D | C | C | C |
| V | L | XXV | V | V | V |
| I | X | V | I | I | I |

2. Gather all intermediate results: MDCCCLXXXVVVVVIII.

3. Simplification yields the final result: MDCCCCVIII.

The table in item 1. above is filled using the single-letter multiplication facts that have to be learned by heart from a multiplication table. Since Roman numerals are on average longer than their decimal place-value counterparts, more intermediate results occur in general. Gathering these and simplifying the resulting numeral is done according to the standard procedure of addition with Roman numerals (see Section 2.3).

We see immediately from this procedure that no difficulties like those of column-wise multiplication arise, since there are no columns to be processed and the format of the representation differs substantially enough from that of addition to avoid any confusions. Nor should the positioning of the intermediate results pose any particular difficulties. Moreover, even the order in which the numerals are written on top and on the side of the table does not effect the result of the computation at all. In other words, switching columns or rows does not influence the final result. Further, since there are no zeros or empty places, confusions relating to these cannot arise. Finally, there are no crutches or carries to be dealt with separately, so that this source of common systematic errors with the decimal place-value system is not present in multiplication with Roman numerals.

In fact, the procedure for multiplication with Roman numerals is very similar to the semantic tool employed by (Wittmann and Müller, 2005) to help students learn multiplication with the decimal place-value system. The difficulties with carries, which do not arise in the Roman multiplication, cannot be completely avoided even in the representation of Wittmann and Müller. When the values in the diagonals in Figure 14 are added, it is possible that carries have to be processed to transgress column (or, in this case, diagonal) boundaries.

---

[32]See (Schlimm and Neth, 2008, p. 2010) for the multiplication table with Roman numerals.

# 5   Conclusion

In the decimal place-value system the meaning of a single symbol depends on the shape of the symbol *and* on the position of the symbol in the numeral. As long as the relation between these two components has not been properly understood by the student, the algorithms for the basic arithmetic operations appear to be arbitrary manipulations that, almost magically, lead to the correct results. In fact, as we have seen above, failure to grasp the semantic content of the notation leads many students to abstract erroneous procedures when learning the algorithms for addition, subtraction, and multiplication. The students do not realize that these algorithms are incorrect, because they haven't grasped the full semantics of place-value numerals.

By analyzing the algorithms for arithmetic operations for an additive numeral system, the Roman numerals, we have shown that many of the mistakes that students make when learning the decimal place-value system would not occur if they were to calculate with the additive system. The underlying reason for this observation is the distinctive feature of additive systems, namely that each symbol encodes a determinate value. Thus, the connection between a symbol and its semantic content is stronger, which facilitates learning the correct manipulation strategies for the notation. An interesting issue that is raised by our discussion concerns the relation between syntactic simplicity and semantic understanding. The stronger connection between syntax and semantics of numerals in an additive system goes hand in hand with algorithms that require simpler, although more frequent, manipulations. This syntactic simplicity might turn out to be an important factor for more error-free computations with the Roman system. After all, the complexity of the decimal place-value algorithms stems from the fact that they have to make sure the places are dealt with correctly.

The conclusion of our analysis is three-fold. On the one hand, we have suggested a new possible criterion for assessing different systems of numerals in terms of their semantic content, which is crucial for learning the symbolic manipulations for arithmetic operations and for a proper understanding of the numerals. In the discussions of different numeral systems, this pedagogical dimension of numeral systems has been largely overlooked. On the other hand, our discussion of the systematic errors that are frequently made by students who learn the decimal place-value system and the comparison of how such erroneous strategies would play out in an additive numeral system can be used to augment the repertoire of semantic tools used by mathematics teachers. This further supports the views of various educators who have emphasized the importance of teaching children different representation systems for numerals.[33] Finally, we have argued that notation can influence the learning of mathematics, and thus, indirectly, also mathematical practice. Consequently, philosophers interested in the latter should also be concerned about the nature of the language of mathematics, i.e., notational systems like numerals, formulas, and diagrams.

# Bibliography

Ashlock, R. B. (1998). *Error Patterns in Computation: Using error patterns to improve instruction.* Merrill, Upper Saddle River NJ, 7th edition.

Becherer, J. and Schulz, A., editors (2007). *Duden Mathematik 3.* Duden Paetec Schulbuchverlag, Frankfurt a.M.

Böttinger, C., Bremer, U., Engel, H., Haubold, H., Jung, T., Mohr, I., Müller-Heidtkamp, M., and Schaper, H. (2008). *Wochenplan Mathematik 4.* Ernst Klett Verlag, Leipzig.

Boyer, C. B. (1944). Fundamental steps in the development of numeration. *Isis*, 35(2):153–168.

Brown, J. S. and Burton, R. R. (1978). Diagnostic models for procedural bugs in basic mathematical skills. *Cognitive Science*, 2(2):155–192.

Cajori, F. (1928). *A History of Mathematical Notations*, volume I: Notations in Elementary Mathematics. Open Court Publishing Company, Chicago IL.

Campbell, J. I. and Epp, L. J. (2005). Architectures for arithmetic. In Campbell, J. I. D., editor, *Handbook of Mathematical Cognition*, pages 347–360. Psychology Press, New York NY.

Chrisomalis, S. (2004). A cognitive typology for numerical notation. *Cambridge Archaeological Journal*, 14(1):37–52.

Cox, L. S. (1975). Systematic errors in the four vertical algorithms in normal and handicapped populations. *Journal for Research in Mathematics Education*, 6(4):202–220.

Davey, P. C. (1975). Review of (Seaborne, 1975). *The Mathematical Gazette*, 59(410):283–284.

---

[33]E.g., (Hughes, 1986; Lengnink, 2006).

De Cruz, H., Hansjörg, N., and Schlimm, D. (2010). The cognitive basis of arithmetic. In Löwe, B. and Müller, T., editors, *PhiMSAMP. Philosophy of Mathematics: Sociological Aspects and Mathematical Practice*, volume 11 of *Texts in Philosophy*, pages 59–106, London. College Publications.

Ferreirós, J. and Gray, J., editors (2006). *The Architecture of Modern Mathematics*. Oxford University Press, Oxford.

Hartshorne, C. and Weiss, P., editors (1932). *Collected Papers of Charles Sanders Peirce. Volume 2. Elements of Logic*. Harvard University Press, Cambridge MA.

Hasemann, K. (2007). *Anfangsunterricht Mathematik*. Spektrum Elsevier, Munich, 2nd edition.

Hughes, M. (1986). *Children and Number. Difficulties in Learning Mathematics*. Basil Blackwell, New York NY.

Ifrah, G. (1981). *Histoire universelle des chiffres*. Seghers, Paris.

Ifrah, G. (1985). *From One to Zero: A Universal History of Numbers*. Viking Penguin Inc., New York NY. Translation of (Ifrah, 1981).

Lengnink, K. (2006). Reflected acting in mathematical learning processes. *Zentralblatt für Didaktik der Mathematik*, 38(4):341–349.

Menninger, K. (1992). *Number Words and Number Symbols: A Cultural History of Numbers*. Courier Dover Publications, New York NY.

Miller, K. F., Smith, C. M., Zhu, J., and Zhang, H. (1995). Preschool origins of cross-national differences in mathematical competence: The role of number naming systems. *Psychological Science*, 6(1):56–60.

Nunes, T., Dias Schliemann, A., and Carraher, D. W. (1993). *Street Mathematics and School Mathematics*. Cambridge University Press, Cambridge.

Padberg, F. (2005). *Didaktik der Arithmetik*. Spektrum Elsevier, Munich, 3rd edition.

Padberg, F. and Thiemann, K. (2002). Alles noch beim Alten? Eine vergleichende Untersuchung über typische Fehlerstrategien beim schriftlichen Multiplizieren. *Sache – Wort – Zahl*, 50:38–45.

Schlimm, D. and Neth, H. (2008). Modeling ancient and modern arithmetic practices: Addition and multiplication with Arabic and Roman numerals. In Love, B., McRae, K., and Sloutsky, V., editors, *Proceedings of the 30th Annual Meeting of the Cognitive Science Society*, pages 2097–2102, Austin TX. Cognitive Science Society.

Seaborne, P. L. (1975). *An Introduction to the Dienes Mathematics Programme*. University of London Press, London.

Spiegel, H. and Selter, C. (2003). *Kinder und Mathematik. Was Erwachsene wissen sollten*. Kallmeyer & Klett, Seelze-Velber.

Van Kerkhove, B., editor (2009). *New Perspectives on Mathematical Practices: Essays in Philosophy and History of Mathematics*. World Scientific Publishing Co., London.

VanLehn, K. (1983). On the representation of procedures in repair theory. In Ginsburg, H., editor, *The Development of Mathematical Thinking*, pages 201–252. Academic Press, London.

Wellman, H. and Miller, K. (1986). Thinking about nothing: Development of concepts of zero. *British Journal of Developmental Psychology*, 4(1):31–42.

Wittmann, E. C. and Müller, G. N. (2005). *Handbuch produktiver Rechenübungen. Band 2: Vom halbschriftlichen und schriftlichen Rechnen*. Klett, Stuttgart.

Zhang, J. and Norman, D. A. (1995). A representational analysis of numeration systems. *Cognition*, 57:271–295.

# Skills and mathematical knowledge

Benedikt Löwe[1,2,3], Thomas Müller[4,5]*

[1] Institute for Logic, Language and Computation, Universiteit van Amsterdam, Postbus 94242, 1090 GE Amsterdam, The Netherlands

[2] Department Mathematik, Universität Hamburg, Bundesstraße 55, 20146 Hamburg, Germany

[3] Mathematisches Institut, Rheinische Friedrich-Wilhelms-Universität Bonn, Endenicher Allee 60, 53115 Bonn, Germany

[4] Departement Wijsbegeerte, Universiteit Utrecht, Heidelberglaan 6, 3584 CS Utrecht, The Netherlands

[5] Institut für Philosophie, Universität Bonn, Lennéstraße 39, 53113 Bonn, Germany

E-mail: bloewe@science.uva.nl, Thomas.Mueller@phil.uu.nl

## 1 Introduction

In (Löwe and Müller, 2008), we discuss the traditional modal view of (propositional) mathematical knowledge that reduces knowledge claims of the form "$S$ knows that $p$" to the ability of $S$ to produce a formal derivation of $p$. We argue that such a modal definition of knowledge cannot be given in a context-insensitive way and that the (now contextually determined) modality will have to be interpreted with respect to *skills* of the subject $S$. When looking at actual knowledge attributions in mathematics, it becomes clear that the access to proof that is purportedly behind mathematical knowledge, has to be of a dispositional nature: nobody has current conscious or physical access to proofs of all, or even of a small fraction of, the items of mathematical knowledge that can be truthfully attributed to her. This modal or dispositional element is present in many other accounts of knowledge, e.g., in Aristotle's conception of knowledge as a ἕξις (*Cat.* 8).

The crucial question is how to make this modalization precise. Our analysis (Löwe and Müller, 2008, p. 104) rests the modalization on the notion of "mathematical skill":

> $S$ *knows* that $p$ iff $S$'s current mathematical skills are sufficient to produce the form of proof or justification for $p$ required by the actual context. $\qquad(‡)$

Skill is both a modal notion (what somebody is able to do even while not doing it) and has an empirical side (skills can be tested). Skill levels can be characterised independently of any conceptual models for mathematical

---

*The authors acknowledge support of the *Wissenschaftliches Netzwerk* PhiMSAMP funded by the *Deutsche Forschungsgemeinschaft* (MU 1816/5-1).

Benedikt Löwe, Thomas Müller (*eds.*). *PhiMSAMP. Philosophy of Mathematics: Sociological Aspects and Mathematical Practice.* College Publications, London, 2010. Texts in Philosophy 11; pp. 265–280.

*Received by the editors:* 1 October 2009; 20 December 2009; 25 February 2010.
*Accepted for publication:* 25 February 2010.

knowledge. Mathematical practice affirms that the concept of mathematical skill is well entrenched as it is customary to comment on students' or researchers' skills, and it is often possible to rank people with respect to their skills. Skills are tested in exams and job talks, and it may well be that the aim of mathematics education is best characterised not as instilling mathematical knowledge, but as teaching mathematical skills. In the mentioned paper we do not discuss this in detail, but instead refer to the Dreyfus-Dreyfus model of skill acquisition as a semi-formal theory of skill levels and relegate a more detailed discussion to future work. In this paper, we provide the necessary background and continue the discussion from (Löwe and Müller, 2008).

In § 2, we shall give a general discussion of the role of skills in epistemology, specializing to the Dreyfus-Dreyfus model of skills in § 3. The original applications of the Dreyfus-Dreyfus model were (relatively) homogeneous skills such as car driving (Dreyfus and Dreyfus, 1986, p. 24) and playing chess (p. 25). Mathematics is much more multi-faceted; in fact, we propose to see mathematics as involving a *professional skill*. There is a well-known treatment of a professional (set of) skills using the Dreyfus-Dreyfus model, viz. Benner's (1984) treatment of nursing skills which we discuss in detail in § 4. After having seen the example of nursing, we return to mathematical skills in § 5, asking a number of questions with very few concrete answers. In our concluding § 6, we summarize the discussion of this paper.

## 2   Skills

It is sometimes claimed that mathematical knowledge is mostly propositional knowledge: knowledge *that*, e.g., a specific mathematical proposition $p$ is true or false; and it is this type of knowledge that we have investigated (Löwe and Müller, 2008). Rav (1999) has argued that mathematics is not really about knowing the truth values of theorems ("knowing that"), but about knowing the techniques and ideas behind their proofs ("knowing how"). Rather than viewing this as a strict dichotomy, we are interested mostly in the role of skills—knowing how—for propositional mathematical knowledge.

Skills aren't new to the philosophical scene. Ryle (1949, Chap. 2) has famously argued for the separation of knowing *how* (which he uses synonymously with "skill") from knowing *that*. Ryle's overall aim is to fight the "intellectualist doctrine which tries to define intelligence in terms of the apprehension of truths, instead of the apprehension of truths in terms of intelligence" (Ryle, 1949, p. 27). He claims that intellectualism is implicit in much of philosophy, but that "'[i]ntelligent' cannot be defined in terms of 'intellectual' or 'knowing *how*' in terms of 'knowing *that*'" (p. 32). If a reduction of one to the other has any chance of success, it should be the other way round—but here also a danger lurks. Skill is a modal notion and thus close to dispositions, but human know-how is different from physical dispositions like the solubility of sugar, which can (arguably) be tested by uniform behavior in specific conditions.[1] This creates a problem for a direct reduction of knowing that to knowing how:

> Epistemologists, among others, often fall into the trap of expecting dispositions to have uniform exercises. For instance, when they recognise that the verbs 'know' and 'believe' are ordinarily used dispositionally, they assume that there must therefore exist one-pattern intellectual processes in which these cognitive dispositions are actualised. Flouting the testimony of experience, they postulate that, for example, a man who believes that the earth is round must from time to time be going through some unique proceeding of cognising, 'judging', or internally re-asserting, with a feeling of confidence, 'The earth is round'. (Ryle, 1949, p. 44)

The role of triggering conditions or, more generally, the role of actual performance for skill assessment is certainly more diverse.

Dreyfus and Dreyfus affirm Ryle's point that "know-how is not accessible to you in the form of facts and rules" (Dreyfus and Dreyfus, 1986, p. 16). From this observation they draw the important conclusion that the genesis of skills contains the key to a better understanding of know-how.[2] Consequently, they choose to focus their investigation on a phenomenologically detailed study of skill acquisition.

## 3   The Dreyfus-Dreyfus model of skills

The philosopher Hubert Dreyfus and the mathematician Stuart Dreyfus propose their five-step skill acquisition model in their book *Mind over Machine* (Dreyfus and Dreyfus, 1986), which itself forms an important contribution to the discussion about symbolic Artificial Intelligence in the 1980s. Their model is grounded in phenomenological observations about the acquisition of various human skills, on the one hand, and in philosophical theories about human practices going back to Heidegger, Merleau-Ponty and the late Wittgenstein (Dreyfus and Dreyfus, 1986, p. 11).

Dreyfus and Dreyfus discern five skill steps in the development of skills in humans—stressing, of course, that not everyone acquiring a skill will necessarily reach the highest, expert level (Dreyfus and Dreyfus, 1986, Chap. 1; summary p. 50):

---

[1] As the discussion about so-called *ceteris paribus* laws and "finkish dispositions" in philosophy of science shows, this assumption about testing may have to be qualified.

[2] Ryle also makes an observation that points in this direction: "Learning *how* or improving in ability is not like learning *that* or acquiring information" (Ryle, 1949, p. 59).

1. *Novice.* Application of context-free rules through information processing.

2. *Advanced Beginner.* Application of rules, also based on perceived similarity with prior examples.

3. *Competence.* Application of a hierarchical procedure of decision-making ("problem solving"; p. 26).

4. *Proficiency.* Deep involvement, experiencing situations from a perspective ("holistic similarity recognition"; p. 28); decisions grounded analytically.

5. *Expertise.* No need for rules. *"[E]xperts don't solve problems and don't make decisions; they do what normally works"* (p. 30f.).

It is part of Dreyfus and Dreyfus's argument against symbolic AI that explicitly rule-based schemes, even if rules include heuristics polled from human experts, will never allow computer programs to advance to proficiency or expertise. According to them, higher skill levels are only reached through repeated *in situ* experience. The Dreyfus-Dreyfus skill model is a situational model offering "no context-free criteria to identify persons as possessing talents or traits indicative of expertise" (Benner, 1984, p. 15).

Note that Hubert Dreyfus (2001) extends the Dreyfus-Dreyfus skill model by two further levels called *Mastery* and *Practical Wisdom.* We shall focus mostly on the original five-level model and only mention some issues of Mastery in our concluding § 6.

We have chosen to focus on the Dreyfus-Dreyfus skill model because it is general, explicit, and empirically grounded. However, despite the fact that Stuart Dreyfus himself is a mathematician, that model has not been applied to the case of mathematics itself.[3] Early applications of the model were skills or skill sets that are clearly delineated such as playing chess and driving a car (or, slightly more complex, the education of airplane pilots: Dreyfus and Dreyfus, 1977). Car-driving and chess are skills that are needed in specific situations; there is a fairly clear distinction to be made between the skills involved in such settings and more general enabling conditions or auxiliary skills. For instance, your car-driving skill can be assessed independently from your competence in using the CD player in the car, even though handling of the CD player is something that is typically done by drivers. Or, in the case of chess, a world class chess player needs to travel to tournaments and many everyday skills (such as booking flights and hotels) are required for this, but we feel confident in completely separating this from the chess-playing skill of an individual: if the world champion of

---

[3] Hubert Dreyfus, personal communication, June 2003.

chess did not know how to book a flight or a hotel, this would not affect the level of his or her skill.

In the case of mathematics, this separation is more difficult: while we think that booking flights or hotels are not parts of the skill set that defines a mathematician (after all, we know research mathematicians who do not go to conferences), this is much less clear for being able to write up proofs intelligibly or to explain proofs to graduate students, etc. (for more discussion, see § 5.1 below).

Later applications of the Dreyfus-Dreyfus skill model have been dealing with skills more complex than car-driving and chess. Examples of this are the famous studies on nursing by Patricia Benner (cf. § 4) and Flevbjerg's studies on social workers (Flyvbjerg, 2001). In order to provide a background for our discussions about mathematical skills, we shall discuss the prototypical case of nursing in more detail in the following section.

## 4 Nursing

Patricia Benner's analysis of the professional skills of nursing in terms of the five-level Dreyfus-Dreyfus model of skills (Benner, 1984) has been called one "among the most sustained, thoughtful, deliberate, challenging, empowering, influential, empirical [...], and research-based bodies of nursing scholarship" (Darbyshire, 1994, p. 760). Her approach has had a substantial influence on the practice of teaching nursing, as witnessed by a number of papers published in the 2001 commemorative edition of the 1984 book (cf. Gordon, 2001; Huntsman et al., 2001; Ullery, 2001; Fenton, 2001; Dolan, 2001). An overview of the impact of Benner's study can be found in (Brykczynski, 2006).

Benner (1984) gave a detailed description of the five stages in the learning of nursing skills based on paired interviews, individual interviews, and participant observation. She covers the stages of *Novice* (Benner, 2001, pp. 20–22), *Advanced Beginner* (pp. 22–25), *Competence* (pp. 25–27), *Proficiency* (pp. 27–31), and *Expert* (pp. 31–36), including implications for teaching of nursing students at the particular levels.[4]

Note that since the Dreyfus-Dreyfus model is a situational model, you cannot expect criteria to determine the skill level of a nurse in an objective, context-independent way. Rather, the study exhibits examples of behaviour and insight indicative of particular skill levels.[5]

---

[4]For instance, for *advanced beginners* she notes that "their nursing care needs to be backed up by nurses" (Benner, 2001, p. 25) and for *proficient* nurses, she concludes that they "are best taught by the use of case studies" and that "the proficient performer is best taught inductively" (p. 30).

[5]Cf. (Benner, 2001, p. 15): "No attempt was made to classify the nurses themselves according to proficiency levels".

As we have argued above, nursing is a good example for what we aim to do in §5 as it is a much richer activity than the one-dimensional examples of chess playing or car driving. Benner identifies a number of skill sets that are all part of the skills of a nurse, including providing comfort (Benner, 2001, §4) and interpretation for patients (§5), diagnostics (§6), situation management (§7), medication (§8), quality control (§9), and organization (§10). The notion of a skilled nurse is related, ultimately, to a nurse's job description, which has developed historically. We are not concerned here with a natural kind of human beings, nurses, of which there are more and less skilled ones. Rather, we are assessing human beings who have chosen a specific profession, as more or less skilled as required by the (historically and sociologically contingent and changing) requirements of that profession. Nursing skills are *professional skills*.

Not every highly skilled nurse will be equally good at all of these skill sets. We can imagine highly accomplished nurses with decades of experience who are not very good at particular parts of the job description. This is why the Dreyfus-Dreyfus model should not be seen as assigning skill levels to individuals but to performance patterns in a given situation.[6] We shall see this phenomenon in more detail in the case of mathematical skills in the next section.

## 5   Mathematical skills

In the mathematics education literature, Ryle's distinction of *knowing that* and *knowing how* has been embraced as a fundamental dichotomy for mathematical epistemology, and has given rise to a number of related (and yet subtly different) dichotomies for mathematical skills. We find examples in Sfard's *structural* vs. *operational* duality,[7] in Anderson's *declarative* vs. *procedural* distinction,[8] and in the notions of *functional* and *predicative thinking* in the work done at the Osnabrück *Institut für kognitive Mathe-*

---

[6]Compare the recent proposal of applying the Dreyfus-Dreyfus model for the profession of infection preventionists by Marx (2009). Here, the skill levels are objective and context-invariant properties of the individuals; e.g., "The Infection Preventionist-Competent would have more experience (> 2 years) and be certified in Infection Control (CIC) OR have a Masters or Doctorate in a healthcare field, > 6 months experience and be certified in Infection Control (CIC)" (p. E157). This is clearly not in line with the set-up of the Dreyfus-Dreyfus skill model and no such attempt should be made for the case of mathematics.

[7]Cf. (Sfard, 1991, p.4): "whereas the structural conception is static, [...], instantaneous, and integrative, the operational is dynamic, sequential, and detailed." Sfard stresses that there is no dichotomy between the structural and the operational approach, but rather a duality of "inseparable, though dramatically different, facets of the same thing" (Sfard, 1991, p. 9).

[8]Cf. (Anderson, 1993, p. 18): "Intuitively, declarative knowledge is factual knowledge that people can report or describe, whereas procedural knowledge is knowledge people can only manifest in their performance."

*matik* by Cohors-Fresenburg, Schwank, and their collaborators (Schwank, 2003). Combining the two sides of the duality, we also find the notion of *procept* (a combination of 'process' and 'concept') in (Gray and Tall, 1994).

A lot of interesting educational and empirical research has come out of these dichotomies and dualities, e.g., proposals for supporting certain talents of students based on the *functional* vs. *predicative* distinction (Cohors-Fresenborg and Schwank, 1992) or studies using eye-tracking as an indicator for such talent focus (Cohors-Fresenborg et al., 2003). However, we do not think that the classification of mathematical skill into a very small number of basic mathematical aptitudes is appropriate for the analysis that we are aiming for here.

We shall approach the issue of mathematical skills and their role in mathematical knowledge by studying three interrelated questions:

1. What kind of skills are mathematical skills? What is their principle of unity? Are they linked to the mathematical profession, or are they rather a type of natural skills?

2. How are mathematical skills individuated? Is it useful to distinguish mathematical skills very finely, in line with the division of the subject itself, or is there an overarching principle of unity?

3. How are mathematical skills measured and assessed? Which indicators are employed in practice; what makes mathematical skills empirically accessible?

We shall pursue these issues, in the above order, in sections 5.1 through 5.3.

## 5.1   What kind of skills are mathematical skills?

When we talk about skills, we group them according to different principles, the spectrum ranging from purely task-related clustering (e.g., when assessing someone's skill at repairing bicycles or driving a car) to clusters that can lay a claim to resonating with some natural subdivision of the activities of a human being (when, e.g., assessing someone's musical skills, or her skills at bringing people together). The latter skills seem to be more strongly associated with the notion of talent than the former ones. Further subdivisions in all these areas are possible and sometimes useful, according to demands set by the context (e.g., even a skilled bicycle mechanic may be poor at some specific task like adjusting the headlights).

Where do mathematical skills lie in the above spectrum? Both extremes of the spectrum have a certain appeal. In a somewhat romantic fashion that pervades the public image of research mathematics, propounded by

popular culture,[9] one may picture mathematical skills as a combination
of different natural talents ranging from an analytical mindset to powerful
visual intuition. At the other extreme, one may view mathematical skills
simply as the skills a research mathematician needs for his or her job, which
would mean, in most cases, that filling in one's travel expense declarations
and LATEX typesetting are as much part of the deal as are finding and
checking proofs.

In this spectrum we lean towards viewing mathematical skills as *professional skills* whose unity is defined by the job that a mathematician is
doing as a researcher. In the process of mathematical research, a lot of
skills are involved in a successful research episode: a mathematician tackles
a research question, asks the right people who give her ideas helping on her
way to the correct proofs, finally finds the proof, writes it up in a way that
she can communicate it to the experts, gives a number of seminar talks on
the proof, receives comments from peers in these talks, fixes a number of
inaccuracies and uncertainties in the proof, types a journal paper, submits
the paper, goes to international conferences reporting on the result, receives
a referee report with revisions, revises the paper, and finally publishes it.
In this overview of the mathematical research process, a number of skills
are needed that are central, others that are less central, and yet others that
are peripheral. The skills involved in finding the proof are certainly central,
but being able to communicate the proof to the experts (i.e., knowing how
the community expects proofs to be communicated) is equally important.
Giving presentations is still relatively important, but probably more dispensable than the earlier mentioned skills. Being able to write a paper in a
form that is acceptable for a referee is slightly further down the scale, and
somewhere at the end of the scale we find skills such as involved in filling
in travel expense forms for the trip to the international conference. While
the extremes on this spectrum (the ones that clearly are part of the mathematical research skills and the ones that clearly aren't) are easily identified,
we do not think that an objective stable core of skills can be identified that
could usefully take the place of the profession as a unifying principle.

To illustrate this, let us give an example from actual practice. A few
years ago, the first author had an autistic student who took several advanced mathematical classes at the graduate level. The student would not
hand in homework exercises, so there was no ordinary means of assessing
the student's understanding of the material. However, even in complicated

---

[9]In a recent episode of the TV series NUMB3RS (episode 6.01, "Hangman", aired 25
September 2009), mathematical skill of the protagonist math professor was manifested by
being able to calculate the position of the attacker while under sniper fire. Similar topoi,
such as being able to detect hidden patterns very quickly in large amounts of unstructured
data, can be found in other mainstream movies such as "A beautiful mind" (2001). For a
detailed discussion of the public representation of mathematicians, cf. (Osserman, 2005).

proofs, the autistic student was able to correct mistakes on the blackboard by shouting corrections, indicating an understanding of the proofs. At the end of the lecture, the student took a personalized exam in which he performed very well on questions that essentially required a binary answer or just an intuitive idea. He performed badly on questions that required the student to give a proper mathematical argument, so for most contexts of research mathematics, the autistic student described would get a low assessment of skill level. We see this as an example corroborating the fact that pure understanding for the mathematical structures under investigation is not enough to have a high level of mathematical skill if this is not paired with the appropriate (historically and culturally determined) skills of communicating why the insights are true.

## 5.2   How are mathematical skills individuated?

The granularity of mathematical skills also leaves open a spectrum of options for analysis. The mathematical community usually puts great emphasis on the unity of the subject, which is indeed one of the special traits of the historical development of mathematics (cf. François and Van Bendegem, 2010). Furthermore, interrelations between seemingly disconnected areas of mathematics are constantly discovered,[10] so there is a strong empirical basis for claiming the unity of the subject and thus, for expecting one unified notion of mathematical skills. On the other hand, mathematicians themselves of course differentiate when it comes to specific aspects of a colleague's skills, and such aspects may also be epistemically important. Classifications of types of mathematicians have been around for a long time and are not a result of the diversification of the mathematical field.[11]

---

[10]To give a famous example, consider Gerhard Frey's 1984 observation that a solution to the Fermat equation would yield a counterexample to the Taniyama-Shimura conjecture, thus linking number theory to the area of elliptic curves.

[11]Cf., e.g., Felix Klein's recommendations for choosing among candidates for a vacant position in Berlin 1892, in the course of which he gives his view of the required balance of skills in a math department: "Bei der Mannigfaltigkeit der Individualitäten kann man ja nicht schematisieren, aber im grossen und ganzen sollten folgende Typen vorhanden sein:

1. Der Philosoph, der von den Begriffen construirt,

2. Der Analytiker, der wesentlich mit der Formel operirt,

3. Der Geometer, der von der Anschauung ausgeht."

(Letter dated 6 January 1892; cited after (Siegmund-Schultze, 1996); translation: "Given the manifold of individualities one cannot press the discussion into a schema, but generally speaking, the following types should be present: 1. The philosopher who constructs conceptually, 2. The analyst who essentially operates with formulae, 3. The geometer who proceeds from intuition.")

Klein mentions Weierstraß and Cantor for type 1, Weber and Frobenius for type 2, and Schwarz and Lindemann for type 3. (We should like to thank to Dirk Schlimm for help related to this reference.)

At the very fine-grained level, there is the division of the subject of mathematics into subfields, e.g., according to the *Mathematics Subject Classification* of the American Mathematical Society (the 2010 version of this classification is a document of 46 pages just listing the names of the subareas). It makes good sense to ask, when confronted with a specific problem in one of these areas, who is an expert in that specific field, i.e., who is a skilled mathematician with respect to that subject. Does that mean that there are as many variants of mathematical skills as there are subfields of mathematics?

With respect to this spectrum, we support a unificationist stance: Mathematics is one subject, and for most purposes, it makes sense to view *general* mathematical skills as the pertinent level of granularity. For purposes of assessing knowledge claims, local dimensions of skill may however also play a role, depending on context.

Suppose that we have a given mathematical theorem $p$ and a given context and would like to know whether a mathematician $S$ satisfies our requirements for knowledge given in (‡) at the beginning of this paper. This will require us to describe a skill level in terms of "mathematical skills" that is sufficient for the task at hand. But if mathematical skills are so diverse, what part of the skill set will be relevant here? We claim that this will be given by the context and the nature of the theorem $p$ in the same sense that a nurse's skill level is not an objectively defined property of a given human being, but situationally determined.

To give an example similar to the situations described in (Löwe and Müller, 2008) as part of the argument for the context-dependency of mathematical knowledge, let us consider a mathematician $S$ and a theorem $p$ that is not from his immediate research area, but a closely related area, and of which he has seen proof sketches, but never a full proof. We assume $S$ to be of lower skill level than *expert* for the relevant area of mathematics.

**Scenario 1.** If we are looking at a context in which only the proof idea matters, then the cognitive access that $S$ has to the proof sketches (by virtue of memory) is enough to satisfy the requirements of our definition (‡), and no skill for transforming the proof sketch into something else is needed. As a consequence, we would conclude that $S$ knows that $p$ in these contexts.

**Scenario 2.** Other contexts (for example, research contexts in which $S$ needs to use parts of the details of the proof in order to generalize the proof to a different setting) need *expert* level skills in order to allow $S$ to bring the proof sketches to which he has cognitive access to the level of detail needed for the context. As we assumed that $S$ is not of *expert* level for the relevant area, in these contexts we would not say that $S$ knows that $p$.

**Scenario 3.** Extending the example a bit, we can consider a context like in the last scenario (i.e., $S$ does not know that $p$); now $S$ asks an expert to give a more detailed proof sketch, gaining cognitive access to a new account of the proof for which a lower skill level than *expert* is enough in order to allow $S$ to transform this detailed account into the level of detail needed for the context, thus creating knowledge.

Comparing the three scenarios in this example, we see that a lot depends on judgments about situations of the type "given cognitive access to $X$, you need skill level $Y$ to produce a proof of type $Z$" or "$S$ has skill level $Y$ with respect to this particular situation". So, in order to make definition ($\ddagger$) useable in practice, we need to be able to make assessments of this type.

### 5.3    How are mathematical skills measured and assessed?

Directly continuing our discussion of the examples in the last section, we consider the question: how do we assess a person's mathematical skills? It is of the essence of a modal or dispositional predicate that while performance in specific circumstances may be a valid indicator, skill also transcends recorded performance. Even a good bicycle rider may fall from her bike, and even a skilled musician can play out of tune (cf. also the long quote from Ryle in § 2 above).

We do not think that mathematical skills are special in this respect. They are dispositional, but performance is an indicator. Typical exam situations show the tension inherent in this: We believe that many mathematicians have experienced a situation in which they wanted to assess the *mathematical skills* of a student, but were forced by exam regulations (and, in some sense, also considerations of fairness) to give marks based solely on performance. When a skilled student underperforms, it is not rare to comment on the less than satisfactory grade by telling the student that one is convinced that she can do better than that. Those are not empty words—rather, such assessment highlights the fact that skill is dearer to most of us than "mere" performance. Of course, performance matters—we all know cases of people who have been promising for just a little too long. Assessment of skill and assessment of performance are not independent. But, as with many other dispositional traits, no strict statistical relationship appears to exist either.

The assessment of mathematical skills and of mathematical knowledge of course go hand in hand. When claiming, as we did above, that mathematical skills can be the key to an analysis of mathematical knowledge, we do not claim that we have access to mathematical skills completely independently of knowledge attributions. This seemingly circular structure is typical of modality—cf., e.g., David Lewis's remarks about the interrelation between formal models giving truth conditions for counterfactuals, on the one hand, and our intuitive assessment of counterfactuals on the other hand (Lewis, 1986, p. 43).

In § 5.1 we have argued that mathematical skills are best viewed as professional skills, i.e., as skills belonging to a specific, historically and sociologically contingent profession, the research mathematician. Thus, performance on the job is certainly another, but again defeasible, indicator of mathematical skills.

It seems that in contexts like the ones discussed at the end of § 5.2, a good test question whether a mathematician has the required skill level is the following: Suppose that I have a certain type $X$ of proof at my cognitive disposal (e.g., a proof by testimony, a proof sketch, a handwritten proof with gaps, etc.) and a certain skill level $Y$ is needed to transform this into the type $Z$ of proof that I need in order to satisfy the definiens of (‡). In order to assess whether a mathematician $S$ has skill level $Y$, I can ask myself the question: "Assuming that $S$ has never heard about my problem before, if I give him the information at my cognitive disposal (of type $X$), will he be able to produce a proof of type $Z$?"

# 6   Conclusion

This paper is a specification of the general ideas starting in (Löwe and Müller, 2008), explaining how a link can be made between the Dreyfus-Dreyfus model of skills and our context-sensitive definition of mathematical knowledge. It raises a large number of questions and answers few of them. The Dreyfus-Dreyfus skill model is a situational model, not allowing for objective characterizations of individuals in terms of levels, but rather describing typical behaviour of individuals at certain levels of expertise in particular situations.

We have proposed that mathematical skill should be seen as a *professional skill*, largely delineated by the skills necessary for being a mathematical researcher, certainly a culturally and historical determined notion.[12] For each given knowledge assessment context, we need to determine which parts of the professional skill set are relevant. Certain parts of the skill set (the ones we called peripheral, such as booking flights to conferences) will almost certainly never of be relevance in mathematical knowledge claims; but we doubt that there is a clear definition of which part of the professional skill set forms the stable epistemological core.

---

[12]One of the consequences of this general perspective is that the set of skills that make a good mathematician can change. Some researchers in automated deduction claim that 25 years from now, proofs will not be checked by referees anymore, but mathematicians will write their proofs in codes checkable by automated theorem checkers. If they are right (witness the December 2008 issue of the *Notices of the American Mathematical Society* on formal proof for some indication of momentum gathering), it may become a central mathematical skill to be versatile in HOL programming, as much as it is now a central skill in mathematical physics to be able to use Mathematica or in statistics to use R.

We have also given examples of how the skill levels would be used in research situations, but the largest part of the empirical project remains: providing an empirical basis (similar to the empirical basis that Benner provided for the area of nursing) that allows us to identify various levels of expertise in research mathematicians. This is a long-term project and will require a lot of observation of research mathematicians in the style of Heintz (2000) and Greiffenhagen (2008). We hope that this paper can serve as a stepping stone for these investigations to come.

To close this paper, we would like to mention the issue of "Mastery". In this paper, we have based the discussion on the five-level Dreyfus-Dreyfus skill model, not on its extension that includes the level of *Mastery* (Dreyfus, 2001). Since mathematics is sometimes closer to an art than to a trade, and issues such as creativity can play a vital role, extending the five-level model by the additional level of *Mastery* seems particularly fitting for the case of mathematics. In mathematics, as in music or art, we run into situations that are difficult for the empirical researcher: there are very few people who understand the most complicated proofs in mathematics; there are very few people who can give us an insight into how the minds of the top-researchers of a field work; some of the best mathematicians claim theorems whose proofs are essentially uncheckable by anyone else.[13] If we want to get behind these epistemological conundra, a theory of *Mastery* could certainly help, in the same sense that it helps to understand singular phenomena in music and art.

However, we believe that currently, we first need to understand the levels of expertise of the non-exceptional research mathematicians (as part of the five-level skill model) before we move on to the more puzzling and exotic realms of exceptional talent and skill.

# Bibliography

Anderson, J. R. (1993). *Rules of the mind.* Lawrence Erlbaum, Hillsdale NJ.

Benner, P. (1984). *From novice to expert: Excellence and power in clinical nursing practice.* Addison-Wesley, Reading MA.

---

[13]Cf. Thurston's discussion of how he struggled in communicating his proof of the geometrization theorem for Haken manifolds: "[S]ome concepts that I use freely and naturally in my personal thinking are foreign to most mathematicians I talk to. [...] At the beginning, this subject was foreign to almost everyone. It was hard to communicate—the infrastructure was in my head, not in the mathematical community" (Thurston, 1994, p. 174–175).

Benner, P. (2001). *From novice to expert: Excellence and power in clinical nursing practice. Commemorative edition.* Prentice Hall, Upper Saddle River NJ.

Brykczynski, K. A. (2006). Benner's philosophy in nursing practice. In Alligood, M. R. and Tomey, A. M., editors, *Nursing theory: Utilization & Application*, pages 131–156, St. Louis MO. Elsevier Mosby.

Cohors-Fresenborg, E., Brinkschmidt, S., and Armbrust, S. (2003). Augenbewegungen als Spuren prädikativen oder funktionalen Denkens. *Zentralblatt für Didaktik der Mathematik*, 35(3):86–93.

Cohors-Fresenborg, E. and Schwank, I. (1992). Die Berücksichtigung des Dualismus von prädikativen und funktionalen mentalen Modellen für das Erfassen und die Förderung von Begabungen im mathematisch-naturwissenschaftlichen Bereich. In Urban, K., editor, *Begabungen entwickeln, erkennen und fördern*, volume 43 of *Theorie und Praxis*, pages 301–309, Hannover. Universität Hannover.

Darbyshire, P. (1994). Skilled expert practice: Is it "all in the mind"? A response to English's critique of Benner's novice to expert model. *Journal of Advanced Nursing*, 19:755–761.

Dolan, K. (2001). Building bridges between education and practice. In Benner (2001), pages 275–283.

Dreyfus, H. L. (2001). *On the Internet (Thinking in Action).* Routledge, London.

Dreyfus, H. L. and Dreyfus, S. E. (1977). Uses and abuses of multi-attribute and multi-aspect model of decision making. Unpublished manuscript.

Dreyfus, H. L. and Dreyfus, S. E. (1986). *Mind over Machine: The Power of Human Intuition and Expertise in the Era of the Computer.* Free Press, New York NY.

Fenton, M. V. (2001). Identification of the skilled performance of master's prepared nurses as a method of curriculum planning and evaluation. In Benner (2001), pages 262–274.

Flyvbjerg, B. (2001). *Making Social Science Matter: Why social inquiry fails and how it can succeed again.* Cambridge University Press, Cambridge.

François, K. and Van Bendegem, J. P. (2010). Revolutions in mathematics. More than thirty years after Crowe's "Ten Laws". A new interpretation. In Löwe, B. and Müller, T., editors, *PhiMSAMP. Philosophy of Mathematics: Sociological Aspects and Mathematical Practice*, volume 11 of *Texts in Philosophy*, pages 107–120, London. College Publications.

Gordon, D. R. (2001). Research applications: Identifying the use and misuse of formal models in nursing practice. In Benner, P., editor, *From novice to expert: Excellence and power in clinical nursing practice. Commemorative edition*, pages 225–243, Upper Saddle River NJ. Prentice Hall.

Gray, E. M. and Tall, D. O. (1994). Duality, ambiguity and flexibility: A proceptual view of simple arithmetic. *Journal for Research in Mathematics Education*, 26(2):115–?141.

Greiffenhagen, C. (2008). Video analysis of mathematical practice? Different attempts to 'open up' mathematics for sociological investigation. *Forum: Qualitative Social Research*, 9(3):art. 32.

Heintz, B. (2000). *Die Innenwelt der Mathematik. Zur Kultur und Praxis einer beweisenden Disziplin*. Springer, Vienna.

Huntsman, A., Lederer, J. R., and Peterman, E. M. (2001). Implementation of Staff Nurse III at El Camino Hospital. In Benner, P., editor, *From novice to expert: Excellence and power in clinical nursing practice. Commemorative edition*, pages 225–243, Upper Saddle River NJ. Prentice Hall.

Lewis, D. K. (1986). *On the Plurality of Worlds*. Blackwell, Oxford.

Löwe, B. and Müller, T. (2008). Mathematical knowledge is context-dependent. *Grazer Philosophische Studien*, 76:91–107.

Marx, J. (2009). Career path for the new infection preventionist using the Dreyfus model of skills acquisition. *American Journal of Infection Control*, 37(6):E156–E157.

Osserman, R. (2005). Mathematics takes center stage. In Emmer, M., editor, *Mathematics and Culture II. Visual Perfection: Mathematics and Creativity*, pages 187–194. Springer, Berlin.

Rav, Y. (1999). Why do we prove theorems? *Philosophia Mathematica*, 7(1):5–41.

Ryle, G. (1949). *The Concept of Mind*. University of Chicago Press, Chicago IL.

Schwank, I. (2003). Einführung in funktionales und prädikatives Denken. *Zentralblatt für Didaktik der Mathematik*, 35(3):70–78.

Sfard, A. (1991). On the dual nature of mathematical conceptions: Reflections on processes and objects as different sides of the same coin. *Educational Studies in Mathematics*, 22:1–36.

Siegmund-Schultze, R. (1996). Das an der Berliner Universität um 1892 "herrschende mathematische System" aus der Sicht des Göttingers Felix Klein. Eine Studie über den "Raum der Wissenschaft". Humboldt-Universität zu Berlin. Institut für Mathematik. Preprint P-96-14.

Thurston, W. P. (1994). On proof and progress in mathematics. *Bulletin of the American Mathematical Society*, 30(2):161–177.

Ullery, J. (2001). Focus on excellence. In Benner, P., editor, *From novice to expert: Excellence and power in clinical nursing practice. Commemorative edition*, pages 258–261, Upper Saddle River NJ. Prentice Hall.

# Proof: Some notes on a phenomenon between freedom and enforcement

Gregor Nickel*

Fachbereich Mathematik, Universität Siegen, Walter-Flex-Straße 3, 57068 Siegen, Germany
E-mail: nickel@mathematik.uni-siegen.de

## 1 Introductory remarks and a disclaimer

Proof lies at the core of mathematics. A large part of every day's work in mathematical research is devoted to searching for provable theorems and their proofs, explaining proofs to students or colleagues, and reading other mathematician's proofs and trying to understand them. Mathematics is special among the other sciences due to the role of proof and insofar as mathematical proof offers the highest possible rigor. Moreover, it is a special feature of mathematics—in contrast to all positive sciences—that it can work on its foundations using its own methods. The mathematical analysis of mathematical proof is systematically developed in 20th century's formal logic and (mathematical) proof theory.[1] Here it was possible to transform proof from the *process* of mathematical arguing to an *object*—e.g., a well-formed sequence of symbols of some formal language—mathematics can argue about. Though this approach offers valuable results I will not concentrate on it.

In fact, there has been a growing interest in 'non-formal' features of mathematics. In fact, a closer look at the 'real existing' mathematics exhibits a much less formal science than the formalistic picture claimed. Here historical studies,[2] arguments from 'working mathematicians',[3] and sociological studies[4] point into a similar direction. Concerning a non-formal 'phenomenology of proof' we refer to the recent discussion initiated by Yehuda

---

*The author would like to thank two anonymous referees for their prompt reports containing instructive and often helpful criticism.

[1] In his comprehensive work on the philosophy of mathematical proof theory, Wille (2008) emphasizes that proof theory is special among the mathematical subdisciplines because it discusses *normative* questions about external justifications of its axiomatic basis. In fact, it can not be denied that the content of the *publications* in proof theory vary between a purely mathematical and a less technical one comprising normative claims and argumentations. However, in *every* mathematical area there are normative discussions—however rarely published—about the 'sense' of axioms, definitions, the fruitfulness of results etc. The only difference seems to be the communication medium.

[2] Cf., e.g., the pioneering work of Imre Lakatos (1976).

[3] Cf., e.g., the polemic in (Davis and Hersh, 1981) against a philosophy of mathematics emphasizing only the formal aspects.

[4] Cf., e.g., (Heintz, 2000).

Benedikt Löwe, Thomas Müller (*eds.*). *PhiMSAMP. Philosophy of Mathematics: Sociological Aspects and Mathematical Practice*. College Publications, London, 2010. Texts in Philosophy 11; pp. 281–291.

*Received by the editors*: 16 December 2009; 4 February 2010; 16 February 2010.
*Accepted for publication*: 22 February 2010.

Rav (1999).[5] With reference to many concrete examples from the mathematical practice it is argued there, that most of the existing proofs are for good reasons non-formal. Only in special situations strictly formalized proofs, called *derivations* occur. We shall contrast this with the classical analytic-synthetic distinction below. A special emphasis is given on the heuristic value of (the semantic side of) proof for the scientific practice. In the related paper (Pelc, 2009), the focus is narrower, concerned only with the function of proof to foster confidence for a claimed theorem. Nevertheless, the importance of non-formal proofs is underpinned, since a transformation of interesting proofs into derivations is hopeless for complexity reasons.

In our note we shall follow a similar, but slightly modified question: 'why and how are we convinced by proofs?' In fact, we are not so much interested in the important question of the genesis and growth of mathematics, but into the question, *how* the validation of its results work. This note, however, does not intend to bring forward any philosophical argument and avoids thus the debates and counter-debates of the professional philosophy of mathematics. Instead, we shall just present a collection of observations from a practitioner's point of view.[6]

## 2  Brian Rotman's semiotic perspective on mathematics

This note will adopt a perspective which is influenced by semiotics;[7] we thus observe proof as a quite special communication process[8] between two or more people or inside one person. Brian Rotman (1988) has offered this semiotic perspective on mathematics in an inspiring paper. Here he presents an integrative view where the basic insights of intuitionism, formalism, and platonism are valuated but also their respective limitations are shown:[9]

> [...] to have persisted so long each must encapsulate, however partially, an important facet of what is felt to be intrinsic to mathematical activity.

For a closer description of the working mathematician's dealing with signs, Rotman splits him into three different acting entities, the *Agent*, the *Mathematician*, and the *Person*:

> If the Agent is a truncated and abstracted image of the Mathematician, then the latter is himself a reduced and abstracted version of

---

[5]Cf. the subsequent discussion in (Azzouni, 2004; Pelc, 2009; Rav, 2007).

[6]Compare the introductory remarks in (Rotman, 1988, p. 97).

[7]A related perspective on proof is presented in (Lolli, 2005).

[8]I will concentrate, however, still on proof as an idealized concept of communication. For sociological observations of the 'real existing' mathematical communications, cf., e.g., (Heintz, 2000) or, more recently, (Wilhelmus, 2008).

[9]Cf. (Rotman, 1988, p. 101).

> the subject – let us call him the Person who operates with the signs
> of natural language and can answer to the agency named by the 'I'
> of ordinary nonmathematical discourse. [...] The Mathematician's
> psychology, in other words, is transcultural and disembodied.

While jointly doing mathematics the Mathematician and the Agent perform
different types of activities, e.g.,

> [...] it is the Mathematician who carries out inclusive demands to
> 'consider' and 'define' certain worlds and to 'prove' theorems in rela-
> tion to these, and it is his Agent who executes the actions within such
> fabricated worlds, such as 'count', 'integrate', and so on, demanded
> by exclusive imperatives.

Solely the Mathematician can communicate the leading idea of a proof to
another Mathematician, while many activities the Mathematician is inca-
pable to perform, e.g., evaluating infinite series, are done by the Agent.
Only the combined activities of these two give a full picture.

We shall follow this perspective and try to emphasize one aspect, Rotman only implicitly touches: The question of freedom and enforcement.
Our leitmotif will thus be, that mathematical proof shows a strange tension
of *freedom* and *enforcement* and can only be understood within this field.
It is thus our aim to pinpoint some characteristic features of mathematical proofs where this tension can be observed. With respect to Rotman's
actors, we might characterize the Agent to be completely determined by
the orders of the Mathematician, who in turn is at least (partially) free to
define certain objects, to suppose certain conditions etc., but restricted to
the rules of the mathematical discourse. The Person, finally, is (completely)
free to join or to leave this discourse.

## 3   Proof: despotism without authorities

One obvious goal of a mathematical proof, though not the only one, is to
give evidence that a certain claim is true. But, how does this takes place?
To characterize an important feature of the pragmatics[10] of proof I would
like to recall an instructive example: the elementary proof of Pythagoras'
theorem in Plato's dialogue *Menon*. Here, I will not summarize the content
of this wellknown dialogue, but just concentrate on some points. The main
subject is a foundational question of ethics namely for the nature of virtue
and whether it could be taught and learned. This question leads to the

---

[10]A completely different pragmatic aspect of proof is the concern of Imre Lakatos
(1976) in his classical work. Here it is shown that giving evidence for the truth of a result
is only *one*—and often not even the most important—goal of a proof. It is much more
important that the argumentation gives hints for understanding the respective roles of
the assumptions and thus for understanding *why* the theorem could hold.

general dilemma of teaching and learning: It seems impossible to learn anything unknown. If we do not know a content, after finding it we can never realize that it was the previously unknown. Socrates solves the puzzle by the concept of *anamnesis*: learning is nothing else than remembering the ideas already seen by the soul before coming to earth. To give evidence for this claim he demonstrates Pythagoras' theorem to an uneducated slave and *at the same time* he 'demonstrates' to Menon, how something previously unknown could be learned. The *aha effect* at the end of the learning process is afterwards interpreted as remembering. After knowing the solution it seems as if it was always known; the question—unsolvable before the proof—is now trivial. The initial state of ignorance appears to be improper, properly spoken, the theorem was always known. Indeed, the phenomenology of *learning* in mathematics and *remembering* is stupendously similar. For the initial question, whether virtue could be learned, Socrates claims at least the chance for a positive answer. Though there is no guaranteed way or indoctrination, we should steadily undertake efforts and hope for finding virtue by divine grace.

It is, however, an ironical feature of the dialogue, that the 'hard' argument for this position is a mathematical proof. The result is thus fixed from the outset and during the course of the proof every alternative route led into obvious errors. So the slave had to follow the necessitating force of proof.[11] There is no fair bargaining about the result, and at the end no freely chosen assent to the theorem; the communication process has thus a **dictatorial** character. On the other hand, the social position or any authority of the dialogue partners is completely irrelevant; thus there is also a **subversive** aspect, the freedom of arguing against any authority. Instead of a dictatorial proclamation of the result mathematical proof argues for it. David Hilbert characterized the value of mathematics for general education mainly in "ethical direction" since it "awakes the confidence in our own intellect, the critical ability of judgment, which distinguishes the truly educated person from the one who merely beliefs in authorities."[12] However, proof is based on strict commands formulated in the definitions and axioms and following the argument of a proof means following the commands of the author; Herbert Mehrtens felicitously describes this basis:

> Die Setzungen der Mathematik haben den Charakter von Befehlen,
> die Theoreme und Schlußfolgerungen sollen immer zwingende Folge
> der Befehlssysteme sein. Das macht den eigenartigen Charakter dieser

---

[11]The winning of dialogues is consequently the intuitive background for a constructive approach to formal logic by Paul Lorenzen and Oswald Schwemmer. A claim is provable, if there exists a definite strategy for winning such a dialogue, cf. (Lorenzen and Schwemmer, 1973).

[12]Cf. Hilbert's 1922/23 lectures edited in (Bödigheimer, 1988, p. 3; translation by the author).

> Sprache aus; sie besteht aus Befehlen, die das Setzen von Zeichen regeln. Die Gewißheit der Mathematik liegt in ihrer befehlsmäßig zwingenden Struktur.[13]

Taking stock, we regard a strange tension in the communication process. On the one hand, a mathematical proof is an ideal communication, which is strictly *herrschaftsfrei* (free of dominion) in the sense of Jürgen Habermas. On the other hand, there is absolutely no tolerance for defending a differing position. Only the predetermined result can be defended with warranty.

## 4    Freely chosen objects—strictly ruled language

There seems to be only one way for escaping the subversive despotism of a mathematical proof—aside from refusing to listen at all. For successfully denying the theorem you could only change the axioms, e.g., switch from plane to spherical geometry. This leads to the question: what are we talking about in mathematical proofs? During the course of the early 20th century we can observe a major change with respect to this question. It is well-known that the result is a switch from external to almost purely internal reference leading to a far reaching autonomy of mathematics. It lies in the *free* choice of the mathematician, which special set of axioms he likes to start with.[14] No external object dictates a certain set. And it is hardly exaggerated when Georg Cantor claims "Das Wesen der Mathematik liegt in ihrer Freiheit" (Purkert and Ilgauds, 1987, p. 121).

Concerning the communication process, we observe that conflicting positions in mathematics can now simply be avoided by subdiscipline branching; if you don't like Euclidean geometry you just change to hyperbolic. If you don't like the axiom of choice you work without it. This is similar in proof theory: you may freely choose from various axiomatic means. This also seems to be *one* reason, why mathematicians will quite fast agree about the validity of an argument.

However, this freedom is restricted in a twofold way. First, there is no freedom of interpretation and no context dependence of the terms. In contrast to all other texts, within a mathematical proof every $x$ must strictly remain the same $x$. There is—so to speak—no hermeneutical problem in a mathematical proof. Of course, this does neither mean that every proof

---

[13] "The determinations of mathematics have the character of commands, the theorems and inferences must be coercive implications of the systems of commands. This constitutes the peculiar character of this language; it is composed of commands ruling the drawing of signs" (translation by the author), cf. (Mehrtens, 1993, p. 101).

[14] Herbert Mehrtens discusses the development of modern mathematics and the disputes during the 'foundational crisis' just under this aspect of creative freedom, cf. (Mehrtens, 1990). Here David Hilbert—following Cantor—stands for a progressive modernity against L. E. J. Brouwer—following Leopold Kronecker—being the representative of reactionary anti-modernity which claims a necessary external reference for mathematics.

is self evident to everybody nor that the inner pictures connected to $x$ are the same for everybody. This strong concept of identity enables and leads to the second restriction: the chosen axioms are not allowed to contain contradictions neither explicitly nor implicitly. The anxious emphasis on this consistency, is the prize we pay for the freedom of choice with respect to the axioms.

## 5   Proof between determined calculation and spontaneous construction

If we cease to refer to given objects 'somewhere outside' there is no other way than to look at the internal structure of mathematical proof to explain its force.

To characterize mathematical inferences I will refer to the classical distinction of *analytic* and *synthetic* judgments[15] due to Immanuel Kant.[16] Though it is disputed, whether this distinction can be applied to all judgments, I think that it still can be used to illustrate important features of mathematical proofs. Thereby, I will postpone the question, whether mathematical proofs *are* empirically based or *a priori*. It is simply observed that mathematicians claim necessity for their inferences, thus try to avoid any empirical flavor. Moreover, the analytic-synthetic distinction can be used independently form the decision about the *a priori* character.

On the one hand, a correct proof can be described as consisting in or being transformable into a—probably very long—chain of identities. This transformation is either a real goal or at least an ideal. Any theorem is thus implicitly contained in the axioms; every judgment is analytic. Here, the assent enforcing power of proof seems to be well captured since nobody could reasonably deny the identity of the identical. We find this position most clearly stated in Leibniz's writings. In his small paper "Zur allgemeinen Charakteristik" he sketches the picture of proof—and broadens the application to all human reasoning—being as simple as a mechanical calculation:

> [E]s müßte sich, meinte ich, eine Art Alphabet der menschlichen Gedanken ersinnen und durch die Verknüpfung seiner Buchstaben und die Analysis der Worte, die sich aus ihnen zusammensetzen, alles andere entdecken und beurteilen lassen [...] Unsere Carakteristik wird alle Fragen insgesamt auf Zahlen reduzieren und so eine Art

---

[15]The usefulness of this distinction has been doubted, e.g., in (Quine, 1951). However, Quine's identification of analytic with "grounded in meanings independently of matters of fact" and synthetic with "grounded in fact" is for our concern much too close to the a priori vs. a posteriori distinction.

[16]It is not our aim to give an interpretation of Kant's complex philosophy of mathematics; for a concise analysis in the context of its functions within his critical philosophy, cf. (von Wolff-Metternich, 1995).

Statik darstellen, vermöge derer die Vernunftgründe gewogen werden können.[17]

Almost the same characterization of a quasi mechanical process of inference without any thinking can be found in Paul Bernays' writings:

> [N]achdem einmal die Prinzipien des Schließens genannt sind, [braucht] nichts mehr überlegt zu werden. Die Regeln des Schließens müssen so beschaffen sein, daß sie das logische *Denken* [emph. G.N.] eliminieren. Andernfalls müßten wir ja erst wieder logische Regeln dafür haben, wie jene Regeln anzuwenden sind. Dieser Forderung der Austreibung des Geistes kann nun wirklich genügt werden.[18]

There are—according to Leibniz—two basic principles for these mechanical proofs,

> die Definitionen oder Ideen und [...] ursprüngliche, d.h. identische Sätze, wie der, daß $B$ gleich $B$ ist.[19]

If we put aside the external reference of the axioms, all mathematical theorems and proofs reduce to identities. There are, however, at least two more drawbacks for this position. The first one on the 'macro-level': no interesting proof has ever been completely broken down to trivialities. Nobody would be able to write or to read and *understand* it.[20] To overcome this problem we could theoretically refer to 'in principle analysis' or to the metaphor of a 'mechanical working' of proof and practically involve computers. In fact, Leibniz's characterization, the concept of analyticity is well illustrated by the mechanical working of a machine. Being just a metaphor for Leibniz, today *computer (assisted) proofs* are technically implemented and became an interesting tool for mathematical research. The status, however, of these proofs is still a question in dispute. Of course, a first question arises, how we shall guarantee its correctness, thus the correct working of the software *and* the hardware of the computer. However, McEvoy (2008) argues against others (e.g., Kitcher (1983) or Resnik (1997)) that computer

---

[17] "I arrived at this remarkable thought, namely that a kind of alphabet of human thoughts can be worked out and that everything can be discovered and judged by a comparison of the letters of this alphabet and an analysis of the words made from them. [...] Our characteristic will reduce all questions to numbers, so that reasons can be weighed, as if by a kind of statics" (translation by the author), quoted after (Cassirer, 1966, pp. 30).

[18] "After designation of the principles of inference, no more thinking is needed. The rules of inference must eliminate the logical thinking. Otherwise we would need new logical rules how these rules can be applied. This demand of an expulsion of spirit can actually be satisfied" (translation by the author), cf. (Bernays, 1976, pp. 9).

[19] "[...] definitions or ideas and [...] identical propositions such as $B = B$" (translation by the author), cf. (Cassirer, 1966, p. 58).

[20] This is the core thesis in (Pelc, 2009).

assisted proofs are as *a priori* as any other 'hand-made' proof. In fact,
the checks and double checks of a proof can always be called 'experiments',
independently on the question, whether it is a machine or a human mathematician who performs it. And in both cases we shall never be absolutely
certain if the proof is sufficiently long. Accepting this view, the grade of
warrant and the aprioricity is therefore not the main difference between a
hand-made and a computer proof. Instead of the *grade* of certainty, it is
rather its *base*, we should look at. A human mathematician performs an
analytical inference by a (freely chosen, but) trusted assent following the
self-posed rule, e.g., of *modus ponens*. The working of a machine, on the
other hand, obtains its reliability by a supposed strict determination. We
thereby replace *necessity* of the result by a *determined contingency*. And at
least on the theoretical level, we must suppose a strong concept of determinism. It is not by chance—I think—that the same Leibniz draws one of
the strongest metaphysical pictures of a completely predetermined course
of the world. It is, however, to question whether we do want to refer to such
a strong metaphysical concept when trying to understand the concept of a
mathematical proof. Second, even on the 'micro-level' there still remains a
question. If we accept a proof to be a chain of tautologies, we should ask
again why or how such a single tautology—thus the principle of identity —
convinces. I do not want to deepen this question here. It may be sufficient
to mention that, e.g., the sociologist Niklas Luhmann[21] again and again
considers this and states a puzzling paradox of identity. Another reference
is Klaus Heinrich (1981) who studies the origin of basic logical principles in
the Greek myth. Both authors show that the seemingly trivial tautologies
are quite complicated phenomena.

On the other hand, it seems strongly that a successful mathematical
proof—as a whole and in every single step—is a non-trivial act of constructive unification. Its basis lies within a spontaneous, non-mechanical, constructive working by the mind of the (human) mathematician. It is Kant
who emphasizes the role of pure intuition,[22] *reine Anschauung*, and the
synthetic character of mathematical theorems and proofs.[23] First, Kant
takes the axioms[24] to be synthetic and *a priori* valid propositions which

---

[21]Cf., e.g., (Luhmann, 1992, pp. 491). Here Luhmann formulates "die Einsicht, daß die
Tautologie letztlich nichts anderes ist als eine verdeckte Paradoxie; denn sie behauptet
einen Unterschied, von dem sie zugleich behauptet, daß er keiner ist."

[22]The emphasis on intuition can also be found in the *Menon* dialogue. The switch
from arithmetic to geometry is precisely the point where intuition is essentially required.
It is important to remark here that for the Greek mathematics there was no arithmetical
solution of the problem, no calculus dealing with irrational 'numbers' as, e.g., $\sqrt{8}$. It was
thus inevitable to switch from the arithmetical calculation to the intuitive geometry.

[23]In (Norman, 2006), this synthetic character is further substantiated when the author
discusses the role diagrams play for mathematical reasoning.

[24]Cf. (Kant, 1781, pp. A732).

could never be known without the construction of their concepts within pure intuition—again we let aside this point in the course of modern internalization. I thus also do not bother with the question of Kant's taking too much structure as *a priori* granted, e.g., the Euclidean space. Second, the proofs of propositions from these axioms can only mediate knowledge if they are also intuitively evident. The requirements for a real demonstration are twofold, it must show that the claim necessarily holds and it must be intuitively (self) evident:

> Nur ein apodiktischer Beweis, sofern er intuitiv ist, kann Demonstration heißen. [...] Aus Begriffen a priori (im diskursiven Erkenntnisse) kann aber niemals anschauende Gewißheit d.i. Evidenz entspringen, so sehr auch sonst das Urteil apodiktisch gewiß sein mag.[25]

Thus only mathematics could offer these demonstrations in the proper sense. For Kant, however, this essentially constructive approach is not restricted to geometry. Even the most 'mechanical' part of mathematics in Kant's horizon, namely algebra, works by using constructions, at least in the form of signs or characters. In fact, the process of proving—not only on the level of heuristics—is always accompanied by jotting down signs and pictures; every single step provokes our own constructions. There is no mathematics without the recognition of the basic signs. Again we also consider the macro-level: Intuition enables us to take a proof or argument as a whole—it is an ability of unification which could be illustrated by the difference between step-by-step calculation and an overall insight. This insight, however, is—as the search for virtue in *Menon*—hardly controllable. Only *after* a successful construction we might be able to give good reasons.

In my opinion it is sensible to keep *both* aspects, analytic and synthetic, as essential, but never completely realized ideals. Just from the outset, mathematical axioms, definitions and proofs are an *invitation* to follow the *free* mental constructions of the author; to repeat his or her synthesis. Though active construction is needed, though there is no proof without synthesis, you *might* always ask for a closer analysis which *enforces* the conclusion. And every proof communicates the *promise* that this further analysis could be done.

---

[25] "An apodictic proof can be called a demonstration, only in so far as it is intuitive. [...] Even from a-priori concepts, as employed in discursive knowledge, there can never arise intuitive certainty, that is, [demonstrative] evidence, however apodictically certain the judgment may otherwise be." Cf. (Kant, 1781, pp. B762).

# Bibliography

Azzouni, J. (2004). The derivation-indicator view of mathematical practice. *Philosophia Mathematica*, 12(2):81–105.

Bernays, P. I. (1976). *Abhandlungen zur Philosophie der Mathematik*. Wissenschaftliche Buchgesellschaft, Darmstadt.

Bödigheimer, C.-F., editor (1988). *David Hilbert. Wissen und mathematisches Denken. WS 1922/23. Vorlesung ausgearbeitet von Wilhelm Ackermann*. Mathematisches Institut, Göttingen.

Cassirer, E., editor (1966). *Gottfried Wilhelm Leibniz. Hauptschriften zur Grundlegung der Philosophie. Band I*. Meiner, Hamburg.

Davis, P. J. and Hersh, R. (1981). *The Mathematical Experience*. Birkhauser, Boston MA.

Heinrich, K. (1981). *tertium datur. Eine religionsphilosophische Einführung in die Logik*. Stroemfeld/Roter Stern, Frankfurt a.M.

Heintz, B. (2000). *Die Innenwelt der Mathematik. Zur Kultur und Praxis einer beweisenden Disziplin*. Springer, Vienna.

Hersh, R., editor (2006). *18 unconventional essays on the nature of mathematics*. Springer, New York NY.

Kant, I. (1781). *Kritik der reinen Vernunft*. Johann Friedrich Hartknoch, Riga. 2nd edition 1787.

Kitcher, P. (1983). *The Nature of Mathematical Knowledge*. Oxford University Press, New York NY.

Lakatos, I. (1976). *Proofs and Refutations*. Cambridge University Press, Cambridge.

Lolli, G. (2005). *QED — Fenomenologia della dimostrazione*. Bollati Boringhieri, Torino.

Lorenzen, P. and Schwemmer, O. (1973). *Konstruktive Logik, Ethik und Wissenschaftstheorie*. BI Verlag, Mannheim.

Luhmann, N. (1992). *Die Wissenschaft der Gesellschaft*. Suhrkamp, Frankfurt a.M.

McEvoy, M. (2008). The epistemological status of computer-assisted proofs. *Philosophia Mathematica*, 16(3):374–387.

Mehrtens, H. (1990). *Moderne Sprache Mathematik*. Suhrkamp, Frankfurt a.M.

Mehrtens, H. (1993). Nachwort. In Barrow, J. D.: *Warum die Welt mathematisch ist*. Campus, Frankfurt a.M.

Norman, J. (2006). *After Euclid. Visual Rasoning and the Epistemology of Diagrams*, volume 175 of *CSLI Lecture Notes*. CSLI Publications, Stanford CA.

Pelc, A. (2009). Why do we believe theorems? *Philosophia Mathematica*, 17(1):84–94.

Purkert, W. and Ilgauds, H. J. (1987). *Georg Cantor 1845–1918*. Birkhäuser, Basel.

Quine, W. V. O. (1951). Two dogmas of empiricism. *Philosophical Review*, 60:20–43.

Rav, Y. (1999). Why do we prove theorems? *Philosophia Mathematica*, 7(1):5–41.

Rav, Y. (2007). A critique of a formalist-mechanist version of the justification of arguments in mathematicians' proof practices. *Philosophia Mathematica*, 15(3):291–320.

Resnik, M. (1997). *Mathematics as a Science of Patterns*. Oxford University Press, New York NY.

Rotman, B. (1988). Towards a semiotics of mathematics. *Semiotica*, 72(1/2):1–35. Page numbers refer to the reprint in (Hersh, 2006, pp. 97–127).

von Wolff-Metternich, B.-S. (1995). *Die Überwindung des mathematischen Erkenntnisideals. Kants Grenzbestimmung von Mathematik und Philosophie*. De Gruyter, Berlin.

Wilhelmus, E. (2008). Socio-empirical epistemology of mathematics. *The Reasoner*, 2(2):3–4.

Wille, M. (2008). *Beweis und Reflexion. Philosophische Untersuchungen über die Grundlagen beweistheoretischer Praxen*. Mentis, Paderborn.

# Are all contradictions equal?
# Wittgenstein on confusion in mathematics

Esther Ramharter*

Institut für Philosophie, Universität Wien, Universitätsstraße 7, 1010 Vienna, Austria
E-mail: esther.ramharter@univie.ac.at

> Der Widerspruch.
> Wieso gerade dieses *Eine* Gespenst?
>
> *Ludwig Wittgenstein*

## 1   Introduction

In this paper I will not focus on contradictions as parts of formal languages (formulae in a logical system) or as an occasion to construct a system of paraconsistent logic, but I will rather study them as phenomena of mathematical practice. I shall do so from a Wittgensteinian point of view, using Wittgenstein's treatment of Cantor's Diagonal Argument as a guideline for most considerations in this paper.[1] There will be no main thesis resulting from my considerations, but I hope to be able to give a partial overview of what may happen, when a mathematician faces a contradiction. I will also discuss Wittgenstein's remarks on the mathematicians' attitudes towards the contradiction, but will leave aside the debate regarding the extent to which Wittgenstein should be understood as anti-revisionistic (cf. Frascolla, 1994; Maddy, 1992; Redecker, 2006; Wright, 1980, and many others).

> But you can't allow a contradiction to stand!—Why not? We do sometimes use this form in our talk, of course not often—but one could imagine a technique of language in which it was a regular instrument. (Wittgenstein, 1956, p. 166e)

Even if one does not call this revisionistic, one has to admit, I think, that adopting Wittgenstein's view would not leave things in mathematical practice as they are. (Of course, contradictions have always been a "regular instrument" in mathematics, in proofs by *reductio ad absurdum*, and also as means to develop and change theories (cf. Byers, 2007, p. 84 and p. 98)—but this is not: allowing a contradiction *to stand*.)

---

*The author would like to thank the *Wissenschaftliches Netzwerk* PhiMSAMP funded by the *Deutsche Forschungsgemeinschaft* (MU 1816/5-1) for travel support.

[1]For a detailed discussion of Wittgenstein's remarks on the Diagonal Argument, cf. (Redecker, 2006).

Benedikt Löwe, Thomas Müller (*eds.*). *PhiMSAMP. Philosophy of Mathematics: Sociological Aspects and Mathematical Practice.* College Publications, London, 2010. Texts in Philosophy 11; pp. 293–306.

*Received by the editors:* 20 September 2008; 8 May 2009.
*Accepted for publication:* 8 May 2009.

## 2   Contradictions which cause confusion (and those which do not)

If we are confronted with a contradiction (in a formal system), why do we not simply eliminate it by just excluding either the contradiction itself from the system or one of the propositions contributing to the contradiction?

> Can we say: 'Contradiction is harmless if it can be sealed off'? But what prevents us from sealing it off? That we do not know our way about in the calculus. Then *that* is the harm. And this is what one means when one says: the contradiction indicates that there is something wrong about our calculus. It is merely the (local) *symptom* of a sickness of the whole body. But the body is only sick when we do not know our way about.
>
> The calculus has a secret sickness, means: What we have got is, as it is, not a calculus, and *we do not know our way about*—i.e., cannot give a calculus which corresponds 'in essentials' to this simulacrum of a calculus, and only excludes what is wrong in it. (Wittgenstein, 1956, III, §80, p. 209)

Wittgenstein's answer is: Such a procedure will not be successful in cases where the contradiction makes us confused, where we cannot find our way ("wir kennen uns im Kalkül nicht aus").

But there are also other cases. Wittgenstein (1956, III, §80, p. 209) gives an example: One could teach Frege's calculus, which includes a contradiction. Nobody might notice it, and everybody would be content. Then: Where is the problem?

We need not even refer to such a thought experiment, mathematical practice provides us with a fairly recent "real life example": Physicists deal with what they call "δ-functions" and live in peace with the contradiction they produce.

This remarkable situation followed from several requirements in physics, especially in connection with solutions to certain differential equations. In theoretical physics, for example, the (ideal) situation of all mass concentrated in one point needs to be modelled. The *desideratum* would be a function $f$ with the properties

$$f(x) = \begin{cases} 0 & \text{for } x \neq 0 \\ \infty \ (\text{or } a \in \mathbb{R}) & \text{for } x = 0 \end{cases} \quad , \quad \int_{-\infty}^{\infty} f(x)\, dx = 1, \text{ and}$$

$$\int_{-\infty}^{\infty} \delta(x) f(x) dx = f(0).$$

Such a function does not exist.

Mathematicians then built the theory of "generalized functions" or "distributions", i.e., linear functionals fulfilling the above requirements such that the "usual" functions can be embedded into the space of these functionals. But what the physicists really wanted were *functions*, not some generalized entities.

If you have a look at notes from a first term course in theoretical physics you will most likely still find something like this:

### III. Theorie der Distributionen

#### 3.1. Die Dirac'sche Deltafunktion

$$Bp: m\nu = p = \int_{t_0}^{t_0+\tau} F(t)\,dt$$

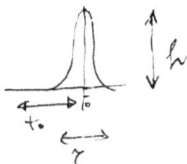

$$Bp: Balken \qquad F = \int dx\, q(x)$$

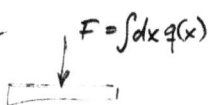 

#### Def:

$$\delta(x) = \begin{cases} 0 & x \neq 0 \\ \infty & x = 0 \end{cases}, \qquad \int_{-\infty}^{\infty} dx\, \delta(x) = 1$$

$$\int_{-\infty}^{\infty} dx\, \delta(x)\, f(x) = f(0)$$

This means that the physicists choose to stick to the self-contradictory objects. In a book on the history of the theory of distributions one finds:

> However, I have not drawn very general philosophical conclusions from the history told in this book, since I suspect the development of the theory of distributions may not be representative of the way mathematics has developed in the 20th century. (Lützen, 1982, p. 2)

Though distributions may not be a paradigmatic example for the history of mathematics in the 20th century, they are at least not un-typical for what becomes of mathematical concepts: From a later point of view we say that the mathematicians of earlier times did not really understand what they were doing,[2] that they did not avoid contradictions (because they did not notice them).

But in what sense did they really have a problem? Wittgenstein argues, that we only need to worry about a contradiction once we lose our understanding of what is going on.

## 3   Contradictions to start and those to end with

One of Wittgenstein's major concerns is Cantor's Diagonal Argument. A mathematician explaining the argument to a student might put it like this:

*We may restrict ourselves to the real numbers between 0 and 1, for if these are uncountable, then the set of all real numbers will be so a fortiori. Now suppose the real numbers between 0 and 1 were countable, then we could make a list of all of them. Let this list of real numbers, in their decimal notations, be:*

$$0, a_{11}a_{12}a_{13}\ldots$$
$$0, a_{21}a_{22}a_{23}\ldots$$
$$0, a_{31}a_{32}a_{33}\ldots$$
$$\vdots$$

*We can build a number between 0 and 1 in the following way: The first decimal place is $a_{11} + 1$, the second decimal place is $a_{22} + 1$, the third is $a_{33} + 1$ etc.—and we get a number, that is different from every number in the list, a contradiction. Therefore such a list does not exist; the real numbers are uncountable.*

Wittgenstein formulates two objections to the Diagonal Argument.

The first one is: We do not know what it means "to be a list of all real numbers (between 0 and 1)". The only conception of a list we possess is that of countable sets of things.[3] In Wittgenstein's words:

---

[2] Think of the convergence of infinite series, for example, or the infinitesimals of the early stages of the development of Calculus.

[3] And even in such cases there is reason for caution: "It is less misleading to say '$m = 2n$ allows the possibility of correlating every number with another' than to say '$m = 2n$ correlates all numbers with others'. But here too the grammar of the meaning of the expression 'possibility of correlation' has to be explained." (Wittgenstein, 1969, p. 466) (The English translation writes "time" for the German word "Zahl". We corrected this to "number".)

Asked: "Can the real numbers be ordered in a series?" the con-
scientious answer might be: "For the time being I can't form any
precise idea of that".—"But you can order the roots and the algebraic numbers for example in a series; so you surely understand the
expression!"—To put it better, I *have got* certain analogous forma-
tions, which I call by the common name 'series'. But so far I haven't
any certain bridge from these cases to that of 'all real numbers'. Nor
have I any general method of trying whether such-and-such a set 'can
be ordered in a series'. (Wittgenstein, 1956, II, §16, p. 130)

The second objection consists of the diagnosis that the outcome of the
argument is a procedure to create a series of numbers rather than a thing
of which we can be sure that it is a real number. Again, Wittgenstein:

If someone says: "Shew me a number different from all these", and
is given the diagonal rule for answer, why should he not say: "But I
didn't mean it like that!"? What you have given me is a rule for the
step-by-step construction of numbers that are successively different
from each of these. (Wittgenstein, 1956, II, §5, p. 126)

Now both objections rèly on this rather colloquial version of Cantor's
argument. And Wittgenstein himself warns us:

The result of a calculation expressed verbally is to be regarded with
suspicion. The *calculation* illuminates the meaning of the expression
in words. It is the *finer* instrument for determining the meaning.
(Wittgenstein, 1956, II, §7, p. 127)

Such a "finer" alternative of the argument would be:

*(We presuppose the real numbers to be given as Dedekind cuts or by some
axioms.) Suppose $\varphi$ were a 1-1-mapping from $\mathbb{N}$ to [0,1]. For each $b \in [0,1]$
there is a decimal notation (this is a theorem), let's say $0, b_1 b_2 b_3 \ldots$. We
define functions $g_i : [0,1] \to [0,1]$ such that*

$$g_i(b) = 0, \underbrace{0 \ldots 0}_{i-1} b_i 00 \ldots.$$

*Then the series $\sum_{n=1}^{N}(g_n(\varphi(n)) + 10^{-n})$ converges (for N to infinity) and
therefore has as its limit a real number d, let's say (between 0 and 1). It
can easily be shown that d is different from all numbers $\varphi(n)$, n = 1, 2, \ldots.
This is a contradiction to the assumption that $\varphi$ is one-to-one.*[4]

What remains of Wittgenstein's objections in the light of this more precise formulation?

---

[4] Instead of working with decimal expansions, this could be done with nested intervals.
I have chosen this option because it is closer to Wittgenstein's considerations.

Let me start with the second objection. This objection vanishes with respect to the "mathematical" version; $d$ is not "less a real number" than any number given as the limit of a sequence (as a Dedekind cut).[5] Not accepting $d$ as a real number would mean to make Analysis impossible all together.

But in spite of his warning Wittgenstein continues to focus on the way we express things in ordinary language. We do in fact explain the Diagonal Argument as it was done in the first version (above), and Wittgenstein is quite right that the listener has not the slightest reason to identify this avoiding ("ausweichende") procedure (Wittgenstein, 1956, II, §8, p. 127) with a real number. The first version is simply not an appropriate way of telling the story. (It hides problems such as that of totality for instance.)

The first objection is more serious with regard to the formal version. Applied to this version it says: we do not know what a one-to-one-function between a countable and an uncountable set "looks like", we do not know any examples. Is this just an objection to any sort of *reductio ad absurdum*?

> How does indirect proof work, for instance in geometry? What is strangest about it is that one sometimes tries tries to draw an ungeometric figure (the exact analogue to an illogical proposition).

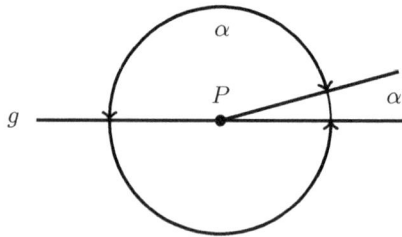

> (But this, of course, only comes from an erroneous interpretation of the proof. It is funny, for instance, to say "assume that the straight line g has two continuations from point $P$". But there is really no need to assume such a thing.)

> Proofs in geometry, in mathematics, cannot be indirect in the real sense of the word because one cannot assume the opposite of a geometrical proposition as long as one sticks to one specific geometry.

> (That proof simply shows that the arcs $\alpha$ and $\alpha+\alpha'$ approximate each other all the more and without limit, the more $\alpha'$ approximates 0.)
> (Wittgenstein, 2000, Item 108, p. 29, transl. E. R.)

---

[5]Wittgenstein could argue that the limits of sequences or series are no more numbers than the result of the diagonal proof, and indeed he is inclined to say that the definition of a limit includes a proof (Wittgenstein, 1956, V, §36, p. 290), but this will not keep us from regarding the result as a number; otherwise the answer to a question could never be a number because a question is never a number.

Obviously Wittgenstein really does not accept indirect proofs, at least not in every case. On the other hand there are remarks like:

> What an indirect proof says, however, is: "If you want *this* then you cannot assume *that*: for only the opposite of what you do not want to abandon would be combinable with *that*." (Wittgenstein, 1956, p. 147e)

This sounds as if Wittgenstein were, in principle, in agreement with the idea of the *reductio ad absurdum*. Consider the following examples:

*Example 1.*

**Theorem.** $\sqrt{2}$ is not a rational number.

*Proof.* Suppose $\sqrt{2}$ were rational, then it could be written as $\frac{a}{b}$ ...
                                                                        Q.E.D.

*Example 2.*

**Theorem.** Every sequence $(x_n)$ in a subset $F$ of $\mathbb{R}$ has a convergent subsequence in $F$ if and only if $F$ is bounded and closed.

*Proof (one direction).* Consider a subset $F$ of $\mathbb{R}$. Suppose it is not closed. Choose $x$ in the closure of $F$, but not in $F$, then for all $n \in \mathbb{N}$ there is an $x_n \in F$ such that $|x_n - x| < \frac{1}{n}$. Then every subsequence of $(x_n)$ converges to $x$, but $x \notin F$, and therefore $(x_n)$ has no convergent subsequence *in F*—a contradiction to the assumption.        Q.E.D.

Obviously the first proof—which is similar in this respect to Cantor's Diagonal Argument—starts with the contradiction, whereas the second proof just uses the contradiction in the end.

We could conclude that Wittgenstein does not accept those proofs by *reductio*, which start with a (hidden) contradiction—but that seems too superficial a distinction. The following section will describe what could be seen as the "deeper" (Wittgenstein does not like this word) distinction behind the "at the beginning/at the end"-distinction.

I agree that the latter distinction may be regarded as superficial, but want to point out that Wittgenstein's understanding of proofs as being fundamentally dependent on their geometry (the "geometry of signs" as he calls it)[6] makes it important *where* the contradiction is placed.

---

[6]The role of visual thinking in symbol manipulation is studied extensively in (Giaquinto, 2007, pp. 191–213 )—not from a Wittgensteinian, but from a cognitivistic point of view; nevertheless there are similarities. Cf. also (Rotman, 2000, pp. 44–70).

# 4   Contradictions which include the impossibility of an intuition or a concept (and those which do not)

In the first example we suppose something that contradicts itself and of which we therefore cannot have any concept or imagination. The Diagonal Argument is similar to the first example in this respect. We do not know what a "list of irrationals" should be—just as we do not know what a "list of water" should be.[7] In the second example everything works well and is coherent *as long as the presupposition is not taken into considera- tion*. (Notice that we can start the proof with "suppose it *is*" instead of "suppose it *were*".) We do not have to start with a contradiction but may decide in the end "if you want *this*—F has no sequences without convergent subsequences—, then you must not assume *that*—F is not closed".[8]

How to deal with the Diagonal Argument now? We might admit that the proof is nonsense—because we started with an absurdity—and therefore give up the idea of proving (in this manner) anything about the size of $\mathbb{R}$ in comparison with $\mathbb{N}$. What we have learnt then is just:

> If it were said: "Consideration of the diagonal procedure shews you that the *concept* 'real number' has much less analogy with the con- cept 'cardinal number' than we, being mislead by certain analogies, are inclined to believe", that would have a good and honest sense. But just the *opposite* happens: one pretends to compare the 'set' of real numbers in magnitude with that of cardinal numbers. The dif- ference in kind between the two conceptions is represented, by a skew form of expression, as difference of expansion. (Wittgenstein, 1956, II, §22, p. 132)

We do not know how to compare them by listing, one-to-one-mapping, etc. And the fact that we do not know this tells us, that they are different. And the proof might be taken as a hint that there is something wrong with any attempt to compare their sizes the way we tried to. So the contradiction destroys parts of the supposed meaning of a concept, it does not render a proposition false.

---

[7]This is, I suppose, a case where the contradiction leads to confusion in the sense of the first section.

[8]Of course one might argue that the proof of Example 1 can easily be restated in the same form: Let $x$ be a rational number, $x = \frac{a}{b}$, etc., and in the end it turns out, that $\sqrt{2}$ would not be a possible candidate for such a number. But this is not how we *do it*! Wittgenstein is right that there are two different ways of proceeding in a proof by *reductio in practice*. In Example 1 we work with the expression $\sqrt{2} = \frac{a}{b}$ throughout the whole proof—and this is clearly a meaningless expression as we very well know from the onset.

# 5 Contradictions between mathematical propositions and contradictions between mathematical and other propositions (empirical ones)

In the early years after his return to philosophy, Wittgenstein thought that the meaning of a mathematical proposition consisted in its proof—at least this seems to be the direction he inclined to. Later on he places greater emphasis on the relation between mathematical propositions and their "roots" in everyday life. But throughout his life the discussion of the status of (certain) propositions is at the centre of his remarks on mathematics. So the topic of the current section would require a detailed analysis of "logical", "grammatical", "empirical" propositions and the propositions called "hinge propositions" in secondary literature. As this goes far beyond the scope of this article I will restrict myself to one little remark.

> Put two apples on a bare table, see that no one comes near them and nothing shakes the table; now put another two apples on the table; now count the apples that are there. You have made an experiment; the result of counting is probably 4. [...] And analogous experiments can be carried out, with the same result, with all kinds of solid bodies.—This is how our children learn sums; for one makes them put down three beans and then another three beans and then count what is there. If the result at one time were 5, at another 7 (say because, *as we would now say*, one sometimes got added, and one sometimes vanished of itself), then the first thing we said would be that beans were no good for teaching sums. But if the same thing happened with sticks, fingers, lines and most other things, that would be the end of all sums.
>
> "But shouldn't we then still have 2+2=4?"—This sentence would have become unusable. (Wittgenstein, 1956, I, §37, p. 51f)

This passage has a harmless reading: We place more trust in mathematics (mathematical propositions) than in our ability to count (correctly).[9] (And only if mathematics were not expedient in the majority of cases, would we consider mathematical propositions to be meaningless.) But there is another reading: We would exclude beans from the teaching of elementary arithmetic if they did not behave well and disappeared—this might still seem harmless, but note that this implicitly states that we would rather believe that a bean has dissolved into nothing than that 3+3 is not 6. So it is not only our possible errors that make us distrust—if we take Wittgenstein seriously in what he says here, we would rather adopt the assumption of a very incoherent "world of experiences" than give up our acceptance of arithmetical theorems.

---

[9] And this is the "main reading" of the passage. I agree with Gierlinger (2008, p. 123) in this point.

At first sight this is not only astonishing *per se*, it also seems to contradict what I presented as Wittgenstein's attitudes in the earlier sections of this article. If mathematicians overestimate the general importance of contradictions, why should we then adopt the deeply unsettling assumption that things can simply disappear, only because 3+3=5 contradicts 3+3=6?

It is not just *any* contradiction we have to deal with here, it is not just the two mathematical propositions contradicting each other, it is the whole system of elementary arithmetic together with all our practices of counting, calculating, paying bills, etc. that is at stake. This situation could be taken as a paradigm for what is meant by a contradiction that causes confusion.

But is not the confusion, that arises from the assumption that things might disappear, even worse? I think the solution of the problem consists in recognizing that we have the wrong picture of what is at issue here. The situation does not, as it might seem, resemble a scale with elementary arithmetic on the one side and the trustworthiness of our daily experience on the other. First, the possibility for things to dissolve is not Wittgenstein's interest in the quoted passage, so we might perhaps better take take his statement *cum grano salis*. And second, in the situation he describes, *only* beans dissolve (if all things can disappear, then mathematics, too, becomes useless—but this would be taking matters one step further). So Wittgenstein contrasts a very local phenomenon with a very global one (arithmetic). And pressed to choose between the two, we would decide in favour of the global regularity.[10]

# 6   Contradictions which we seek and those which surprise us

This distinction does not coincide with a distinction between "contradictions in indirect proofs" and "real" contradictions (those which "harm" the mathematicians). In practice something can *turn out* to be an indirect proof. You try to prove something and suddenly you realize that you have found a contradiction. So "contradictions of indirect proofs" are not a subspecies of "contradictions we seek".

The comparison I have in mind here is rather between contradictions we are suddenly confronted with, ones that shock us, and those we search for. Of course (some of the) indirect proofs are examples for the second sort. Another example from Wittgenstein:

> For might we not possibly have *wanted* to produce a contradiction?
> Have said—with pride in a mathematical discovery: "Look, this is
> how we produce a contradiction"? Might not e.g. a lot of people

---

[10]I nevertheless perceive Wittgenstein's argument as problematic: He wants us to imagine a world where beans have the ability to dissolve, but nothing else has. We have to regard this as a thought experiment and be very charitable to engage in it.

possibly have tried to produce a contradiction in the domain of logic
[...]? These people would then never actually employ expressions of
the form $f(f)$, but still would be glad to lead their lives in the *neighbourhood* of a contradiction. (Wittgenstein, 1956, III, §81, p. 211)

There is something curious about the first sort of contradictions: Wittgenstein once states that "if you are surprised, then you have not understood it
yet".[11] Applied to contradictions this means: if the contradiction surprises
you, then you do not understand what you do not understand. On the
other hand it is quite natural to see a contradiction as the point at which
it becomes manifest that you do not understand what is going on. So if the
result of a proof surprises you, you are expected to think it through again,
until the surprise gives way to understanding. But what if you are surprised
by a contradiction? Should you try to understand the contradiction better? This does not seem to be very sound or reasonable. I will not follow
this line of thinking here, but there is any case a seamless transition from
trying to understand the steps that led to a contradiction to searching for
a contradiction.

Now we are in the very centre of Wittgenstein's objections against the
mathematicians' exaggerated fear of contradictions. He doubts that it is of
any use to try to avoid contradictions mechanically in cases where we do
not have any reason to distrust the system we are using (cf. Wittgenstein,
1956, III, §83, pp. 214ff).

The working experience of mathematicians gives them no motive to
suppose that there could be a contradiction in elementary arithmetic and
therefore—according to Wittgenstein—no reason to busy themselves with
proofs of soundness of this system. (Note that this is exactly what mathematicians nowadays do—and did when Wittgenstein wrote his remarks,
which was just after Gödel had completed his paper and the foundations of
mathematics constituted a central topic—, however, they are doing so for
pragmatic reasons, whereas Wittgenstein offers a philosophical background
for this attitude vis-à-vis contradictions.)

# 7   Concluding remarks

Wittgenstein wonders why it is just "this *one* bogy" (Wittgenstein, 1956,
IV, §56, p. 254)—the contradiction—that mathematicians seem to fear so
much. An answer could be given by following his own considerations: There
is not just one kind of contradiction, there are many. And they can hamper
the mathematician's work in many different ways. But this does of course
not answer the question Wittgenstein actually had in mind: Why no *other*
bogies? And again we find hints for an answer in the very passages I have

---

[11](Wittgenstein, 1956, App. II, §2, p. 111); for Wittgenstein on surprise in mathematics
cf. (Floyd, 2008, 2010; Mühlhölzer, 2001).

discussed: Many different threats become visible when we analyse what happens when contradictions occur. For example the contradiction we face when we try to compare the natural with the real numbers with respect to their magnitude teaches us to be careful with expansions of concepts to other areas. A much simpler example: There is nothing to be said against talking of the "length" of a desk and also of the "length" of the distance between the earth and the moon—but we must not forget that we measure them in totally different ways and that "length" therefore has a different meaning in the two cases (Wittgenstein, 1956, III, §4, p. 147). The same holds for expansions of concepts in mathematics in general. Wittgenstein also discusses the transition from finite numbers to infinite cardinal numbers. He argues that a proof concerning finite numbers consists in ascertaining that it holds for every number of the finite totality. Whereas—as there is no infinite totality—a proof for all numbers has to yield a procedure. Stuart Shanker sums up:[12].

> The word 'class' means *totality* when it is used in the context of a finite *Beweissystem* (a group of objects all sharing the same property); but in its 'infinite' framework 'class' signifies a rule-governed series (the possibility of constructing a series *ad infinitum* by the reiteration of an operation)." (Shanker, 1987, p. 165)

So thoughtless expansion of concepts could be considered another bogy, we should beware of.

As we have seen, a contradiction may be, but need not be a symptom of a state of confusion. *If* it is a symptom of a state of confusion, then it can be seen within a more general context: Wittgenstein says, "a philosophical problem has the form: I don't know my way about" (Wittgenstein, 1953, § 123). And this is exactly what a contradiction in the "serious" sense says. The contradiction then tells us that we do not know our way about, but it *only* tells us *that*. It is the mere expression of perplexity; it tells us where the problem is, not what the problem is. If we want to understand the reason for the perplexity—if we want to see the philosophical problem—, we must have a look around, investigate the context of the contradiction. My proposal to see a confusing contradiction in a more general context can therefore supplemented by the "opposite" statement: A more specific context (a philosophical tradition, for example) is needed to understand what has happened.

As I said at the beginning of this text, I will not state a thesis, but I am now able to formulate a suggestion, which is not entirely unfounded:

---

[12]Wittgenstein's interest in "system", "class" and "totality" has one of its origins in Waismann's considerations and discussions with him; cf. (Waismann, 1986).

it could be useful to search for "other bogies" in the neighbourhood of contradictions.[13]

## Bibliography

Byers, W. (2007). *How Mathematicians Think: Using ambiguity, contradiction, and paradox to create mathematics.* Princeton University Press, Princeton NJ.

Floyd, J. (2008). Wittgenstein über das Überraschende in der Mathematik. In Kross, M., editor, *Ein Netz von Normen. Ludwig Wittgenstein und die Mathematik*, pages 41–77. Parerga Verlag, Berlin. Translated by Joachim Schulte.

Floyd, J. (2010). On being surprised. Wittgenstein on aspect perception, logic and mathematics. In Day, W. and Krebs, V. J., editors, *Seeing Wittgenstein Anew. New Essays on Aspect Seeing*, pages 314–337. Cambridge University Press, Cambridge.

Frascolla, P. (1994). *Wittgenstein's Philosophy of Mathematics.* Routledge, New York NY.

Giaquinto, M. (2007). *Visual Thinking in Mathematics. An Epistemological Study.* Oxford University Press, Oxford.

Gierlinger, F. (2008). Wright, Wittgenstein und das Fundament des Wissens. In Hieke, A. and Leitgeb, H., editors, *Reduktion und Elimination in Philosophie und den Wissenschaften. Beiträge des 31. Internationalen Wittgenstein Symposiums*, pages 122–124, Kirchberg am Wechsel. Österreichische Ludwig Wittgenstein Gesellschaft.

Lützen, J. (1982). *The Prehistory of the Theory of Distributions.* Springer, Heidelberg.

Maddy, P. (1992). Wittgenstein's anti-philosophy of mathematics. In Puhl, K., editor, *Philosophy of Mathematics. Proceedings of the 15th International Wittgenstein Symposium*, pages 52–72, Kirchberg am Wechsel. Österreichische Ludwig Wittgenstein Gesellschaft.

Mühlhölzer, F. (2001). Wittgenstein and surprises in mathematics. In Haller, R. and Puhl, K., editors, *Wittgenstein and the Future of Philosophy. A Reassessment after 50 Years. Proceedings of the 24th International Wittgenstein-Symposium*, pages 306–315, Kirchberg am Wechsel. Österreichische Ludwig Wittgenstein Gesellschaft.

---

[13] I would like to thank the referee for very helpful remarks on this concluding section.

Redecker, C. (2006). *Wittgensteins Philosophie der Mathematik. Eine Neubewertung im Ausgang von der Kritik an Cantors Beweis der Überabzählbarkeit der reellen Zahlen.* Ontos Verlag, Frankfurt a.M.

Rhees, R. and Kenny, A., editors (1978). *Ludwig Wittgenstein. Philosophical Grammar.* University of California Press, Berkeley CA.

Rotman, B. (2000). *Mathematics as Sign. Writing, Imagining, Counting.* Stanford University Press, Stanford CA.

Shanker, S. (1987). *Wittgenstein and the Turning-Point in the Philosophy of Mathematics.* Croom Helm, London.

Waismann, F. (1986). The nature of mathematics: Wittgenstein's standpoint. In Shanker, S., editor, *Ludwig Wittenstein: Critical Assessments*, volume III, pages 60–67. Croom Helm, London.

Wittgenstein, L. (1953). *Philosophische Untersuchungen. Philosophical Investigations.* Blackwell, Oxford. Edited by G. E. M. Anscombe and R. Rhees, translated by G. E. M. Anscombe.

Wittgenstein, L. (1956). *Remarks on the Foundations of Mathematics. Bemerkungen über die Grundlagen der Mathematik.* Basil Blackwell, Oxford, 3rd edition. Edited by G. H. von Wright, R. Rhees and G. E. M. Anscombe; translated by G. E. M. Anscombe; page numbers refer to the 1978 edition.

Wittgenstein, L. (1969). *Philosophische Grammatik.* Basil Blackwell, Oxford. Edited by R. Rhees; page numbers refer to the English translation in (Rhees and Kenny, 1978).

Wittgenstein, L. (1998–2000). *Wittgenstein's Nachlass. The Bergen Electronic Edition.* Oxford University Press, Oxford.

Wright, C. (1980). *Wittgenstein on the Foundations of Mathematics.* Duckworth, London.

# Fraenkel's axiom of restriction: Axiom choice, intended models and categoricity

Georg Schiemer*

Institut für Philosophie, Universität Wien, Universitätsstraße 7, 1010 Vienna, Austria
E-mail: georg.schiemer@univie.ac.at

## 1 Introduction

A recent debate has focused on different methodological principles underlying the practice of axiom choice in mathematics (cf. Feferman et al., 2000; Maddy, 1997; Easwaran, 2008). The general aim of these contributions can be described as twofold: first to clarify the spectrum of informal justification strategies retraceable in the history of mathematical axiomatics. Second, to evaluate and to philosophically reflect on the actual reasoning involved in the introduction of new axioms in mathematical practice such as large cardinal axioms in set theory.

The most extensive treatment of these matters for the case of set theory can be found in (Maddy, 1997). Her philosophical discussion of axiom choice covers both of the mentioned approaches, i.e., it is both descriptive in reconstructing the justification types in early axiomatic set theory as well as normative in devising "methodological maxims" for the evaluation of present set theoretic axiom candidates. More specifically, Section 1.3 of her book provides a historical survey of the arguments given for ZFC by Zermelo, von Neumann, and Fraenkel (among others) with the intention "to explicate and analyze its distinctive modes of justification" given there (Maddy, 1997, p. 72). In the final two sections of her book (Sections 3.5 and 3.6), Maddy in turn devises "a naturalistic program" for discussing more recent axiom candidates (starting from Gödel's axiom of constructibility to large cardinal or determinacy axioms) intended to model the "justificatory structure of contemporary set theory" (Maddy, 1997, p. 194).

In this paper I attempt to take up Maddy's historical discussion by drawing attention to a historical episode from early axiomatic set theory centered on Abraham Fraenkel's *axiom of restriction* (*"Beschränktheitsaxiom"*) (in the following AR). The axiom candidate was first introduced by Fraenkel in the early 1920s and can be considered as a minimal axiom devised to

---

*The translations in this paper are mostly my own; if not I refer to the source from which they are quoted. I would like to thank Erich Reck, Richard Heinrich, Johannes Hafner, Matt Glass and Miles MacLeod for helpful remarks and suggestions. I am also thankful to the members of the audience at the PhiMSAMP-3 conference in Vienna 2008, where a preliminary version of this paper was presented, as well as the members of the SoCal HPLM Group and two anonymous referees for useful comments and criticism.

Benedikt Löwe, Thomas Müller (*eds.*). *PhiMSAMP. Philosophy of Mathematics: Sociological Aspects and Mathematical Practice.* College Publications, London, 2010. Texts in Philosophy 11; pp. 307–340.

*Received by the editors:* 1 September 2008; 13 May 2009.
*Accepted for publication:* 12 August 2009.

express a restriction clause, more specifically a minimal condition for any set model satisfying ZFC. His attempts to devise such a restriction clause on set models varied in the course of his intellectual development and eventually led to different, partly independent versions of restrictive axioms for set theory. Now, all of these axiom candidates are considered in retrospect as "ad hoc devices" due to their "vague, metatheoretic" character without any real, remaining significance in modern axiomatic set theory (cf., e.g., Kanamori, 2004, p. 515) (cf. Section 5). Nevertheless, I will argue that Fraenkel's attempts to introduce such a minimal axiom remain interesting from an historical point of view since the axiom takes a central and so far neglected place in a broader discussion on the (non-)categoricity of set theory and its role as a foundational discipline in mathematics. Moreover, closer study of it will also prove to be instructive for the general methodology of axiom choice given the specific justifications that Fraenkel provides for his axiom candidate.

The paper has two main aims. The first is to reconstruct the different arguments for AR in light of Maddy's account of extrinsic (in contrast to intrinsic) justification. Given Fraenkel's case, I show that one can expand Maddy's analysis by a new type of extrinsic argument that concerns the metatheoretic property of categoricity of the resulting axiomatization. The second aim is to analyze AR in terms of Maddy's "methodological maxims", namely UNIFY and MAXIMIZE, devised for axiom choice in contemporary set theory (cf. Maddy, 1997, pp. 208–215). I argue that AR deserves closer attention since—being a minimizing principle for the set theoretic domain— it is *prima facie* diametrically opposed to Maddy's second principle that calls for a maximization of the set theoretic domain. This—given the overall viability of Maddy's principles - could be taken as an additional argument against the legitimacy of AR in axiomatic set theory. However, I argue that a direct evaluation of Fraenkel's axiom candidate in terms of Maddy's maxims is problematic since they are motivated by different conceptions of set theory as a foundational enterprise.

The paper is organized as follows: I present a brief overview of different types of axiom justification described in (Maddy, 1997), focusing on types of extrinsic, non-epistemic arguments she identifies in the early axiomatization of set theory (Section 2). Her account of extrinsic evidence is compared to the different lines of argumentation Fraenkel develops for his axiom candidate (Section 3). His main motivation for AR is a metatheoretic consideration, i.e., to restrict the set theoretic universe to his intended model of ZFC and thereby to render his axiom system categorical (Section 3.1). I discuss different proposed versions of AR intended to achieve this categorical axiomatization (Section 3.2). Further, I suggest that Fraenkel seems to develop his views on minimal models and the intended effect of

AR in close analogy to Dedekind's approach to defining sets via closure principles (Section 4). In Section 5 a number of objections directed against AR by Baldus, von Neumann, and Zermelo from the late 1920s will be discussed that eventually resulted in a fundamental shift in Fraenkel's own understanding of his axiom candidate. Fraenkel's response to these objections and the resulting new versions of AR will be discussed (Section 6). Finally, I compare Fraenkel's different versions of AR with Maddy's naturalistic method of axiom choice, specifically with her maxims UNIFY and MAXIMIZE, and discuss whether such an evaluation can be justified (Section 7).

## 2   Maddy on extrinsic justification

Maddy's historical survey of the axiomatization of set theory and the different motivations for the axioms of ZFC is based on the distinction (anticipated in Gödel, 1964) between two types of justification. In the case of extrinsic justification, an axiom is assessed in terms of its theoretical fruitfulness, i.e., with an eye to its intended consequences for the resulting theory. In the case of intrinsic justification, an axiom is defended in terms of the intuitive nature of the properties it is supposed to express (cf. Maddy, 1997, pp. 36–37). The two kinds of arguments are commonly associated with different types of mathematical axioms: *structural axioms* of the "working mathematician" (e.g., the axioms of rings, groups etc.) and *foundational axioms* concerning structures that "underlie all mathematical concepts" (e.g., the Peano axioms for arithmetic and ZFC for set theory) (cf. Feferman, 1999, p. 3). Structural axioms are often considered to be justifiable on extrinsic grounds comparable to the experimental testing of hypotheses in the natural sciences. Foundational axioms in turn are attributed an entirely different status. Their justification is often based on intrinsic considerations, either by reference to certain epistemic norms (such as those of intuitiveness, obviousness, immediacy, and naturalness) or by reference to a pre-axiomatic conception of the subject matter, i.e., the mathematical structure the axiom in question is supposed to capture.[1]

However, Maddy succeeds in showing that the assumed link between foundational axioms and intrinsic arguments is not exclusive. Her survey of the practice of axiom choice in set theory identifies a number of genuinely extrinsic considerations laying the ground for the foundational axioms of

---

[1]Numerous examples from the history of axiomatic mathematics suggest such a relation between foundational axioms and intrinsic arguments. In discussing the (epistemological) primacy of the Peano axioms over ZF, Skolem argues that in contrast to the latter the former are "immediately clear, natural and not open to questions" (Skolem, 1922, p. 299). In opposition to this, Gödel—in a well known passage—describes a faculty of "intuition" as a sufficient "criterion of truth" for the set theoretic axioms. (Gödel, 1964, p. 271)

ZFC.[2] Moreover, she points out that the extrinsic arguments given there
are explicitly non-epistemic in character. In several cases (most prominently
in Zermelo's defense of the axiom of choice, cf. (Maddy, 1997, p. 56) the
motivation for the acceptance of an axiom does not depend on its intuitiveness but rather on its theoretical consequences for mathematics.[3] Naturally,
since ZFC is primarily considered a foundational theory, these consequences
have to be closely tied to what Maddy describes as its "foundational goal"
within mathematics. Her specific understanding of this goal is clearly non-
epistemological: in contrast to a stronger "foundationalist" reading of ZFC
in terms of an ontological reduction that "reveals the true identities of [...]
mathematical objects" or the reduction to an epistemologically secure basis, set theoretic axioms in her "modest" version of foundations share no
"preferred epistemological status" (cf. Maddy, 1997, pp. 24–25).[4] Instead,
they provide a fruitful codification of all other branches of mathematics by
allowing a set theoretic "representation" of all other mathematical entities
and structures (cf. Maddy, 1997, pp. 25–26). By this,

> [...] vague structures are made more precise, old theorems are given
> new proofs and unified with other theorems that previously seemed
> quite distinct, similar hypotheses are traced at the basis of disparate
> mathematical fields, existence questions are given explicit meaning,
> unprovable conjectures can be identified, new hypotheses can settle
> old problems, and so on. (Maddy, 1997, pp. 34 35)

For Maddy, it is the sum of these theoretical virtues that amount to the
foundational goal of ZFC. Concerning the question of axiom justification,
she argues that the capacity of a particular axiom to contribute to these
theoretical objectives (and thus to the overall success of the foundational
discipline) can be taken as direct extrinsic evidence for it: "[...] I see the
effectiveness of an axiom candidate at helping set theoretic practice reach its
foundational goal as a sound extrinsic reason to adopt it as a new axiom".[5]

---

[2]Cf. (Feferman et al., 2000) for an interesting discussion between Feferman and Maddy
on the intrinsic/extrinsic distinction and its bearing on Feferman's classification of mathematical axioms mentioned above (Feferman et al., 2000, pp. 416–419).

[3]Maddy refers to Russell (1973) for an early methodology of axiom choice along similar
lines. In fact, in his lecture, Russell proposes a "regressive method" for justifying logical
axioms (such as the axiom of reducibility) without any direct intrinsic support by a
kind of probabilistic confirmation through its "obvious" consequences: "Hence we tend
to believe the premises because we can see that their consequences are true, instead of
believing the consequences because we know the premises to be true." (Russell, 1973,
pp. 273-274)

[4]For a comparable account of (higher-order) logic as a non-foundationalist foundation
for mathematics, cf. (Shapiro, 1991).

[5](Feferman et al., 2000, p. 418); there is a certain tension between Maddy's account
of the "foundational goal" of ZFC and Feferman's account of foundational axioms. For
her own evaluation of this relation (cf. Feferman et al., 2000, pp. 417–419).

One central criterion of success that allows the assessment of an individual axiom concerns set theory's unifying power. Its strong unifying role is due to the creation of a single domain of discourse to which all of mathematics is reducible:[6]

> The force of set-theoretic foundations is to bring (surrogates for) all mathematical objects and (instantiations of) all mathematical structures into one arena—the universe of sets—which allows the relations and interactions between them to be clearly displayed and investigated. (Maddy, 1997, p. 26)

Note that one central implication of this picture of set theoretic unification through a "unified arena" is that a specific conception of the domain of set theory, i.e., the intended universe of sets, becomes a central issue in the foundational goal. Now, Maddy's historical discussion of extrinsic argument types for the axioms of set theory focuses on the standard axioms of ZFC. What is not mentioned in her survey, however, is that there was already a strong and ongoing debate throughout in 1920s on how to conceive this universe of sets and characterize it axiomatically. In the course of different attempts to fix a domain of set theory that is capable of providing such a "unified arena" for mathematics, one specific axiom candidate, namely Fraenkel's axiom of restriction stands out as the most prominent contribution. In the remaining sections of the paper I will focus on this specific episode in the early history of the axiomatic set theory in general and Fraenkel's axiom candidate in particular. It will be shown that one can identify an extrinsic argument in his remarks on set theoretic restriction based on a similar motivation for unification not discussed in (Maddy, 1997).

## 3   Fraenkel's axiom of restriction

In the early 1920s, Fraenkel suggested two axioms to be added to the axiom system presented in (Zermelo, 1908b): the axiom of replacement, now a standard axiom of ZF, as well as the lesser known axiom of restriction (AR). The latter was basically devised to express a restriction clause, more specifically a minimal condition for any set model satisfying the axioms set up by Zermelo. In what follows I will give a brief reconstruction of the evolution of Fraenkel's thought on the notion of restriction during this period.

The first mention of AR can be found in an article titled "Zu den Grundlagen der Cantor-Zermeloschen Mengenlehre" in *Mathematische Annalen*

---

[6]Compare also the following passage: "One methodological consequence of adopting the foundational goal is immediate: if your aim is to provide a single system in which all objects and structures of mathematics can be modeled or instantiated, then you must aim for a single, fundamental theory of sets." (Maddy, 1997, pp. 208–209)

from 1922 (Fraenkel, 1922b). Fraenkel's motivation for adding the axiom
candidate is mainly pragmatic and explicitly concerns set theory as a foundational discipline. He states that "Zermelo's concept of set is more comprehensive than seems to be necessary for the needs of mathematics [...]."
(Fraenkel, 1922b, p. 223) Fraenkel goes on to mention two types of possible
sets in the "domain" ("*Grundbereich*") of set theory that are consistent with
the existing axioms, but are irrelevant for mathematical purposes. The first
are "non-conceptual" sets consisting of physical elements. The second are
non-well-founded sets, i.e., sets with infinite membership chains originally
specified by Mirimanoff (1917). From their possibility within Zermelo's
axiomatization, Fraenkel draws an interesting consequence for its general
status:

> Whereas sets of the first as of the second kind are not necessary for set
> theory considered as a mathematical discipline, it in any case follows
> from the fact that they have a place within Zermelo's axiomatization
> that the axiom system [...] does not have a "categorical character",
> that is to say it does not determine the totality of sets completely.
> (Fraenkel, 1922b, p. 234)

Categoricity is understood here as a "complete" characterization of the domain of sets. In an added footnote, Fraenkel refers to one of his earlier
works on number theory, more specifically on different axiomatizations of
$p$-adic numbers, for an informal definition of categoricity based on Veblen's
notion of a "categorical set of postulates" (Fraenkel, 1911, p. 76).[7] A more
structured presentation of his arguments for AR can be found in the second
edition of his monograph *Einleitung in die Mengenlehre* (Fraenkel, 1924).
Here, the introduction of the additional axiom leads to a "simplification of
the set theoretic edifice" by ruling out non-well-founded numbers without
losing its significance for mathematics due to the fact that "all mathematically relevant sets can [...] be saved with such a restricted axiomatization."
(Fraenkel, 1924, p. 218) As a second independent argument the property
of categoricity is mentioned: "Moreover, without such a restriction it is
not within reach that our axiom system captures the totality of admissible
sets *completely* as is desirable for the construction of every axiomatization."
(Fraenkel, 1924, p. 218) Two short remarks are in order here. First, one
can identify at least two related but non-identical objections against Zermelo's original axiomatization here: the non-eliminability of extraordinary
sets that are redundant for the formalization of mathematics on one hand.
On the other hand, the non-categoricity of Zermelo's proposed axiom system is considered as a general theoretical deficiency of any axiomatization.[8]

---

[7]On Veblen's understanding of categoricity and a closer comparison to the modern
notion, cf. (Awodey and Reck, 2002, pp. 22–25).

[8]This second point is further highlighted in a passage in his published lectures from

Note second that the two mentioned issues, i.e., the applicability of set theory to mathematics and the axiomatic property of categoricity, are treated independently here. One can find no explicit remark about the possible implications of the categoricity of the extended axiom system ZF + AR for its foundational goal in mathematics. I will return to this point in the last section.

## 3.1   The (non-)categoricity of set theory

In the second edition of *Einleitung* we also find an explicit definition of the notion of categoricity as one type of completeness of an axiom system referred to in the argument above:[9]

> According to it an axiomatic system is called complete, if it determines uniquely the mathematical objects governed by it, including the basic relations between them, in such a way that between any two interpretations of the basic concepts and relations one can effect a transition by means of a 1–1 and isomorphic correlation. (Fraenkel 1924, quoted from Awodey and Reck 2002, p. 30)

For the specific case of set theory the following explication is given:

> If the axiom system is complete and one has chosen in two distinct ways, each in accord with the axioms, an interpretation of the concept of set—in particular also its extension—and of the basic relation $a \in b$, then it has to be possible to maintain a correlation between the sets of the one interpretation and those of the other such that first, to each set of the first interpretation corresponds one and only one [...] set of the other interpretation and vice versa and that secondly, if $a \in b$ is a valid relation in the first interpretation [...] then the relation $a' \in b'$ also holds for the sets $a'$ and $b'$ that have been assigned to $a$ and $b$ in the other interpretation and vice versa. (Fraenkel, 1928, p. 228)

This is probably the first application of the concept of categoricity via isomorphism to axiomatic set theory. Nevertheless, his presentation remains sketchy compared to modern standards. The central concept used in these remarks about the conditions of categoricity for set theory is the notion of an isomorphic correlation between set models. In modern terminology such

---

1925: "It means more than a mere flaw of our axiom system that the totality of all possible sets is not unequivocally fixed but that instead there are always narrower and more comprehensive interpretations of the concept of set that remain compatible with our axiom system." (Fraenkel, 1927, p. 101)

[9]For Fraenkel's treatment of three types of completeness, i.e., semantic completeness, syntactic completeness and categoricity, as well as their relationship and later reception, mainly by Carnap, cf. (Awodey and Reck, 2002).

In the third edition of *Einleitung* the equivalence of the type of completeness given in the quotation with the notions of "categorical" (Veblen) and "monomorph" (Feigl-Carnap) is explicitly stated (cf. Fraenkel, 1928, p. 349).

a correlation is taken as a 1–1 mapping between two set models $M$ and $N$ that is structure-preserving, i.e., as a function $f$ mapping $M$ one-to-one onto $N$ such that for two binary relations $F$ and $G$ (on $M$ and $N$ respectively), for all members $a$ and $b$ of $M$, $F(a, b)$ iff $G(f(a), f(b))$. However, in the 1920s, Fraenkel does not yet provide a comparable notion of isomorphism for set theory.[10] Nor does he get more explicit on the kind of the formal background language in which his axiom candidate and the notion of isomorphism should be cast. It should be stressed here that Fraenkel's remarks on AR and the categoricity of axiom systems from this period are in general presented informally. There is no attempt to provide a formulation of the axiom candidate in a formal symbolism such as Russell's type theoretic language commonly used at that time. This lack of formalization makes Fraenkel's claim about the categoricity of his axiomatization ZF + AR debatable from a modern point of view. If the axiom system ZF (specifically the axiom of replacement) is thought to be presentable in first order logic, then the expanded ZF + AR fails to be categorical due to the Löwenheim- Skolem theorems. Fraenkel's claim is only valid if a second-order axiomatization is assumed.[11] However, this fact is simply not noticed in his writings on the categoricity of set theory from that time. Despite discussing the Skolem paradox in the second and third edition of *Einleitung*, Fraenkel seems to be simply ignorant of its impact on his own project of providing a categorical axiomatization.[12] (I will return to this point in Section 6.2)

It is also in the third edition of *Einleitung* (Fraenkel, 1928) that one can find an interesting remark concerning his understanding of the concept of isomorphism. Following a more general discussion of the categoricity of axiom systems, he adds in a footnote:

> The expression "isomorphic" has a considerably more general sense than is usually common [. . .]. In fact the isomorphism is applicable to arbitrary relations, not only to those tertiary and $n$-ary relations denoted as "operations".[13]

---

[10] An alternative notion of isomorphism for sets had already been introduced some years before Fraenkel's version in (Mirimanoff, 1917). His definition is based on the simple notion of equivalence between sets and does not take into account a correlation between set models (cf. Mirimanoff, 1917, p. 41). For an early discussion of this definition, cf. (Sierpiński, 1922).

[11] Compare (Shapiro, 1991, pp. 85–86).

[12] van Dalen and Ebbinghaus (2000) retrace the different receptions of the Skolem paradox by Zermelo, von Neumann, and Fraenkel in the 1920s. For the latter's case they state that "the role of logic in set theory was not quite clear to Fraenkel". They in my mind correctly conclude that the impact of logical formalization on his categoricity claim transcended his "expertise" on logical matters in that period (van Dalen and Ebbinghaus, 2000, p. 148).

[13] (Fraenkel, 1928, p. 349); it was due to Rudolf Carnap who seems to have followed Fraenkel's informal remarks on a generalized concept of isomorphism to develop a formal

Irrespective of this, Fraenkel holds that Zermelo's axiomatization from 1908 is non-categorical in the sense specified above. This was a commonly acknowledged position in the 1920s shared by such eminent figures such as Skolem, von Neumann and Zermelo himself. Subject to debate were the possible reasons for this fact and whether Zermelo's original axiomatization could be rendered categorical by adding additional axioms.[14] As we have seen, according to Fraenkel's view anno 1924, the non-categoricity of Z is mainly due to the non-eliminability of "extraordinary sets" by the existing axioms. This in turn is due to the fact that the existential axioms, i.e., the empty set axiom and the axiom of infinity, do not restrict the domain of sets whereas the restrictive axioms like the axiom of separation are not restrictive enough to yield an "unequivocal specification" of the concept of set. As a solution to this Fraenkel proposes to introduce his AR which is described in analogy to Hilbert's completeness axiom in geometry:

> [...] as is the case there, the mentioned deficiencies can be remedied by setting up a [...] last axiom, the "axiom of restriction" that imposes on the concept of set, or more appropriately the domain [of sets], the smallest extension compatible with the remaining axioms. (Fraenkel, 1928, p. 234)

An alternative definition of the axiom can be found in (Fraenkel, 1924): "Aside from the sets imposed by the axioms [of Zermelo (1908)] there exist no further sets." (Fraenkel, 1924, p. 219) Now, the underlying motivation for introducing AR is clearly extrinsic in Maddy's more general spirit. The intention behind both versions of the axiom is evident: to rule out non-intended and non-well-founded sets by restricting either the interpretation of the concept of set or the domain of set and, by doing so, to render the axiom system categorical.

## 3.2    Versions of restriction

Fraenkel's early elucidations of the intended effect of AR do not go beyond the level of informal remarks. The most detailed exposition can be found in the article "Axiomatische Begründung der transfiniten Kardinalzahlen" (Fraenkel, 1922a) in which he develops an axiomatization for cardinal numbers. Here Fraenkel formulates two versions of AR that prove to be instructive for the case of standard set theory. According to the first, restriction is considered as a minimality condition on sets: There exist no sets apart from the sets implied by the given axioms. The second reading is more interesting, since it sketches the intended effect of the axiom. According to

---

definition of a "$n$-stage isomorphism correlator" for a type-theoretic language in his works on a general methodology of axiomatics. Cf. (Bonk and Mosterín, 2000; Carnap and Bachmann, 1936).

[14]Compare, e.g., (Kanamori, 2004, pp. 515–516) and (Shapiro, 1991, pp. 184–189).

it AR can be viewed as imposing a minimal model for the axiom system: "If the domain (*Grundbereich*) $B$ contains a smallest submodel (*Teilbereich*) $B_0$ satisfying the axioms [...], then $B$ is identical with such a smallest submodel $B_0$." (Fraenkel, 1922a, p. 163) This in effect rules out the existence of any possible submodel of $B_0$ that also satisfies the axiom system. The second definition is followed by a footnote concerning the method of constructing such a minimal model:

> As is usual, a smallest submodel of the indicated character is to be understood as a model that is the intersection of all submodels of $B$ with the property in question and that also possesses the property itself. (Fraenkel, 1922a, p. 163)

Two claims are made here: first that a minimal model for Z can be conceived as the intersection of all possible models satisfying the axioms. Second, if such a minimal model exists, the extended axiom system ZF + AR, i.e., the Zermelo axioms plus replacement and restriction, is categorical. Now, Fraenkel does not get more explicit about his conception of the domain or the models of set theory. How are these notions conceived? In approaching this question it will prove to be fruitful to take into consideration Fraenkel's intellectual background. Specifically, a closer look at Richard Dedekind's methodological innovations concerning set formation and mapping in *Was sind und was sollen die Zahlen* from 1888 will be helpful for the understanding of how Fraenkel's ideas behind restriction might have evolved.

## 4   Dedekind's heritage

Two interpretive issues concerning AR are in need of further consideration. First, how exactly did Fraenkel conceive the intended restrictive effect of his axiom on the possible set models satisfying ZF? Second, how is it supposed to constitute the categoricity of the axiom system? I argue that in order to address both questions a closer look at Dedekind's methodological work in the foundations of arithmetic will be instructive. More specifically, I show that there is a striking similarity between Fraenkel's scattered remarks about his understanding of AR and Dedekind's theory of chains (*"Kettentheorie"*) introduced in 1888 that suggests that Fraenkel actually modeled his idea of restriction based on Dedekind's approach.

   Concerning the first question, we can find an insightful remark in *Einleitung* from 1928 about the "special character" of the axiom compared to the "existential" and "relational axioms" of ZF. Here AR is described as similar in effect to Peano's induction axiom. Fraenkel states that "in both versions [of AR], the inductive moment is essential." (Fraenkel, 1928, p. 355) What is his intuition about this "inductive character" underlying AR? As we have already seen, the concept of intersection plays a central role for the

intended effect of the axiom. It is supposed to impose a minimal model as
the intersection of all possible models satisfying ZF. From a methodological point of view, this is essentially a paring down approach of defining a
specific minimal structure by taking the intersection of all closed subsets
of a given set. This method was first introduced by Dedekind in 1888 and
used for fixing the standard model of arithmetic. One could conjecture that
Fraenkel's idea of a minimal model for set theory was shaped in direct analogy to Dedekind's strategy of defining the natural numbers as a minimal
set closed by induction. Now, there is no immediate textual evidence that
Fraenkel was directly guided by Dedekind's method in his thinking about
set theoretic restriction. However, I will present a number of points that
strengthen the plausibility of this relation of influence. In the next section
a short presentation of the central concepts developed in (Dedekind, 1888)
will be presented that seem of relevance for Fraenkel's axiom candidate.

## 4.1   The theory of chains

Dedekind's project of developing an "unambiguous foundational conception" of the natural numbers in 1888 is based on a number of methodological
results concerning the central concepts that allow the reduction of numbers
to a logical basis (Dedekind, 1888, p. 351). Here the idea of an isomorphism
based on a 1–1 mapping (*"ähnliche Abbildung"*) between elements of two
systems is expressed formally for the first time. Systems that are isomorphic
in this sense are terminologically fixed as "classes of similar systems".[15] A
second newly introduced concept allowing Dedekind to devise the sequence
$\mathbb{N}$ of the natural numbers is that of a *chain* (relative to mapping function
$\varphi$ and a system $S$): in modern terminology, a subsystem $B$ of $S$ is called
a chain if it is closed under a mapping $\varphi$ (Dedekind, 1888, p. 352). Subsequently, a system $A_0$ is defined as the *chain of $A$* (*"Kette des Systems $A$"*)
if and only if $A_0$ is the intersection of all chains containing $A$ (Dedekind,
1888, p. 353). The way Dedekind conceives $A_0$ as the intersection of closures implies that it is also the smallest chain containing $A$, i.e., the smallest
subset of $S$ closed under $\varphi$. Again, in modern terminology, this effectively
says that $A_0$ is the minimal closure of $A$ under $\varphi$.[16]

---

[15](Dedekind, 1888, p. 351); compare (Sieg and Schlimm, 2005) for a systematic presentation of the evolution of the concept of mapping in Dedekind's foundational work.

[16]Compare (Sieg and Schlimm, 2005) on this fact: "$A_0$ obviously contains $A$ as a
subset, is closed under the operation $\varphi$; and is minimal among the chains that contain
$A$, i.e., if $A \subseteq K$ and $\varphi(K) \subseteq K$ then $A_0 \subseteq K$." (Sieg and Schlimm, 2005, p. 145)

Dedekind (1888) himself is not explicit about the minimality property of *chains of A*.
There exists, however, as Sieg and Schlimm have pointed out, a note in Dedekind's earlier
manuscript "Gedanken über Zahlen" from the Nachlass in which this issue is explicitly
mentioned: "$(A)$ [i.e., the chain of $A$] is the "smallest" chain that contains the system
$A$". (Quoted from Sieg and Schlimm, 2005, p. 144). I would like to thank Dirk Schlimm
for drawing my attention to this passage.

There is an obvious similarity between the idea of minimal chains developed here, i.e., the method of building minimal closures of a given base set and a specific operation via intersection, and Fraenkel's remarks on AR throughout the 1920s. A number of additional points can be mentioned that further highlight this affinity. First, both positions are strikingly similar in their motivations for imposing a minimal condition on the intended model. In Fraenkel's case, as we have seen, the aim is to restrict the model to well-founded and abstract sets, thereby keeping out all types of non-standard and extraordinary sets. A comparable account can also be found in Dedekind's writings, most explicitly in his famous letter to Keferstein from 1890. After a short discussion of his basic concepts used for expressing $\mathbb{N}$ he states:

> [...] however, these facts are still far from being adequate for completely characterizing the nature of the number sequence $\mathbb{N}$. All these facts would hold also for every system $S$ that, besides the number sequence $\mathbb{N}$, contained a system $T$, of arbitrary additional elements $t$, to which the mapping $\varphi$ could always be extended while remaining similar and satisfying $\varphi(T) = T$. [...] What, then, must we add to the facts above in order to cleanse our system $S$ again of such alien intruders $t$ as disturb every vestige of order and to restrict it to $\mathbb{N}$. (Dedekind, 1890, p. 100)

To exclude such non-standard elements from the interpretation in question can thus be considered a common motivation behind the method of devising a minimal model. In Fraenkel's case this restriction is imposed by his AR. In Dedekind's proto-axiomatic presentation of the natural numbers it is required by his clause four of a "simple infinite system" stating, in modern terms, that $A_0$ is the smallest set containing $A$ and closed under $\varphi$ (cf. Dedekind, 1888, p. 352).[17]

This immediately leads to a second observation concerning Fraenkel's original conception of the intended model of $\mathbb{Z}$ which seems to be modeled based on this idea of closure. In Dedekind's account of the natural numbers 1 is the base element and the sequence $\mathbb{N}$ the intersection of all sets containing 1 and closed under the successor operation. Accordingly, Fraenkel's intended set model is understood as the intersection of all set models that share the properties of (a) containing the empty set and the infinite set $\mathbb{Z}$ and (b) being closed under the operations specified in the Zermelo axioms, i.e., pairing, union, power set, etc.. This is essentially an understanding of models as "algebraic closures" (cf. Kanamori, 2004, p. 515). One can find additional textual evidence for this conception in Fraenkel's work from that time, mainly in the context of building different set models satisfying certain

---

[17]As pointed out by Awodey and Reck (2002, pp. 8–9), the latter effectively corresponds to Peano's (second order) axiom of induction.

restricted versions of Z—e.g., as sets closed under the operations of power set or union—used for independence proofs (cf., e.g., Fraenkel 1922b, p. 233; also Fraenkel 1922a, pp. 165–171). Here, as well as in the third edition of *Einleitung*, he gives an informal sketch of his account of the standard model (*"Normalbereich"*) of Z as a system closed under all operations specified in the axioms. Adding the AR to ZF would impose the following effect:

> This will probably result in the fact that only the empty set func-
> tioning as the primary building block for all sets is set up as the
> initial point. Then only those sets are admissible which emerge from
> the empty set and the sets imposed by [the axiom of infinity] by an
> arbitrary but certainly finite application of the individual axioms.
> (Fraenkel, 1928, p. 355)

Even though Dedekind's notion of chains is not explicitly mentioned in Fraenkel's remarks on model building, it seems obvious that AR can be understood here as a "restriction clause for closures" (Kanamori, 2004, p. 515), i.e., for a universe of sets conceived in direct analogy to Dedekind's method of constructing minimal systems.

## 4.2   Categoricity results

As I have mentioned before, there is no direct indication in Fraenkel's writings of Dedekind's influence on his conceptualization of models and AR. In the first edition of *Einleitung* from 1919, Dedekind is mentioned only for his existence proof of infinite systems and his definition of a finite system given in 1888. In the concluding remarks of the second edition there is a single reference to his theory of chains that, as Fraenkel writes, has received a "general and fundamental significance in set theory." [18] No connection is made to his concept of restriction. There exists, however, a passage in his lectures from 1925 that allows one to draw a direct link between Dedekind's minimal closures and his own approach of devising a minimal model for set theory. In a section on the "non-predicative" methods in mathematics, more specifically the debate between Poincaré and Zermelo on the indispensability of non-predicative proofs in mathematics, there is an interesting footnote mentioning Dedekind's theory:

> In a series of important and thoughtful proofs in set theory especially
> due to Dedekind and Zermelo [. . .], deductions of the following kind
> take center stage: a set $M$ is considered whose elements are all sets of
> a specific property $E$ exclusively characteristic for it; $M$ is thus the
> set of all sets sharing the property $E$. For the cases in question it is

---

[18](Fraenkel, 1924, p. 244); most important in this respect is of course Zermelo's adaptation of Dedekind's "closure approach" in terms of chains in his second proof of the well-ordering theorem (as well as in his proof of the Schroeder-Bernstein theorem) in (Zermelo, 1908a). For further details compare (Kanamori, 2004, pp. 501–503 and pp. 510–511).

then shown that the sum $s$ and the intersection $d$ of all elements of $M$ themselves share the property $E$; therefore $s$ and $d$—which exist by virtue of the definition as *sum* and *intersection* respectively— also belong to the set $M$ and can be characterized as the sets the *most comprehensive* and the *most limited in size* sharing the property $E$. Due to this characterization $s$ and $d$ play a decisive role in the concerned proof. (Fraenkel, 1927, p. 29; notation slightly changed)

The approach described here essentially follows the proof strategy introduced by Dedekind in 1888 to prove the categoricity of arithmetic. And it is precisely this idea—here formulated in Fraenkel's own words—that also seems to lie behind Fraenkel's own understanding of AR. To interpret his tacit assumptions underlying restriction in this way also sheds further light on the second unsolved issue mentioned above, namely how to understand the claim that the addition of AR to Z would render the resulting axiomatization categorical. Fraenkel's remarks alone are not conclusive on this intended effect. Here a glance at Dedekind's categoricity proofs will be instructive to show how Fraenkel might have conceived a similar categoricity result for set theory.

Dedekind's well-known metatheoretic results (cf. Dedekind, 1888, §10) can be qualified as instances of a categoricity based on minimal models according to which a theory is categorical if and only if it has a minimal model and any two minimal models are isomorphic.[19] His proofs (in remarks 132 and 133) that the simple infinite system $N$ can be captured completely, i.e., up to isomorphism by the conditions equivalent to the Peano axioms, strongly depends on his idea of minimal chains (cf. Dedekind, 1888, pp. 376–377). This connection follows from Dedekind's definition of a *mapping of a number sequence by induction* used in his subsequent proofs.[20] The equivalence $\psi(\mathbb{N}) = \theta_0(\omega)$ proved in remark 126 makes explicit the central link between minimal chains and the mapping of a simple infinite system via induction that plays a central role in Dedekind's categoricity proofs.

Now, unlike Dedekind, Fraenkel did not develop an actual proof of the categoricity of ZF + AR nor does he make any remarks how such a proof based on AR might be built. Besides this fact, the presentation of the central concepts of his theory (most importantly those of restriction, the

---

[19](cf. Grzegorczyk, 1962, p. 63); a minimal model can be defined as a model $M$ satisfying a theory T such that for every submodel $N$ of $M$ that also satisfies T, $N$ is isomorphic to $M$. Cf. (Grzegorczyk, 1962, p. 63).

[20]In remark 126 he shows that there is one and only one mapping of $\mathbb{N}$ into any system $\Omega$ via a function $\psi$ that satisfies the conditions that (i) the closure of $\mathbb{N}$ is a subset of $\Omega$, that (ii) $\psi(1) = \omega$, where $\omega$ is an element of $\Omega$ and that (iii) for any number $n$, $\psi(n') = \theta\psi(n)$, where "$n'$" stands for the successor of "$n$" and $\theta$ is a function on $\Omega$ (cf. Dedekind, 1888, pp. 370–371). In remark 128, Dedekind then proves that there exists an equivalence between such an inductive mapping $\psi(\mathbb{N})$ and a minimal closure $\theta_0(\omega)$ of $\Omega$ that contains $\omega$, i.e., $\psi(\mathbb{N}) = \theta_0(\omega)$ (Dedekind, 1888, p. 372).

set universe and minimal models) is not comparable in technical rigor to Dedekind's foundational work in arithmetic. Nevertheless, given the textual evidence above as well as his various informal remarks on the effect of the axiom candidate as imposing a minimal model, on its "inductive character" as well as on his conception of the intended set model as a minimal closure, it seems at least a plausible interpretation that AR was conceived by Fraenkel in close analogy to Dedekind's method developed in 1888.

# 5 Objections to AR

A number of serious objections were raised against Fraenkel's axiom candidate shortly after its first presentation in print in 1922 that led to general skepticism concerning the validity of AR as a set theoretical axiom and eventually prevented it from being added to the canonical list of ZF. In what follows I will first briefly present the main arguments adduced against AR. Then, Fraenkel's reaction to the objections and its subsequent impact on his own conception of set theoretic restriction will be discussed.

## 5.1 Baldus on meta-axioms

One point of criticism against AR was first put forward by the German mathematician Richard Baldus during a discussion of Hilbert's completeness axiom in geometry (Baldus, 1928). It concerns the general metatheoretic character of Hilbert's and related axioms, later terminologically clustered as "extremal axioms" (cf. Carnap and Bachmann, 1936). Baldus argues that unlike the other axioms in Hilbert's axiomatization of geometry (e.g., the axioms of order), the completeness axiom makes an assertion *"not only over the thought things* [of an interpretation] *but actually over all conceivable things"* (Baldus, 1928, p. 331). This assumption of the *non-extensibility* (*"Nicht-Erweiterungsfähigkeit"*) of the basic elements of the domain involves generalizing over the individuals in all models. Baldus correctly indicates a methodological doubt about the validity of such quantification over models:

> In order to preserve the completeness axiom's status as an axiom, one would have to allow as axioms also assertions over other things than those thought in the respective interpretation of the axiom system, which would extend the concept of axioms in geometry in a precarious and superfluous way. (Baldus, 1928, p. 331)

In an attached footnote, Baldus explicitly mentions Fraenkel's AR in this respect expressing a direct critique of it based on similar grounds:

> At a meeting in Kissingen Mr. A. Fraenkel has suggested that set theory can in no other way be rendered monomorphic than by a "postulate" [...], namely by an axiom of restriction, against which similar objections can be raised as against the axiom of completeness. (Baldus, 1928, p. 331)

Baldus' criticism of the problematic semantic character of the axiom has meanwhile become a standard argument against extremal axioms in general and Fraenkel's axiom candidate in particular. The basic objection is that AR imposes no condition on sets as the individuals of set theory, but on set models, thus conflating "formal languages with their model-theoretic semantics".[21]

## 5.2   von Neumann's subsystems

A second and somewhat related objection raised specifically against Fraenkel's axiom candidate is found in (von Neumann, 1925). Von Neumann presents an alternative axiomatization of set theory based on the primitive notions of functions ($II$-objects), arguments ($I$-objects), and objects that can be both arguments and functions ($I$-$II$-objects).[22] Furthermore, two primitive operations $[x, y]$ denoting the value of a function $x$ for an argument $y$ and $(x, y)$ expressing an ordered pair of arguments are given (von Neumann, 1925, pp. 397–398). Given his specific axiomatization of set theory based on these terms, von Neumann provides the first formalized version of AR intended to capture Fraenkel's original intention of imposing a minimal model for the resulting theory. In von Neumann's terminology, a subsystem of a given system is minimal if and only if it contains no subsystem that also satisfies the axioms:

> Let $\Sigma$ be the system of $I$-objects and $II$-objects. Let $\Sigma'$ be a subsystem of $\Sigma$. Let $I_{\Sigma'}$-objects and $II_{\Sigma'}$-objects be the $I$-objects and $II$-objects, respectively, that are in $\Sigma'$. Let $[x, y]_{\Sigma'}$ (where $x$ is an $II_{\Sigma'}$-object and $y$ an $I_{\Sigma'}$-object) mean $[x, y]$; let $(x, y)_{\Sigma'}$ (where $x$ and $y$ are $I_{\Sigma'}$-objects) mean $(x, y)$; let $A_{\Sigma'}$ be $A$ and let $B_{\Sigma'}$ be $B$. Now if these $I_{\Sigma'}$-objects and $II_{\Sigma'}$-objects, the operations $[x, y]_{\Sigma'}$ and $(x, y)_{\Sigma'}$ and the objects $A_{\Sigma'}$ and $B_{\Sigma'}$ also satisfy our axioms, we say for short that $\Sigma'$ satisfies our axioms. Then the axiom of restriction just mentioned simply requires that besides $\Sigma$ itself no other subsystem $\Sigma'$ of $\Sigma$ shall satisfy Axioms I–V. (von Neumann, 1925, p. 404)

$A$ and $B$ (and their respective correlates $A_{\Sigma'}$ and $B_{\Sigma'}$) are both arguments, i.e., $I$-objects. A subsystem in this sense is thus a collection of $I$-objects and $II$-objects resulting from a restriction of the original system. The operations of $[x, y]$ and $(x, y)$ in turn are restricted to the (two types of) elements of the subsystem. Given this formal presentation of AR, von Neumann then

---

[21](Shapiro, 1991, p. 185); compare also (Ferreirós, 1999) who states that: "Formulated as above [as a minimal condition on set models], the axiom is unacceptable it is no condition on sets but on models of set theory, i.e., it is not an axiom but a metaaxiom." (Ferreirós, 1999, p. 369)

[22](von Neumann, 1925, pp. 399–402); von Neumann's later class/set distinction is clearly anticipated here: those $II$-objects that are not $I$-$II$-objects have to be treated as classes. Compare (von Neumann, 1925, p. 401).

presents two "serious objections" against the axiom that are "equally true in Fraenkel's system." (von Neumann, 1925, p. 404). According to the first, AR presupposes notions of "naive set theory", most importantly that of a submodel that is not precisely definable in his own theory of sets.[23] The resulting regression to informal set theory would make the whole process of axiomatizing set theory circular. A possible remedy for this is to assume a "higher set theory" and a corresponding expanded domain $P$ in which the original domain $\Sigma$ can be properly defined as a class of $P$ (and the subsystems $\Sigma'$ of $\Sigma$ as subclasses of $P$).[24] However, this additional "hypothesis" implies a second, even greater difficulty for expressing a restriction clause for his axiomatization. Von Neumann argues that Fraenkel's proposed method of devising a model via the intersection of all possible models need not necessarily lead to a single, unique minimal model satisfying the other axioms (and thus to a categorical axiomatization) (von Neumann, 1925, p. 405). This is due to the fact that the range of the generalization over all (sub-)models of a theory involved in the intersection approach strongly depends on which higher background set theory is assumed. Different higher systems might allow different ranges of submodels. Since Fraenkel's minimal model is defined via the intersection of all possible models (as in Dedekind's approach), the method may lead to different results when different systems are assumed as the background theory.[25]

A second and more technical argument against the intersection approach presented by von Neumann is closely related to his own axiomatization and the satisfaction conditions he devises for subsystems of a given system. One problem about relativizing functions and arguments of $\Sigma$ to $\Sigma'$ concerns the fact that in order for $\Sigma'$ to satisfy von Neumann's axioms, additional satisfaction conditions stronger than the axioms have to be devised. For example, axiom III.2 states that there is a $II$-object $a$ as such that for all $I$-objects $x$: $[a, x] = x$. Von Neumann argues that this imposes the undesirable effect for a possible subsystem $\Sigma'$ of $\Sigma$ that there is a $II$-object $a$ in $\Sigma'$ such that for all $II$-objects $x$ in $\Sigma'$: $[a, x] = x$. In order to evade this problem, a stronger condition is added that the subsystem has to satisfy in order to

---

[23] (von Neumann, 1925, p. 404); for his distinction between sets and classes that plays a central role in his argumentation (cf. von Neumann, 1925, p. 403).

[24] (von Neumann, 1925, p. 404); this remark can in fact be taken as a first and informal expression of the conception of the set theoretic universe as a infinite progression of "higher set theories" in which lower systems are submodels of the higher theories. This is eventually fully spelled out in (Zermelo, 1930) (cf. the next section). One can also find a reference in von Neumann's remarks on the set theoretic hierarchy to Russell's type theory that seems to anticipate Gödel's later work on set theory as a cumulative and transfinite hierarchy of types (cf. Section 6): "The idea is partly the same as the one upon which Russell's "hierarchy of types rests"" (von Neumann, 1925, p. 405).

[25] Compare (Shapiro, 1991, p. 186) on this objection to Fraenkel's paring down approach for AR.

satisfy the axiom system: condition 3 states that there is a $II$-object $a$ in $\Sigma'$ such that for all $I$-objects $x$ in $\Sigma$: $[a, x] = x$ (von Neumann, 1925, p. 406). Here (as in similar cases for other axioms postulating the existence of sets) the quantification over all objects of the subsystem that is considered to be improper is substituted by the quantification over all objects of the original system. In the case of axiom III.2 the improper quantification $\forall x_{II} \in \Sigma'$ is replaced by $\forall x_I \in \Sigma$.[26] The resulting conditions are thus stronger than the axioms with the effect that they are sufficient, but no longer necessary conditions for $\Sigma'$ to satisfy the axioms. In addition, von Neumann points out that there exists a smallest submodel that satisfies the given additional conditions and that can be constructed via intersection, but not necessarily a minimal model for the axioms (cf. von Neumann, 1925, pp. 405–408). Given this set of objections, the following strong conclusion is drawn for Fraenkel's axiom candidate:

> For these reasons we believe that we must conclude, first, that the axiom of restriction absolutely has to be rejected and, second, one cannot possibly succeed in formulating an axiom to the same effect. (von Neumann, 1925, p. 405)

According to von Neumann, this fact, together with the existence of "inaccessible sets" (such as "descending sequences of sets") that lie "outside the system" in question are the main sources of the non-categoricity of set theory (von Neumann, 1925, p. 405).

## 5.3   Zermelo on set models

A third and more general objection against Fraenkel's axiom candidate was expressed in Zermelo's seminal paper "On boundary numbers and setdomains" (Zermelo, 1930). Zermelo introduces a new conception of the set universe as an open and unbounded sequence of set models (*Normalbereiche*) of increasing size that satisfy his axiomatization.[27] We have mentioned that a comparable view of a sequence of larger and larger set models had already been presented but not further developed in (von Neumann, 1925). Zermelo, in giving a formal explication of a cumulative hierarchy of sets, also provides a definitive clarification of the semantic notions of set models and submodels that can be found both in Fraenkel and in von Neumann. According to Zermelo, each set is decomposable into *layers* and cumulative *sections* that include all sets formed at earlier layers in the set theoretic hierarchy.[28] Set models $\mathbf{V}_\kappa^Q$ in turn are treated as sets that can be specified by

---

[26] For a comparable notion of relativization (cf. also Gödel, 1940, p. 76).

[27] Zermelo's proposed axiom system can be considered as a second order version of ZF since it contains a second order formulation of the axiom of replacement.

[28] (Zermelo, 1930, pp. 32–33); for the technical details of this early version of an iterative conception of sets (cf. Kanamori, 2004, pp. 521–524).

two numbers, a *base Q*—the cardinality of its first rank, i.e., the base set of individuals—and a *characteristic* or *boundary number* $\kappa$ as the least ordinal greater than all ordinals contained in the model. From this it follows that each model can act as a submodel of a set model with a higher boundary number (cf. Zermelo, 1930, p. 31). Thus, the universe **V** is composed of a "boundless progression" of set models (Zermelo, 1930, p. 29).

This conception of the set theoretic universe as an unlimited sequence of models obviously differs substantially from Fraenkel's static conception of a closed and fully describable universe of sets. This divergence also results in an opposing view on the issue of the (non-)categoricity of set theory. Whereas Zermelo shows that his axiomatization captures set models of a given base and boundary number up to isomorphism—the main results of his article in fact are three "isomorphism theorems" and their respective proofs[29]—an absolute concept of categoricity in Fraenkel's meaning of capturing a unique model is not possible due to the boundlessness of the set theoretic universe, i.e., the "existence of a unlimited sequence of boundary numbers" (cf. Zermelo, 1930, pp. 40–41).

Given Zermelo's picture of **V**, talk about the single intended model captured by ZF is shown to be inadequate. This insight also underlies Zermelo's more general critique of restrictive axioms. We have seen that von Neumann holds the assumption that for set theory there always exists a larger domain, a higher set theory in which the original model is definable as a set and in which a restriction for the lower theory yielding categoricity could at least in principle be formulated. Zermelo's theory of relative or quasi-categoricity essentially conforms to this view. Nevertheless, for him a domain restriction will never be desirable from a practical point of view, because it decisively delimits the functional role of set theory as a foundational discipline:

> Our axiom system is non-categorical which in this case is not a disadvantage but rather an advantage, for on this very fact rests the enormous importance and unlimited applicability of set theory. (Zermelo, 1930, p. 45)

Here lies the central objection to Fraenkel's account of restriction. Its effect is not considered a theoretical virtue of the axiomatization, but rather as a deficiency in a practical sense: it restricts set theory in its proper foundational goal, i.e., in the task of formalizing mathematics. Zermelo explicitly

---

[29]Briefly, the first holds that two models $A$ and $B$ with the same base cardinality and the same characteristic are isomorphic to each other. The second theorem states that a model $A$ that shares the same base cardinality with model $B$ but has a different characteristic is isomorphic to a certain cumulative rank of $B$. Theorem 3 holds that if $A$ and $B$ share the same characteristic but have different bases, then one is isomorphic to a submodel of the other model. Cf. (Kanamori, 2004, p. 527) for a more detailed account.

refers to Fraenkel's axiom candidate in order to underline the difference
between their conceptions. He remarks that,

> Naturally one can always force categoricity artificially by the addition
> of further 'axioms', but always at the cost of generality. Such postu-
> lates, like those proposed by Fraenkel [...] do not concern set theory
> as such, but rather only characterize a quite special model chosen by
> the author concerned. [...] the applicability of set theory has to be
> given up. (Zermelo, 1930, p. 45)

Note that Zermelo's actual objection against restrictive axioms like Fraenkel's
primarily concerns the fruitfulness of set theory as foundational discipline.
Any deliberate restriction of the set universe negatively affects the "full
generality" of set theory, i.e., its "unlimited applicability" to mathematics
(Zermelo, 1930, p. 45). As we shall see in Section 7, this objection antici-
pates Maddy's recent critique of what she terms "restrictiveness" concerning
set theoretic structures in questions of axiom choice. For present purposes,
it suffices to point to the fact that Zermelo, in his motivation for a cumulative universe claims to be more attentive to this pragmatic ideal of the
foundational goal than Fraenkel in his call for a categorical axiomatization.

## 6   Fraenkel's reaction

Fraenkel's reaction to the presented objections against the axiom of restriction in his subsequent work is instructive in several ways. First, it better
illustrates his own tacit understanding of the concepts involved in his earlier
presentation of the axiom. Second, it also highlights substantial shifts in
his understanding of the axiom as a direct result of these criticisms.

### 6.1   Vindications of AR

As far as I know Fraenkel never responded in print to Baldus' legitimate
doubts about the metatheoretic character of extremal axioms and their semantic implications. Even though he acknowledged the "special character"
of the axiom in comparison to the other axioms of ZF he never seemed to
become aware of the problem that the axiom in fact requires a generalization over set models.[30] More generally, as mentioned above, Fraenkel—in
his writings on set theory in the 1920s—seems to have been indecisive concerning questions of the adequate logical presentation of his axiomatization,
and in particular of his AR. The question of the proper formalization of the
axiom candidate is eventually taken up by Fraenkel almost three decades
later in his *Foundations of Set Theory* (Fraenkel and Bar-Hillel, 1958). Here

---

[30] As Shapiro points out this fact is probably due to the circumstance that a clearly
delineated syntax/semantic-distinction was far from being standard by the time Fraenkel
developed his theory. Compare (Shapiro, 1991, p. 184).

the authors refer to different attempts to formalize AR as a minimal axiom in a higher order logical language by Carnap:

> True, recently [...] Carnap proposed a vindication of this axiom of restriction, and Carnap formulated it symbolically, as an axiom of a minimal model [...]. (Fraenkel and Bar-Hillel, 1958, p. 90)

Carnap, in his (Carnap, 1954) formally expressed Fraenkel's account of a minimal property codified in AR by the use of higher-level binary relations (representing membership relations) and the notion of a partial relation: "There exists no proper partial relation of $E$ that also satisfies the properties stated in axioms A1 to A8 [i.e., ZF]." (Carnap, 1954, p. 154). His formal version depends on the (higher order) universal quantification over (partially defined) membership relations (Carnap, 1954, p. 154):

$$(H)[(x)(y)(Hxy \supset Exy).\mathrm{Kon}(H) \supset (x)(y)(Hxy \equiv Exy)]$$

"Kon" stands for the union of all axioms of ZF; $x$ and $y$ are individual variables ranging over sets. The axiom in this logical formulation basically states that $E$ is the minimal interpretation for set theoretic membership consistent with ZF.[31] Fraenkel, in an attached footnote to the passage cited above stresses this notion of a restriction condition in Carnap's formal presentation as an adequate version of his own informal treatment of AR: "The pith of the axiom is then the demand that no "partial relation" $\varepsilon$ should fulfill the conditions expressed by the other axioms".[32]

Whereas comments on the logical status of AR are limited to his later writings, Fraenkel immediately reacted to the objections leveled against the axiom by von Neumann. This might seem surprising at first sight because it is far from obvious that the latter's critique actually meets Fraenkel's informal presentation of the restriction on set models. First, it seems more reasonable that the technical objections against the AR rather concern von Neumann's own non-standard axiomatization of set theory (and specifically his formalization of the AR) than Fraenkel's preliminary and informal ideas. Second, the general validity of his critical remarks against restriction can be challenged by drawing attention to a number of inconsistencies in his own treatment of set models. As Shapiro (1991) has shown, von Neumann's set theory does not allow a consistent presentation of models (as classes containing its subclasses) due to the fact that proper classes cannot be conceived as

---

[31] Compare (Carnap and Bachmann, 1936) for a slightly different formalization of AR as a minimal model axiom.

[32] (Fraenkel and Bar-Hillel, 1958, p. 90); it remains debatable—in light of modern model theoretic semantics—if Carnap's approach actually vindicates Fraenkel's AR from objections along the lines of Baldus' remarks. For interesting related remarks concerning the axiom (cf. Shapiro, 1991, p. 186).

elements of either sets or classes in his theory.[33] Nevertheless, Fraenkel in
grosso modo seems to have acknowledged von Neumann's critique. Already
in 1927 he considers it "very doubtful" whether his version of restriction
can be attributed "a sound meaning". He states that,

> One seriously has to take the eventuality into consideration that the
> possible realizations of the axiom system that differ in their size do
> not have a smallest common subpart that would also satisfy all the
> axioms. Also the previously given instruction for a "construction" of
> such a smallest model [...] need not lead to a definite result since the
> axioms IV–VI [i.e., the axioms of power set, separation and choice]
> themselves do have a purely constructive character. This is a seri-
> ous and so far not satisfactorily solved problem from which possibly
> the natural necessity of a certain "boundlessness" and also a certain
> vagueness (so to speak at the boundaries) of the yet legitimate con-
> cept of set will follow. (Fraenkel, 1927, p. 102)

The first remark essentially rephrases von Neumann's critique. The second
remark concerning the "boundlessness" of the set concept seems already—
i.e., three years before Zermelo (1930)—to indicate doubts about his conception of the set universe as an (algebraic) closure.[34] Despite the acknowledged criticism, Fraenkel remains optimistic about the practical usefulness
and general correctness of a restrictive axiom.[35]

## 6.2  New axioms of restriction

Fraenkel's subsequent discussion of the axiom candidate is marked by a
number of substantial modifications. By 1958, in *Foundations of Set Theory*, both the intended character of AR and its extrinsic justification change
substantially. One can witness here a more tolerant attitude concerning
different additional axioms candidates for set theory, including stronger infinity axioms that are—as Fraenkel puts it—"antithetical" to AR (Fraenkel
and Bar-Hillel, 1958, p. 87). Fraenkel acknowledges that, in contrast to prior
belief, inaccessible numbers have "significance not only for the foundations
of set theory but also for some applications." (Fraenkel and Bar-Hillel,
1958, p. 87) Following a short presentation of different kinds of axioms that

---

[33]Cf. (Shapiro, 1991, p. 186); Shapiro suggest a modification of this system in order to
vindicate the axiom of restriction, to the effect that "sets and proper classes of the original
theory (can be treated as) as elements, i.e., as sets" thereby allowing to treat models for
a theory $T$ as the subclasses of a higher-level theory $T'$. The effect on restriction would
then be that "[...] one can state in the higher theory that a given class has no proper
subclasses that are models of ordinary set theory." (Shapiro, 1991, p. 186)

[34]Similar remarks along these lines can be found in (Fraenkel, 1928).

[35]In (Fraenkel, 1928) he concludes his discussion of the axiom by stating that: "Nevertheless I like to believe that the mentioned doubts can be resolved and that the axiom of
restriction can be maintained—and then considered as a very central part of the axiomatization! if only its formulation can be made more precise." (Fraenkel, 1928, p. 355).

postulate inaccessible cardinalities, he returns to a discussion of his axiom of restriction, however, not without providing a new motivation for it. The main intention behind AR is no longer to rule out non well-founded sets but to be restrictive in another sense, namely to "yield just all ordinals less than the least inaccessible ordinal number" (Fraenkel and Bar-Hillel, 1958, p. 88). Whereas the well-foundedness of the sets composing the set theoretic universe is now secured by the axiom of foundation (as proposed in von Neumann 1925 and Zermelo 1930), the main rationale for introducing the axiom of restriction is to secure the "non-existence of inaccessible numbers" (Fraenkel and Bar-Hillel, 1958, p. 88). The former version of AR is thus functionally divided into an independent axiom of foundation and a kind of accessibility axiom introduced to exclude all inaccessible numbers. Interestingly, Fraenkel's original paring down method for imposing a minimal model as the intersection of all "realizations" is still upheld as an adequate means of achieving this new and structurally different set theoretic restriction: "Then the non-existence of inaccessible numbers or of the corresponding sets, as well as of extraordinary sets [...] could be proved." (Fraenkel and Bar-Hillel, 1958, p. 89)

Fraenkel's change in perspective concerning the nature of AR calls for two comments. First, recall Zermelo's critical remarks on restrictive axioms like AR and the fact that inaccessible cardinals ("*Grenzzahlen*") were first introduced in his 1930 paper as a direct consequence of his dynamic account of the set theoretic universe as a progression of set models (delimited by such boundary numbers). Given this, it is natural to interpret the shift in Fraenkel's conception of set theoretic restriction as a direct reaction to the theory of set models proposed in (Zermelo, 1930). Second, notice also the underlying change from Fraenkel's original understanding of the set theoretic universe as a static structure that becomes evident here. In *Foundations of Set Theory*, Fraenkel must, at least to a certain point, have acknowledged Zermelo's conception of the set theoretic universe. Otherwise this new reading of AR as an accessibility axiom that rules out higher natural models would simply make no sense. Nevertheless, he still argues for a restriction of the universe to a minimal model. His motivation outlined in 1958 seems to be a somewhat unjustified disapproval of Zermelo's presentation of the set models in general and the idea of relative categoricity results for set theory in specific. In a remark he critically comments on Zermelo's results from 1930:

> The cardinal of the basis and the ordinal α together are an invariant characteristic of the intended domain of sets. The first leads to the domain of finite sets, the second to the domain of sets up to the first inaccessible number. However, Zermelo's proof that this invariant guarantees the monomorphism (categoricalness) of the domain

can hardly be considered stringent, and even the concepts used, e.g., "cardinal of the basis" are objectionable. (Fraenkel and Bar-Hillel, 1958, p. 92).

Zermelo's different categoricity results for his axiomatization, especially his version of categoricity in a given power seems to be in conflict with Fraenkel's principal aim of providing a single categorical axiomatization of set theory. This adherence to the axiomatic ideal of absolute categoricity is probably the main reason why Fraenkel not only did not accept Zermelo's 1930 objection against his axiom candidate but in effect redefined AR with the motivation to rule out the existence of Zermelo's higher boundary numbers. The extrinsic motivation of gaining a categorical axiomatization via this new version of a "limitative axiom" is again clearly stated in 1958:

> A suitable axiom of restriction should enable us to prove that all models of the axiom system are isomorphic, i.e., admit a one-to-one mapping which preserves the ∈-statements. (Cf. also Gödel's postulate of constructibility [...]). (Fraenkel and Bar-Hillel, 1958, p. 89)

Fraenkel's reference to Gödel's axiom $V=L$ (or alternatively *axiom A*) in the passage above deserves closer attention here. As has been pointed out by Maddy (1997), the axiom $V=L$ also imposes a kind of minimal model on ZFC. Without going into further details here, one can understand the constructible universe $L$ as the minimal inner model of ZFC,[36] since for all inner models $M$ of ZFC, the constructible model $L^M$ (as the class of all constructible sets in $M$) is identical to $L$, i.e., $L^M = L$ due to the absoluteness of $L$.[37] In Maddy's own terms: "[...] if $M$ is a transitive model of ZFC containing all ordinals, then the constructible sets of $M$ are the real constructible sets, and thus, $L \subseteq M$." (Maddy, 1997, p. 73) Thus any extension of $L$ "will contain sets different from all relevant constructible sets." (Maddy, 1997, p. 73)

Now, it is worth noting that for both accounts, $V=L$ and AR, the respective minimal models can be constructed—despite von Neumann's original objections—by the kind of intersection approach first outlined by Fraenkel. One can understand the model $L$ as the intersection of all inner models of ZFC.[38] In contrast, the model satisfying ZF plus Fraenkel's AR (as an accessibility axiom) can be constructed via the intersection of all natural models with an empty base. Concerning the question of the categoricity of

---

[36]The layers of $L$ are constructed in similar fashion to the cumulative hierarchy in $V$, however, as elements of a level $\alpha + 1$ only of the (first order) definable subsets of $L_\alpha$ are admitted. Compare (Maddy, 1997, p. 65).

[37]Cf. (Gödel, 1940, pp. 68–78) for a formal presentation of $L$ and on the notion of absoluteness. Compare also (Maddy, 1997, p. 73) and, for a more detailed discussion of Gödel's set theory, cf. (Kanamori, 2007).

[38]Compare (Jech, 2002, p. 187).

axiomatic set theory, there is also a point of continuity between Fraenkel's AR and Gödel's axiom. What is especially worth noting in this respect is that on both accounts categoricity can be achieved by (different types of) set theoretic minimization (again given that ZF is second-order). In a recent paper, Kanamori explicitly points out this methodological affinity between Fraenkel and Gödel:

> Gödel's axiom A, that every set is constructible, can be viewed as formally achieving this sense of categoricity [i.e., "Fraenkel's idea of a minimizing, and hence categorical axiomatization"] since [...] in axiomatic set theory $\mathbf{L}$ is a definable class, containing all the ordinals, that, together with the membership relation restricted to it, is a model of set theory, and $\mathbf{L}$ is a submodel of every other such class. (Kanamori, 2004, p. 539)

Obviously, the (minimal) set models characterized up to isomorphism differ substantially in the two cases. In Fraenkel's 1958 version of restriction, ZFC + AR yields the model $\mathbf{V}^0_\kappa$, where 0 signifies the empty basis and $\kappa$ the first inaccessible rank of $\mathbf{V}$. In contrast, ZFC + $\mathbf{V}=\mathbf{L}$ does not rule out any limit numbers but characterizes the least inner model of ZFC.

Beside the reference to Gödel's axiom in the passage cited above, Fraenkel does not get more explicit on how his AR relates to the axiom of constructibility. However, there exists an interesting aftermath to Fraenkel's own conceptualization of AR that can be found in the second and revised edition of *Foundations*, (Fraenkel et al., 1973), published after Fraenkel's death in 1966. In an expanded section devoted to axioms of restriction, Lévy proposes two new candidates, both reflective of Fraenkel's own considerations given in 1958 (Fraenkel et al., 1973, pp. 113–119). It is argued that both new versions of AR can be "equated with" Fraenkel's original conception but can be formulated "within our axiomatic theory", thus evading the objections against the former axiom's metatheoretic character (Fraenkel et al., 1973, p. 114). The first axiom candidate, $AR_1$, is conceived as the conjunction of an axiom of foundation and an accessibility axiom (as already suggested in the first edition of *Foundations* by Fraenkel himself). For the second and stronger restrictive axiom candidate, $AR_2$, Lévy seems to have taken into account Fraenkel's analogy with Gödel's axiom of constructibility. $AR_2$ is thus defined as the conjunction of $\mathbf{V}=\mathbf{L}$ and a principle ruling out transitive sets as models of ZF (Fraenkel et al., 1973, p. 116).

### 6.3   An "intuitive" consideration

Overall, the presentation of the restrictive axioms in (Fraenkel et al., 1973) is close in spirit to Fraenkel's considerations in the first edition. In particular the extrinsic motivations stated for the two axiom candidates, namely to rule out non-well-founded sets and sets of inaccessible ordinals, are identical to his arguments from 1958. Lévy, in 1973, also mentions the fact that the addition of either one of the two axioms to a second order version of ZF would lead to a categorical axiomatization.[39] However, one can also detect a significant change in the second edition compared to Fraenkel's earlier remarks. We have seen that Fraenkel, in his writings from the 1920s onwards, provides exclusively extrinsic motivations for AR. There is to my knowledge not a single remark in his work indicating a kind of intrinsic motivation for the axiom candidate based on mere reflection on his underlying concept of set universe.[40] In contrast, in the second edition of *Foundations* a discussion of "the desirability of restriction in general" as a basis for evaluating the axioms $AR_1$ and $AR_2$ is given. The extrinsic arguments for the axiom candidates mentioned above are complemented here by a new type of reflection. Lévy makes the following remark concerning a necessary "intuitive justification" of the axioms of restriction:

> In the case of the axiom of induction in arithmetic and the axiom of completeness in geometry, we adopt these axioms not only because they make the axiom systems categorical or because of some metamathematical properties of these axioms, but because, once these axioms are added, we obtain axiomatic systems which perfectly fit our intuitive ideas about arithmetic and geometry. In analogy, we shall have to judge the axioms of restriction in set theory on the basis that the set theory obtained after adding these axioms fits our intuitive ideas about sets. (Fraenkel et al., 1973, p. 117)

This passage documents a substantial shift in the methodology of axiom choice compared to Fraenkel's original approach. Part of the new evaluation strategy proposed here is clearly intrinsic in character in Maddy's sense: not in terms of an allusion to a set-theoretic intuition or self-evidence, but explicitly based on a prior explanation of the set universe underlying the axiomatization (cf. Fraenkel et al., 1973, pp. 87–88). What is effectively called for here is that the justification of an axiom has to depend on a combination of extrinsic and intrinsic considerations. The latter are necessary to confirm (so to speak in a second loop) the choice originally made on extrinsic grounds. An acceptance solely based on extrinsic considerations—such as on "metamathematical properties" of the resulting axiomatization—do not provide sufficient evidence unless backed up by intrinsic reflection whether

---

[39] In contrast to Fraenkel's own perception, Lévy makes clear in the revised version that the possible categoricity results for set theory are those "essentially contained in Zermelo [1930]." (Fraenkel et al., 1973, p. 115).

[40] It is usually Zermelo that is credited with providing the first genuinely intrinsic justification of his axioms based on the cumulative hierarchy of set. Compare Kanamori on this point in Zermelo's 1930 paper: "In a notable inversion, what has come to be regarded as the underlying iterative conception became a heuristic for motivating the axioms of set theory generally." (Kanamori, 2004, p. 521)

the axioms actually fit or comply with the underlying conception of the universe they are intended to characterize. Turning to the cases of AR (without focusing on a specific version), Lévy draws a rather pessimistic conclusion for Fraenkel's axiom candidate:

> To restrict our notion of set to the narrowest notion which is compatible with the axioms of ZFC just for the sake of economy is appropriate only if we have absolute faith that the axioms of ZFC (and the statements which they imply) are the only mathematically interesting statements about sets. It is difficult to conceive of such absolute faith in the sufficiency of the axioms of ZFC [. . .]. Even if one had such faith in the axioms of ZFC, it is likely that he would settle rather for something like an axiom of completeness, if there were some reasonable way of formulating it. (Fraenkel et al., 1973, p. 117)

This can only be understood as an intrinsically based claim against minimizing principles in set theory. Unfortunately, no supporting argument showing exactly why AR does not match the cumulative hierarchy of sets is given. However, the underlying idea here seems to mirror Zermelo's objection in his 1930 paper that any kind of minimizing principle artificially restricts the cumulative hierarchy of sets, thereby ruling out potentially mathematically interesting structures.[41]

# 7   Maddy's MAXIMIZE

Fraenkel's axiom candidate is not mentioned in (Maddy, 1997). Nevertheless, Lévy's negative conclusion concerning "the desirability of restriction" as a set theoretic principle also seems to be in line with her more systematic approach of evaluating axiom candidates in terms of two "methodological maxims", UNIFY and MAXIMIZE (cf. Maddy, 1997, pp. 208–212). We have seen that Fraenkel shares with Maddy an understanding of set theory as a foundational discipline and thus the motivation for assuring its "foundational goal" through unification, i.e., through providing a single arena in which mathematics can be presented.[42] Therefore, AR fits perfectly with

---

[41]Note that one can also identify an interesting affinity to Gödel's changing attitude towards his axiom V=L as a minimizing principle here. In a footnote in (Gödel, 1939), an intrinsic motivation for his axiom A is presented: "In order to give A an intuitive meaning, one has to understand by sets' all objects obtained by building up the simplified hierarchy of types on an empty set of individuals (including types of arbitrary transfinite orders)." (Gödel, 1939, p. 29) By 1964, in contrast, Gödel provides a direct intrinsic argument against a "minimum property" such as expressed by V=L: "Note that only a maximum property would seem to harmonize with the concept of set [. . .]." (Gödel, 1964, pp. 262–263) Compare also (Maddy, 1997, p. 84) on this shift in Gödel's view.

[42]Compare Maddy on this point: "One methodological consequence of adopting the foundational goal is immediate: if your aim is to provide a single system in which all objects and structures of mathematics can be modeled or instantiated, then you must aim for a single fundamental theory of sets." (Maddy, 1997, pp. 208–209)

Maddy's first maxim, UNIFY. The tension with her program clearly arises in relation to the more extensively discussed second principle of axiom choice, MAXIMIZE. The general idea underlying this principle, namely to defend axioms on the basis whether they maximize the set theoretic structure, i.e., secure new isomorphism types not reachable from the given axioms, looks antithetical to Fraenkel's intentions behind AR.[43] In fact, Maddy proposes a notion of MAXIMIZE based on a formal "criterion of restrictiveness" for axiom systems (cf. Maddy, 1997, III.6). According to it, a theory "$T$ is restrictive iff there is a consistent $T'$ that strongly maximizes over $T$" (Maddy, 1997, p. 224). $T'$ in turn strongly maximizes over $T$ iff (i) it delivers new isomorphism types not provable from $T$ , (ii) it is inconsistent with $T$ and (iii) there is no $T''$ that maximizes over $T'$ (cf. Maddy, 1997, p. 224). Given this explication, we can see in what sense the different versions or AR are restrictive in terms of this formal criterion. Maddy is clear in pointing out that Fraenkel's first version of the axiom candidate as a kind of axiom of foundation is not restrictive (in relation to an axiom system including an "anti-foundation axiom") since the addition of non well-founded sets does not provide any additional isomorphism types compared to the models of ZFC (Maddy, 1997, p. 217). What about Fraenkel's 1958 understanding of AR as an accessibility axiom? Here again, Maddy's criterion would not rule out the axiom candidate. As Maddy shows in Section III.6., the axiom system ZFC + "there exists an inaccessible cardinal" (IC) does not maximize over ZFC in a proper sense, even though it clearly allows new isomorphism types. The reason is that it simply does not satisfy condition (ii) of being inconsistent with ZFC. Given this, ZFC is not restrictive in relation to ZFC + IC (Maddy, 1997, p. 222). By the same token, ZFC + "there exist no inaccessible cardinals" is not restrictive relative to ZFC since the latter does not "inconsistently maximize" over the former. However, compared to higher cardinal axioms like IC, the AR is in fact restrictive in the formal sense. The case remains where AR is conceived as a kind of axiom of constructibility (as pointed out by Lévy in Fraenkel et al. 1973). Here again, Maddy shows that ZFC + $\mathbf{V}=\mathbf{L}$ is restrictive in the strong sense relative to different axiom systems imposing higher isomorphism types like ZFC + "$0^{\#}$ exists", ZFC + $\mathbf{V}\neq\mathbf{L}$ or ZFC + MC (a measurable cardinal axiom) (Maddy, 1997, pp. 223–224).

One can interpret these results (provided that one accepts Maddy's formal criterion of restrictiveness as a viable principle for axiom choice) as a direct and strong extrinsic argument against Fraenkel's AR. However, a

---

[43]Cf. again Maddy: "[...] if set theory is to play the hoped-for foundational role, then set theory should no impose any limitations of its own: the set theoretic arena in which mathematics is to be modeled should be as generous as possible; the set theoretic axioms [...] should be as powerful and fruitful as possible." (Maddy, 1997, pp. 210–211)

direct evaluation in terms of Maddy's second maxim would seem somewhat misplaced, for several reasons. First, it would be insensitive to the historical context of axiomatic set theory prior to 1930 in which the axiom candidate was originally devised. It would fail to take into account the general understanding of the axiomatic project in mathematics around that time that guided Fraenkel in setting up his axiom candidate in the first place. We have seen that one of the main motivations of introducing AR was to yield the "completeness of an axiom system" in the sense of its categoricity which was conceived as a central theoretic ideal for any axiomatization, irrespective of its subject matter.[44] Second, modern higher set theory in general and the various higher cardinal axioms in specific which form Maddy's actual field of investigation for her methodological principles certainly transcend Fraenkel's original horizon of set theory as a foundational discipline.[45] His attempt to meet the foundational goal was to propose an axiomatization that would allow a "clean and secure construction of the foundations of all mathematical sciences", without further detailing the spectrum of mathematics referred to here (Fraenkel, 1928, p. 393). In his discussion of AR he explicitly states the belief that "likely all mathematically significant sets, e.g., sets of numbers and points, can [. . .] be secured within the thus restricted axiomatic [system]." (Fraenkel, 1928, p. 355) Fraenkel assumed— and correctly so—that the axiomatization ZFC + AR sufficiently fulfilled the foundational goal for all relevant mathematics faced with in this period.[46]

Given this, one could be inclined to argue that even though Fraenkel's project might have been consistent with the mathematical knowledge of his time, it is simply outdated as a foundational enterprise from a modern point of view given the progress in mathematics since 1930 (and specifically the recognition of higher set theory as a proper and autonomous mathematical discipline). But here again, a final judgment depends on the specific role ascribed to set theory. If it is considered as a proper mathematical discipline, Maddy's maximizing maxim is a perfectly reasonable principle for the choice of its axioms. Note, however, that set theory then can hardly be called a foundational discipline in the original strict sense.[47] If, in contrast,

---

[44] Compare (Corcoran, 1980).

[45] Fraenkel, in 1928, already discusses the possibility of "special existence axioms" postulating higher cardinals (than those secured by the axioms of infinity and replacement) and then concludes: "However, theses problems lie in the most remote regions of the theoretical science and have so far barely a connection to the questions raised by the scientific demands of the present." (Fraenkel, 1928, p. 310)

[46] The model characterized by $ZFC_2 + AR_1$ is sufficiently strong to represent the real and complex numbers, as well as function over real numbers etc. used in classical mathematics. For different models of second order ZF compare, e.g., (Uzquiano, 1999, p. 290).

[47] In this case, it is difficult to provide a sound argument why unification should be a desirable maxim (in contrast to allowing a series of possibly incompatible set theories designed for different tasks).

it is primarily understood as a foundational enterprise used for modeling "everyday mathematics" (and not as a proper mathematical field), it seems less clear that MAXIMIZE is actually a reasonable methodological guideline, since it is simply not necessary from a mathematical point of view. One can sense here a certain tension in Maddy's naturalistic account between the two principles she attributes to axiomatic set theory.[48] A similar point was recently stressed by Friedman in (Feferman et al., 2000). He argues that MAXIMIZE as a maxim for set theoretic axiom choice is not relevant for the working mathematician exclusively interested in set theory as a tool for "a more or less standard set theoretic interpretation of mathematics, with ZFC generally accepted as the current gold standard for rigor".[49] Not only is ZFC then perfectly sufficient for supplying a foundation for ordinary mathematics, one can also quite naturally limit it to more restricted versions along the lines of Fraenkel's suggestions. Concerning Maddy's objections to V=L, Friedman remarks: "However, for the normal mathematician, since set theory is merely a vehicle for interpreting mathematics so as to establish rigor, and not mathematically interesting in its own right, the less set theoretic difficulties and phenomena the better." (Feferman et al., 2000, p. 436) He puts forward an informal principle inverse to Maddy's MAXIMIZE for foundational set theory: "more is less and less is more" (Feferman et al., 2000, p. 437)

One is inclined to view Fraenkel's motivation for introducing his AR in exactly this sense, i.e., in securing a streamlined concept of set (or of the set theoretic universe) that allows for the reconstruction of standard mathematics and rules out anything nonessential for this task. Given this, Fraenkel's motivations for AR have to be considered as fully rational (in Maddy's own naturalistic sense). Moreover, as Friedman's remarks underline—given a modest understanding of set theory's foundational task—it remains far from evident that minimizing principles like AR are less reasonably grounded in mathematical practice than Maddy's principle for set theoretic maximization.

---

[48]Maddy explicitly mentions a possible tension between UNIFY and MAXIMIZE if taken as complementary maxims for some cases of axiom choice (cf. Maddy, 1997, pp. 211–212). The point I want to make here is more general and concerns the rationality of MAXIMIZE if one presupposes a foundational goal for set theory.

[49](Feferman et al., 2000, pp. 434–435); a similar point is made by Feferman in his discussion of Maddy's naturalistic project. He states that "there is not a shred of evidence so far that we will need anything beyond ZFC—or even much weaker systems—to settle outstanding combinatorial problems of interest to the working mathematician." (Feferman et al., 2000, p. 407)

# Bibliography

Awodey, S. and Reck, E. H. (2002). Completeness and categoricity, Part 1: 19th century axiomatics to 20th century metalogic. *History and Philosophy of Logic*, 23:1–30.

Baldus, R. (1928). Zur Axiomatik der Geometrie. Über Hilberts Vollständigkeitsaxiom. *Mathematische Annalen*, 100:321–333.

Bonk, T. and Mosterín, J., editors (2000). *Rudolf Carnap. Untersuchungen zur allgemeinen Axiomatik.* Wissenschaftliche Buchgesellschaft, Darmstadt.

Carnap, R. (1954). *Einführung in die Symbolische Logik.* Springer, Vienna.

Carnap, R. and Bachmann, F. (1936). Über Extremalaxiome. *Erkenntnis*, 6:166–188.

Corcoran, J. (1980). Categoricity. *History and Philosophy of Logic*, 1:187–207.

Dedekind, R. (1888). *Was sind und was sollen die Zahlen.* Vieweg, Braunschweig. Page numbers refer to the edition in (Fricke et al., 1932); English translation in (Ewald, 1996, pp. 787–833).

Dedekind, R. (1890). Letter to Keferstein (1890a). Edited in (van Heijenoort, 1967, pp. 98–103).

Easwaran, K. (2008). The role of axioms in mathematics. *Erkenntnis*, 68(3):381–391.

Ewald, W. (1996). *From Kant to Hilbert: A Source Book in the Foundations of Mathematics.* Clarendon Press, Oxford.

Feferman, S. (1999). Does mathematics need new axioms? *American Mathematical Monthly*, 106:99–111.

Feferman, S., Dawson, J. W., Goldfarb, W., and Parsons, C., editors (1990). *Kurt Gödel: Collected Works*, volume 2. Oxford University Press, New York NY.

Feferman, S., Friedman, H. M., Maddy, P., and Steel, J. R. (2000). Does mathematics need new axioms? *Bulletin of Symbolic Logic*, 6:401–446.

Ferreirós, J. (1999). *Labyrinth of Thought: A History of Set Theory and its Role in Modern Mathematics.* Birkhäuser, Basel.

Fraenkel, A. A. (1911). Axiomatische Begründung von Hensels $p$-adischen Zahlen. *Journal für die reine und angewandte Mathematik*, 141:43–76.

Fraenkel, A. A. (1922a). Axiomatische Begründung der transfiniten Kardinalzahlen. *Mathematische Zeitschrift*, 13(1):153–188.

Fraenkel, A. A. (1922b). Zu den Grundlagen der Cantor-Zermeloschen Mengenlehre. *Mathematische Annalen*, 86:230–237.

Fraenkel, A. A. (1924). *Einleitung in die Mengenlehre*. Springer, Berlin, 2nd edition.

Fraenkel, A. A. (1927). *Zehn Vorlesungen über die Grundlegung der Mengenlehre*. Teubner, Leipzig.

Fraenkel, A. A. (1928). *Einleitung in die Mengenlehre*. Springer, Berlin, 3rd edition.

Fraenkel, A. A. and Bar-Hillel, Y. (1958). *Foundations of Set Theory*. North Holland Publishing Company, Amsterdam.

Fraenkel, A. A., Bar-Hillel, Y., and Levy, A. (1973). *Foundations of Set Theory*. North Holland Publishing Company, Amsterdam, 2 edition.

Fricke, R., Noether, E., and Ore, O., editors (1932). *Richard Dedekind: Gesammelte mathematische Werke*, volume 3. Vieweg & Sohn, Braunschweig.

Gödel, K. (1939). Consistency proof for the generalized continuum hypothesis. *Proceedings of the National Academy of Sciences of the United States of America*, 25(4):220–224. Quoted from the edition in (Feferman et al., 1990).

Gödel, K. (1940). *The Consistency of the Axiom of Choice and of the Generalized Continuum Hypothesis with the Axioms of Set Theory*. Princeton University Press, Princeton NJ. Quoted from the edition in (Feferman et al., 1990).

Gödel, K. (1964). What is Cantor's continuum problem? *The American Mathematical Monthly*, 54:515–525. Quoted from the edition in (Feferman et al., 1990).

Grzegorczyk, A. (1962). On the concept of categoricity. *Studia Logica*, 13:39–66.

Jech, T. (2002). *Set Theory*. Springer Monographs in Mathematics. Springer, Berlin, 3 edition.

Kanamori, A. (2004). Zermelo and set theory. *Bulletin of Symbolic Logic*, 10(4):487–553.

Kanamori, A. (2007). Gödel and set theory. *Bulletin of Symbolic Logic*, 13(2):153–188.

Maddy, P. (1997). *Naturalism in Mathematics*. Oxford University Press, Oxford.

Mirimanoff, D. (1917). Remarques sur la théorie des ensembles et les antinomies cantoriennes. *L'enseignement mathématique*, 19:209–217.

Russell, B. (1973). *The regressive method of discovering the premises of mathematics*, pages 272–283. Georg Braziller, New York NY.

Shapiro, S. (1991). *Foundations without Foundationalism. A Case for Second-Order Logic*. Oxford University Press, Oxford.

Sieg, W. and Schlimm, D. (2005). Dedekind's analysis of number: Systems and axioms. *Synthese*, 147(1):121–170.

Sierpiński, W. (1922). Sur la notion d'isomorphisme des ensembles. *Fundamenta Mathematicae*, 3:50–51.

Skolem, T. (1922). Einige bemerkungen zur axiomatischen begründung der mengenlehre. In *Matematikerkongressen i Helsingfors den 4–7 Juli 1922. Den femte skandinaviska matematikerkongressen. Redogörelse*, pages 217–232, Helsinki. Akademiska Bokhandeln. Quoted after the English translation in (van Heijenoort, 1967, p. 290–301).

Uzquiano, G. (1999). Models of second-order Zermelo set theory. *Bulletin of Symbolic Logic*, 54(3):289–302.

van Dalen, D. and Ebbinghaus, H.-D. (2000). Zermelo and the Skolem Paradox. *Bulletin of Symbolic Logic*, 6(2):145–161.

van Heijenoort, J., editor (1967). *From Frege to Gödel. A Source Book in Mathematical Logic, 1879–1931*. Harvard University Press, Cambridge MA.

von Neumann, J. (1925). Eine Axiomatisierung der Mengenlehre. *Journal für die reine und angewandte Mathematik*, 154:219–240. Quoted after the English translation in (van Heijenoort, 1967, pp. 393–413).

Zermelo, E. (1908a). Neuer Beweis für die Möglichkeit einer Wohlordnung. *Mathematische Annalen*, 65:107?–128. Quoted after the English translation in (van Heijenoort, 1967, pp. 183–198).

Zermelo, E. (1908b). Untersuchungen über die Grundlagen der Mengen-lehre I. *Mathematische Annalen*, 65:261–?281. Quoted after the English translation in (van Heijenoort, 1967, pp. 199–215).

Zermelo, E. (1930). Über Grenzzahlen und Mengenbereiche: Neue Unter-suchungen über die Grundlagen der Mengenlehre. *Fundamenta Mathemat-icae*, 16:29–47.

# Exploratory experimentation in experimental mathematics: A glimpse at the PSLQ algorithm

Henrik Kragh Sørensen*

Institut for Videnskabsstudier, Aarhus Universitet, C. F. Møllers Allé 8, 8000 Århus C, Denmark

E-mail: hks@ivs.au.dk

## 1 Introduction

From a philosophical viewpoint, mathematics has traditionally been distinguished from the natural sciences by its formal nature and emphasis on deductive reasoning. Experiments—one of the corner stones of most modern natural science—have had no role to play in mathematics. However, in the past two to three decades, a mathematical subdiscipline has been forming that describes itself as "experimental mathematics", and it is the purpose of this paper to investigate and discuss the ways in which experimental mathematics is *experimental*.

Since the 1990s, many domains of knowledge production have witnessed a "computational turn" during which the wide use of computers has influenced established ways of thinking.[1] In mathematics, computers have been utilized since their first construction, but in the 1990s, their use led to a new subdiscipline of experimental mathematics in which computers were central to most—if not all—the experiments that give the subdiscipline its name. Using high speed computers and software packages such as Maple and Mathematica, mathematicians can now manipulate data and structures of immense complexity through real-time interaction with computers, and these practices are at the heart of experimental mathematics, I will argue. Thus, computers—and the "experiments" that they seem to carry with them—have entered into wide areas of traditional mathematics ranging from combinatorics to partial differential equations.

It is no coincidence that the name of the subdiscipline under consideration is often given as *experimental mathematics* or sometimes as *computer-based* or *computer-assisted* mathematics. Thus, it does *not* refer to a partic-

*The author wishes to thank participants at the conferences PhiMSAMP-3 (Vienna, May 2008) and ECAP-08 (Montpellier, June 2008), colleagues at the *Institut for Videnskabsstudier* (Aarhus), Jonathan Borwein (University of Newcastle, Australia), and two anonymous referees for valuable comments that helped shape and sharpen the argumentation in this paper.

[1]This "computational turn" was noticed also in the philosophy of knowledge, cf., e.g., (Burkholder, 1992, p. vii).

Benedikt Löwe, Thomas Müller (*eds.*). *PhiMSAMP. Philosophy of Mathematics: Sociological Aspects and Mathematical Practice.* College Publications, London, 2010. Texts in Philosophy 11; pp. 341–360.

*Received by the editors:* 28 September 2008; 19 May 2009.
*Accepted for publication:* 1 July 2009.

ular subject matter within mathematics, but rather to a specific technology (namely that of computers) and a specific, yet largely unspecified, methodology, namely that of *experimental* mathematics. The subdiscipline has consolidated itself around these two central notions with a set of key research questions and tools, a considerable infrastructure including journals and institutional embeddings,[2] and a philosophical legitimization brought about in articles and introductions to books such as *Mathematics by Experiment: Plausible Reasoning in the 21st Century* written by Borwein and Bailey (2004), two of the leading figures in experimental mathematics.

The philosophical literature on experimental mathematics and the discussions of computer-assisted mathematics have often revolved around a number of cardinal examples starting with the proof of the *Four Color Theorem* by Appel and Haken in 1976.[3] This prototypical example of a computer-based proof of a mathematical theorem involved computer-assisted 'number crunching' of a finite, but large, number of configurations to verify the claim. The admittance of such examples of 'number crunching' is discussed in the philosophical literature because they are perceived to involve a loss of surveyability that could challenge the *apriority* of mathematics.[4] Although these examples of 'number crunching' succeed in testing *all* the instances that fall under a general hypothesis, criticism can be levelled against them for not providing any *explanation* for the results obtained.[5] Another use of computers has been to perform searches within infinite domains in order to lend *inductive* support for a general hypothesis. An example of a distributed 'number crunching' of this sort is the search for huge *Mersenne primes*.[6] Similarly, the *Goldbach Conjecture* serves as a typical unsolved problem in number theory that has been subjected to immense 'number crunching' to establish that every even number less than $10^{18}$ can be written as the sum of two primes.[7] Hales' computer-assisted proof of

---

[2]Cf. (Gallian and Pearson, 2007, p. 14).

[3]Cf., e.g., (Tymoczko, 1979; Wilson, 2002; Bassler, 2006).

[4]Cf. (Bassler, 2006; McEvoy, 2008).

[5]Cf. also (Baker, 2007, 2008; Van Kerkhove and Van Bendegem, 2008; Van Kerkhove, 2005, pp. 287–307) and references in these for some of the themes of the discussion. The fact that 'number crunching' has an interesting history going back at least to the nineteenth century has been documented for the famous cases of the *Prime Number Theorem* (Echeverría, 1992), the *Goldbach Conjecture* (Echeverría, 1996), and *Fermat's Last Theorem* (Corry, 2008); cf. also (Goldstein, 2008).

[6]*Mersenne primes* are primes of the form $2^p - 1$. In September 2008, the distributed *Great Internet Mersenne Prime Search* (GIMPS) reported finding the 45th and 46th *Mersenne prime* each of which has more than $10^7$ decimal digits.

[7]As of July 14, 2008, all even numbers less than $12 \times 10^{17}$ have been verified to be expressible as the sum of two primes by the *Goldbach conjecture verification* project coordinated by Oliveira e Silva. The philosophical implications to be drawn from this form of inductive support for a mathematical statement has been discussed in, e.g., (Baker, 2007; Van Kerkhove and Van Bendegem, 2008).

the *Kepler Sphere Packing Problem* can be seen as a modern combination of these techniques, and it embodies most of the philosophical challenges posed by them.[8]

In the present paper, I go beyond these examples by bringing into play an example that I find more *experimental* in nature, namely that of the use of the so-called PSLQ algorithm in researching integer relations between numerical constants. It is the purpose of this paper to combine a historical presentation with a preliminary exploration of some philosophical aspects of the notion of experiment in experimental mathematics. This dual goal will be sought by analysing these aspects as they are presented by some of the protagonists of the field and discussing them using notions from contemporary philosophy of science.

Thus, in the following, I will introduce some of the most important philosophical discussions pertaining to experimental mathematics. I will then go on to illustrate how another feature of experiment may come into play in mathematics: namely that of exploration. In so doing, I describe a recent approach to experiments in science that focuses on their exploratory aspects and their importance in concept formation. Then, by combining this with a presentation of the PSLQ algorithm, I suggest to incorporate *experiments* in the informal portion of mathematics in a way that is loyal to mathematical practice while seeking to integrate the contexts of discovery and justification that are customarily separated.

## 2    Central themes: Induction and the role of computers

In 1998, when experimental mathematics was still in its infancy, Van Bendegem argued that mathematicians were using the term "experiment" in two essentially different ways, either as referring to computations or to real-world experiments. He found, that "it seems that it is very difficult to make any sense out of the idea of considering a computation (whether aiming for a numerical result or a visual image) as a form of mathematical experiment" (Van Bendegem, 1998, p. 178). A decade later, philosophical interest in experimental mathematics has only increased, and the impact of computers on traditional questions within the philosophy of mathematics is beginning to be investigated more. When Van Bendegem took up the philosophical analysis from (Van Bendegem, 1998) again together with Van Kerkhove, they noticed that experimental mathematics had "become established in [the] philosophical literature as an epistemic concept," (Van Kerkhove and Van Bendegem, 2008, p. 423) although the confusion over its precise meaning had still not been resolved. Consequently, they took up examples—some

---

[8]Cf., e.g., (Hales, 1994; Aste and Weaire, 2008).

of them exemplary ones—to discuss an "irreducible role in mathematics
for genuine [as opposed to mathematical] induction (whether it be considered truly experimental or not)" (Van Kerkhove and Van Bendegem, 2008,
p. 424).

In another recent paper, Baker continues the philosophical task of clarifying the meaning of experimental mathematics. First, he argues that "a
literal reading of 'experiment,' in the context of clarifying the nature of experimental mathematics, is unfruitful" (Baker, 2008, p. 339). In particular,
he questions whether computers are really *essential* to experimental mathematics and whether experimental mathematics is essentially about gathering
inductive support for hypotheses. He finds flaws with both suppositions and
instead suggests that the central feature of experimental mathematics is the
calculation of instances of general hypotheses. However, Baker observes,

> [c]onfusingly, both computer use and inductive reasoning also have
> links to aspects of experimentation in science. What I [i.e., Baker]
> have argued, however, is that neither is an essential feature of experi-
> mental mathematics. There is experimental mathematics that makes
> no use of computers, and there is experimental mathematics that in-
> volves no inductive relations—claimed or actual—between evidence
> and hypothesis. (Baker, 2008, p. 343)

Surveying the field of computers in mathematical inquiry for an important recent volume on the philosophy of mathematical practice, Avigad also
takes up the philosophical discussion of the role of computers in mathematics. Avigad's conclusion is that

> issues regarding the use of computers in mathematics are best under-
> stood in a broader epistemological context. [...] What we need now
> is *not* a philosophy of computers in mathematics; what we need is
> simply a better philosophy of mathematics. (Avigad, 2008, p. 315)

In particular, the use of the computer in mathematics and the impact it
may have on so-called *experimental* mathematics needs to be described and
understood within a framework sensitive to mathematical practice and to
broader epistemological discussions.

Before I take up the discussion of these issues by extending and reshap-
ing the notion of experiment, it is time to analyse the formation of the
subdiscipline of experimental mathematics in the 1990s.

# 3  Experimental mathematics in the late twentieth century

Within the mathematical community, experimental mathematics became
a debated topic in the 1990s, when the *American Mathematical Society*
(AMS) and its journals devoted resources to discussing the implications of

computers for mathematics. In retrospect, a column on "Computers and Mathematics" run in the *Notices of the AMS* from the summer of 1988 to the end of 1994 may be seen as the immediate context of experimental mathematics in the form discussed here (Devlin and Wilson, 1995, p. 248). Among reviews of software packages and technical results, some discussions evolved around visions of the impact of the computer (and digital storage and correspondence) for the communication of mathematical results—and conjectures. In connection with this, a perceived loosening of the standards of rigour was also discussed in the context of the enormous potential of the computer. In 1993, this discussion hit the mathematical community with force when Jaffe and Quinn published their suggestion for a separation between two branches of mathematicians, one that was *speculative* and one that was *rigorous*.[9] Among the debaters, it was argued that the mathematical community should be *inclusive* when it came to computers. From a variety of points, mathematicians argued that mathematics ought to embrace computers, even if this would lead to changes in the means of doing and communicating mathematics, although any loosening of epistemic standards was hard to accept for most. These issues since became important for discussions of experimental mathematics.

In the mainstream views, experiments could be useful as *heuristics*, but according to these deductivist and formalist conceptions of mathematics that Lakatos has identified as the *Euclidean myth*, such heuristics were confined to the informal spheres of discovery and teaching. Most certainly, they could not be allowed to aspire to anything like the status of proof. To these mathematicians, deductive proof is the exclusive mode of knowledge production in mathematics. However, in the 1990s, a new wave of "experimental" mathematicians was about to challenge these views.

The number of mathematicians actively identifying themselves with the new experimental programme in the early 1990s was rather limited, as would be expected. But soon, it came to associate also with prominent mathematicians such as Fields-medalists Thurston, who has a strong interest in the impact of computers on mathematics.[10] Despite its size, the group of experimental mathematicians was a heterogeneous one. Individuals committed to experimental mathematics held differing conceptions of the scope of experiment, and in particular of the kind of use that computers could

---

[9](Jaffe and Quinn, 1993); cf. also (Atiyah et al., 1994; Jaffe and Quinn, 1994). Taking their lead from the division of labour within physics, Jaffe and Quinn used the terms "theoretical" and "rigorous". However, as used, I find that what they called "theoretical" is better captured under the heading "speculative"; cf. also (Thurston, 1994, p. 163).

[10](Bown, 1991). Thurston had received the Fields medal in 1980 and was recognized as a leading figure within the mathematical community in the 1990s. Cf. also (Thurston, 1994) for Thurston's reactions to Jaffe and Quinn and (Horgan, 1993) for some of the controversy involved in the issue at the time.

be put to in experimenting with mathematics. One cluster formed around
the journal *Experimental Mathematics* when it was founded in 1991 and
began appearing the following year with Epstein and Levy as editors (Epstein and Levy, 1992). A different cluster formed in Canada at *Simon
Fraser University* around a group of individuals including J. M. Borwein,
P. Borwein, and Bailey. In geographical proximity and sharing a common
research agenda and a powerful set of computer routines and algorithms,
this group institutionalised as *Centre for Experimental and Constructive
Mathematics* (CECM) in November 1993. A third group to be mentioned
here centers around Zeilberger whose research programme is related to that
of the CECM-group but exhibits some remarkable philosophical differences.

## 4   Experimental mathematicians philosophizing

The protagonists of experimental mathematics have been quite explicit
about the philosophical problems involved in their line of mathematical
research. In a central article in the early history of experimental mathematics, "Making Sense of Experimental Math" (published in the *Mathematical Intelligencer* in 1996), the brothers J. M. Borwein and P. Borwein
and two collaborators describe the new field, its possibilities, and its challenges (Borwein et al., 1996). They offer the following definition of the field
drawn from a characterization of four roles of scientific experiments given
by the philosophizing immunologist Medawar (1979):

> *Experimental Mathematics* is that branch of mathematics that con-
> cerns itself ultimately with the codification and transmission of in-
> sights within the mathematical community through the use of ex-
> perimental (in either the Galilean, Baconian, Aristotelian or Kantian
> sense) exploration of conjectures and more informal beliefs and a care-
> ful analysis of the data acquired in this pursuit. (Borwein et al., 1996,
> p. 17)

This definition emphasises the changes in infrastructure required for a shift
towards more experimental methods in mathematics. In other (later) publications, the same authors expand on the particular roles for the use of
computers in mathematics (Borwein and Bailey, 2004, pp. 2–3), as they
see the "utilization of modern computer technology as an active tool in
mathematical research" in the style of experimental mathematics (Bailey
and Borwein, 2005, p. 502). The role of computers in mathematical experimentation will include (cf. Figure 1) *heuristics* [1–3] (gaining insight
and intuition and discovering new patterns using symbolic or graphical experiments), *refining and evaluating conjectures* [4–5] (testing and falsifying
conjectures and exploring the conjecture to see if it is worth attempting
a formal proof), and *aiding in the procedure of proving conjectures* [6–8],

either by suggesting strategies for formal proof or by allowing computer-based derivations or confirmations of, e.g., intricate identities (Borwein and Bailey, 2004, pp. 2–3).

---

1. Gaining *insight and intuition*.

2. *Discovering new patterns* and relationships.

3. Using graphical displays to *suggest underlying mathematical principles*.

4. Testing and, especially, *falsifying conjectures*.

5. Exploring a possible result to see if it is *worth formal proof*.

6. *Suggesting approaches* for formal proof.

7. Replacing lengthy hand derivations with *computer-based derivations*.

8. *Confirming* analytically derived results.

---

FIGURE 1. Roles for computers in mathematics, according to (Borwein and Bailey, 2004, pp. 2–3).

Among the types of experiments extracted from Medawar's classification, "Baconian experimentation" includes "trying things out" and observing "things as they really are" (Medawar, 1979, pp. 69–70). This type of experimentation could seem to come close to the use of computers in visualizing and exploring mathematical structures and problems. However, compared to Medawar's four types of scientific experiments, the authors argue, experimental mathematics is only a "serious enterprise" insofar it resembles the critical (or even crucial) experiments that Medawar calls "Galilean" which "discriminate between possibilities and, in doing so, either gives us confidence in the view we are taking or makes us think it in need of correction" (Borwein and Bailey 2004, p. 6; cf. also Medawar 1979, p. 71). Although superficial, Medawar's presentation of Baconian experiments highlights their ideal characteristic as unbiased recordings of contrived facts of nature that can subsequently be subjected to inductive arguments. Thus, the authors contrast the *inductive* Baconian experiments with the crucial, and thereby essentially *deductive*, experiment that Medawar ascribes to Galilei. By their emphasis on the latter, these authors therefore also argue for a deductive and justificatory role for experiments in mathematics that goes beyond the heuristics of fact (number) gathering.

However, as I will discuss below, I find that the discussions of so-called "new experimentalism" and the notion of "exploratory experimentation" bring a new and refined meaning to Baconian experimentation that has consequences for the understanding of mathematical experiments (Steinle, 1997).

## 5   Contexts of experiments in the sciences and in mathematics

In the second half of the twentieth century, a standard view of the role of experiments in the sciences has been to test hypotheses or theories. Such *theory-driven* experimentation resembles Medawar's Galilean experiments and serves to justify theories. In mathematics, on the other hand, experimentation has been accepted as a (powerful) heuristic that can aid in the *discovery* of plausible conjectures. But, in mathematics, experiments have been confined to the realm of discovery while deductive *proofs* remained the exclusive means for justifying claims in mathematics.

What has happened in the past decades has been that these opposite confinements for experimentation have begun to be loosened in the philosophical literature. In the sciences, "exploratory experimentation" has been analysed as an important means of concept formation. Thus, experiments in the sciences have found a place in the context of discovery together with the heuristic of gathering data. In mathematics, where experiments were typically confined to the context of discovery, experiments have started to blur the distinction in the other direction. Some proponents of experimental mathematics—in particular Zeilberger—have claimed that experiments possess powers of justification that are not those of proofs, but could and should be allowed into mathematics. Thus, the notion of experiment is changing both in the sciences and in mathematics, but because the traditional views differ in science and mathematics, the new roles for exploratory experimentation also differ.

Thus, when the concept of experimental mathematics made its dramatic entry onto the scene of the mathematical community in the first half of the 1990s, the image of experimentation, itself, was undergoing refinement in the philosophical literature.[11] In the following, I suggest that "exploratory experimentation" offers a framework for understanding some of the epistemological claims of experimental mathematics.

---

[11]Cf., e.g., (Steinle, 2002, 1997; Franklin, 2005).

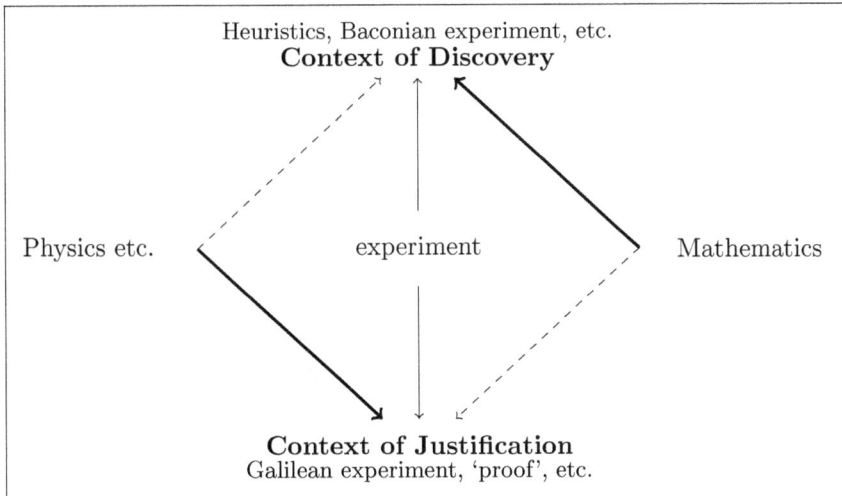

FIGURE 2. The uses of experimentation is changing both in the sciences and in mathematics. The thick arrows indicate traditional embeddings of experiments in science (justification) and mathematics (discovery). The dashed arrows indicate new roles for experiments, and the vertical arrows are meant to indicate that changes in the notion of experiment are blurring the distinctions between the contexts of discovery and justification. Importantly because the traditional views of the two types of science are different, the new roles for exploratory experiments also differ.

## 6    Exploratory experimentation and wide instrumentation

The philosopher of (physical) science Steinle describes the differences between theory-driven and exploratory experimentation in terms of the possible empirical outcomes:

> Theory-driven experiments are typically done with quite specific expectations of the various possible outcomes in mind. [...] Exploratory experimentation, in contrast, is driven by the elementary desire to obtain empirical regularities and to find out proper concepts and classifications by means of which those regularities can be formulated. (Steinle, 1997, p. S70)

Steinle goes on to explain that exploratory experimentation typically takes place in phases of scientific development in which no well-formed conceptual framework is available (Steinle, 1997, p. S70). Thus, Steinle's exploratory experiments in science are *open-ended* and highly important and influential in the processes of concept formation.

Drawing on examples from research in molecular biology during the last decades, the philosopher Franklin adds an interesting dimension to the notion of "exploratory experimentation", namely that of *wide instrumentation*. The availability of high-throughput instruments that can simultaneously measure many features or repeat measurements very quickly has, so Franklin argues, made it feasible (again) to address the enquiry of nature without local theories to guide the experiments. In the process, experiments have gained another quality to be measured by, namely efficiency in bringing about new results (Franklin, 2005, p. 895).

These aspects of exploratory experimentation and wide instrumentation originate from the philosophy of (natural) science and have not been much developed in the context of experimental mathematics. However, I claim that, e.g., the importance of wide instrumentation for an exploratory approach to experiments that includes concept formation also pertain to mathematics. However, it could seem that experimental situations with no well-formed conceptual framework do not occur in mathematics. Therefore, I first give a short outline of an example to illustrate that this can and does occur. I then go on to discuss the idea of exploratory experimentation based on another example from experimental mathematics, namely the so-called PSLQ algorithm.

## 7   In the absence of a well-formed conceptual framework

Experimental mathematicians have taken an interest in studying polynomials with coefficients from a finite set. For instance, a group at the CECM has studied zeroes of polynomials with coefficients 1 or $-1$ and degree up to 18.[12] Extensive data mining was used to produce graphical illustrations of the sensitivities of the zeroes of these polynomials. From these images, two observations can be made that are important in relation to Steinle's notion of exploratory experimentation. One point is that the images were found to exhibit a remarkable behaviour near roots of unity, and indeed, that behaviour has now been rigorously proved (Borwein et al., 2008). The other point is that the images were capable of interactive experimentation such as measuring sensitivities relative to one of the coefficients. Figure 3 illustrates the sensitivities of zeroes relative to the $x^9$ term. These new figures, and in particular the banded features that appear, are not yet fully understood. However, it is widely believed among the experimental experts that the features are stable and not merely programmatic artifacts. To understand these new phenomena, it is most likely that some new conceptual

---

[12]Polynomials with coefficients $\pm 1$ are called *Littlewood polynomials*. This example and Figure 3 below is presented in (Borwein and Jörgenson, 2001) and briefly described in various publications on experimental mathematics, such as (Borwein, 2008).

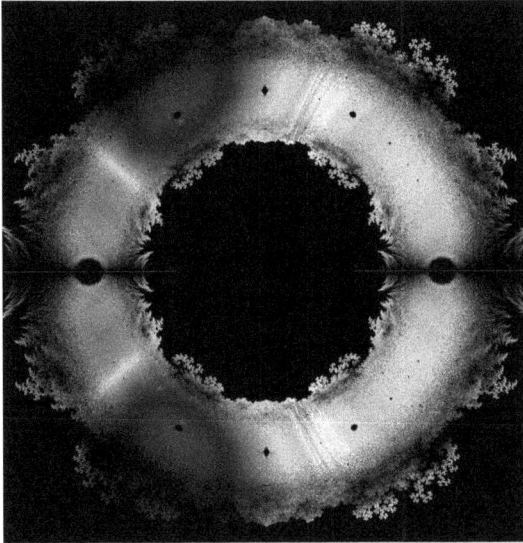

FIGURE 3. Sensitivies of zeroes of *Littlewood polynomials* of degree up to 18 relative to the $x^9$ term; image from Loki Jorgensen's webpage.

developments are required. Thus, this is a case in which computer-assisted methods—both in the form of data mining (or 'number crunching') and as open-ended, exploratory and interactive experimentation—are at work in a situation when no well-formed conceptual mathematical framework for explaining the phenomena has yet been devised. A variation on this situation will appear below in the case of the PSLQ algorithm.

## 8    The PSLQ algorithm as an example

The group of experimental mathematicians centred around the CECM and the brothers J. M. Borwein and P. Borwein have made use of an algorithm developed by Ferguson, Bailey and others, known as the PSLQ algorithm.[13] This algorithm is a good example of a powerful tool in experimental mathematics and therefore merits some attention and analysis in the present context. First, the domain of the algorithm needs to be clarified. The PSLQ algorithm is a so-called "integer relation detection algorithm" that can be used to search for integer relations between given numerical constants expressed in finite, but high-precision decimal representation. By a special application, the PSLQ algorithm can also be used to investigate

[13](Ferguson et al., 1999). The PSLQ algorithm has also been discussed in (Corfield, 2003, pp. 64–66).

whether given numerical constants are algebraic numbers or not. Thus, the
algorithm has found its use, e.g., in summing various series in closed form
(Bailey et al., 1994). Second, the special nature of the PSLQ algorithm that
makes it *experimental* needs to be emphasized. The algorithm takes as its
input a vector of high-precision real (or complex) numbers $(x_1, \ldots, x_n) \in \mathbb{R}^n$
and after a bounded number of iterations either produces 1) a vector of in-
tegers $(m_1, \ldots, m_n) \in \mathbb{Z}^n \setminus \{0\}$ such that the linear integer combination
$\sum_{k=1}^{n} m_k x_k$ is very close to zero with high precision, or 2) a lower bound
$R$ such that for all vectors of integers bounded by $R$, the linear integer
combination $\sum_{k=1}^{n} m_k x_k$ is provably different from zero. The two potential
outcomes of the algorithm, therefore, represent an unproven suggestion or a
provable non-existence result: The result is either a suggestion for an integer relation or a lower bound showing that any such relation has to include
very large coefficients.

## 9   Consequences and applications of the PSLQ algorithm

In case the PSLQ algorithm provides a lower bound for the size of any
possible integer relation between the specified real constants, the result is
exact (and provable). However, the community of experimental mathematicians have come to hold different—some of them rather radical and
provocative  views on the need for making the proof explicit. The most
outspoken radicalist among them, Zeilberger, has suggested that when such
exact results are provided by an established algorithm, it suffices to anno-
tate the announcement with "QED" as a seal that the author has, indeed,
performed the computation and observed the result as indicated.[14] In some
of his papers and preprints, of which his software package "Ekhad" is listed
as co-author, Zeilberger has even implemented this (cf. Figure 4).[15] The
provocative nature of the views of the most radical experimentalists in mathematics is even clearer when it comes to the status of a result such as an
output of the first type from the PSLQ algorithm: a statement of a high-
precision verification of some relationship. Following up on a provocative
suggestion by Chaitin and in line with some of the arguments behind the
suggestions of a "theoretical mathematics" discussed above, Zeilberger has
suggested to affix statements of this kind with a (consistently derived) estimate of how difficult or time-consuming it would be to produce a formal
proof of a relationship that is verified to great numerical precision.[16]

What lies behind Zeilberger's suggestion is that mathematics is presently
in a phase of wide expansion, cultivating 'new lands'. During such a phase,

---

[14]Cf., e.g., (Zeilberger, 1993, p. 31).
[15]Cf., e.g., (Ekhad and Zeilberger, 1996).
[16]For Zeilberger's elaboration, cf. (Zeilberger, 1994).

*Proof.* While this is unlikely to be new, it is also irrelevant whether or not it is new, since *such things are now routine, thanks to the package qEKHAD*, accompanying [PWZ]. Let's call the left side divided by $q^{m(m-1)/6}$, $Z(m)$. Then we have to prove that $Z_0(m) := Z(3m)$ equals $(-1)^m$, $Z_1(m) := Z(3m+1)$ equals $(-1)^m$, and $Z_2(m) := Z(3m+2)$ equals 0. It is directly verified that these are true for $m = 0, 1$, and *the general result follows from the second order recurrences produced by qEKHAD*. The input files inZ0, inZ1, inZ2 as well as the corresponding output files, outZ0, outZ1, outZ2 *can be obtained by anonymous ftp* to ftp.math.temple.edu, directory pub/ekhad/sasha. The package qEKHAD can be downloaded from http://www.math.temple.edu/~zeilberg. Q.E.D.

FIGURE 4. A "proof" according to Ekhad and Zeilberger in the *Electronic Journal of Combinatorics*, (Ekhad and Zeilberger, 1996, p. 2, emphasis added).

efficiency in establishing results is to be valued more, they think, than absolute rigor according to the traditional standards. Therefore, Chaitin has suggested assuming unproved hypotheses such as the Riemann Hypothesis as axioms if they are experimentally justified and lead to fruitful research.[17]

Unsurprisingly, Zeilberger's suggestion was met with criticism from proponents of a more classical view of mathematical epistemology.[18] The CECM-group also takes a different view of the situation. As the nicely titled paper "Strange Series and High Precision Fraud" argues (Borwein and Borwein, 1992), even high-precision verification is *not* a substitute for formal proof.

Although the field of experimental mathematics is considerably broader than applications of the PSLQ algorithm and similar searches for symbolic identities using numerical methods, the PSLQ algorithm is a nice example from which to discuss features of the new experimental approach to mathematics. Integer relation detection is located on the border between rigorously provable theorems and experimentally obtained conjectures. The PSLQ algorithm allows mathematicians to interactively navigate open problems in many fields of mathematics in search of either rigorous lower bounds or suggestions for relations. If the algorithm provides a rigorous lower bound, mathematicians may continue pursuing a proof that no such relations exist at all. On the other hand, if the algorithm suggests a relation,

---

[17]Cf., e.g., (Chaitin, 1993, pp. 326–327).
[18]Cf., e.g., (Andrews, 1994).

mathematicians may seek traditional proof of this suggestion. In both cases, the situation is likely to be further explored using either experimental techniques or more traditional proof, or a combination thereof.

## 10    Fact-gathering or interactive exploration

A traditional (post-foundationalist) framework for discussing the philosophy of mathematical practice has been Lakatos' conception of mathematics as a quasi-empirical science directed by thought-experiments (also known as proofs) and refutations through counter examples (Lakatos, 1976). Thus, Lakatos' philosophy as expressed in the *Proofs and Refutations* deals mainly with concept formation presupposing phases of heuristic conjecturing. If Lakatos' philosophy is to be fully applied, another aspect of experimental mathematics has to be considered. Gauss' calculations (with the help of *human* computers) of prime number tables to a large extent fell into the category of fact-gathering for subsequent conjecturing. As such it can be isolated as a most fundamental form of experiment, but not (yet) as exploratory experimentation. Similarly, many of the efforts that have gone into computer visualisation and 'number crunching' of mathematical problems have served these roles of compiling and making accessible data on which to form hypotheses. However, the interactive use of systems such as Maple and Mathematica have opened the door for an integrated process of experimentation, concept formation, and conjecturing. This can be illustrated using an example relating to so-called *Euler sums* and using the PSLQ algorithm that has been discussed by members of the CECM-cluster in (Bailey et al. 1994; cf. also Borwein et al. 1996, pp. 13–15).

As recorded in (Borwein et al., 1996, pp. 13–15), an undergraduate student observed the curious numerical fact that

$$\sum_{k=1}^{\infty} \left(1 + \frac{1}{2} + \cdots + \frac{1}{k}\right)^2 k^{-2} = 4.59987\cdots \approx \frac{17}{4}\zeta(4) = \frac{17\pi^4}{360}$$

and brought it to the attention of his professor (J. M. Borwein). At first a serendipitous discovery, the experimental mathematicians began to explore the relationship to greater numerical precision. When the relation was confirmed to a precision of 100 digits, they undertook to generalise the setting and produce a framework for experimentally exploring similar conjectures using the PSLQ algorithm. Among many similar so-called *Euler sums*, they investigated sums of the form

$$s_a(m, n) = \sum_{k=1}^{\infty} \left(1 - \frac{1}{2} + \cdots + \frac{(-1)^{k+1}}{k}\right)^m (k+1)^{-n} \quad \text{for } m \geq 1, n \geq 2$$

and

$$s_h(m,n) = \sum_{k=1}^{\infty} \left(1 + \frac{1}{2} + \cdots + \frac{1}{k}\right)^m (k+1)^{-n} \quad \text{for } m \geq 1, n \geq 2.$$

These new expressions are thus part of a conceptualisation introduced in order to formulate and generalise the outcomes of exploratory experimentation.

The integer relation detection of the PSLQ algorithm functioned as an important tool in finding potential integer relations involving other known mathematical constants. When such relations are found using high precision computations, the experiment is concluded, and it may either (according to Zeilberger's view) be accepted as an experimentally derived and supported result, or be subjected to formal proof. For some of the many identities obtained by Bailey and his coauthors, formal proofs have been found, whereas others remain open conjectures. For instance,

$$s_h(2,2) = \frac{3}{2}\zeta(4) + \frac{1}{2}\zeta^2(2) = \frac{11\pi^2}{360}$$

has been formally proved (1996), whereas

$$s_h(3,2) = \frac{15}{2}\zeta(5) + \zeta(2)\zeta(3)$$

remained a conjecture (1996). As part of the process of exploratory experimentation, new connections between various constants (and, therefore, presumably between various sets of theories) are suggested and need to be explored.

Experimental mathematicians repeatedly stress the importance of *interactive* experimentation using computers, for which high-speed processors as well as a functional user interface is imperative. For this, they often profess their debt to *Moore's Law* that semiconductor performance doubles every 18 to 24 months.[19] In doing so, they repeat some of the arguments for wide instrumentation discussed by Franklin. Only with interactive experimentation and multi-purpose software—so it seems—could mathematical experimentation make a transition from 'number crunching' to exploration. Thus, high-speed computers and interactive software packages seem to be the mathematical equivalents of wide instrumentation in the sciences. In particular, this suggests an answer to why experimental mathematics *only* became a discussion in the 1990s when both hardware and software had been developed to meet these demands.

---

[19]Cf., e.g., (Borwein and Bailey, 2004, pp. 3–4). However, for the notion of *Moore's Law* and technological determinism, cf. (Ceruzzi, 2005).

Thus, perhaps as a result of not referring to the most recent literature that includes exploratory experimentation, experimental mathematics has so far missed out on an obvious and yet very powerful symmetry with the use of experimentation in the sciences. It is becoming increasingly clear that experiments have an important role in the concept formation in the sciences—a realization that is a reaction to the exclusive focus on theory-driven experiments in much of the twentieth century. Similarly, an important role is to be played by computational experimentation in mathematics, when it is performed in the interactive, exploratory ways discussed above.

## 11   Is (experimental) mathematics special?

This paper has described aspects of the emerging discipline forming around the use of computer experimentation in mathematics and institutionalising as experimental mathematics. This subdiscipline has challenged the usual deductivist philosophy of mathematics by arguing for a role for experiments in mathematics. Thereby, its proponents not only blur the separation between the context of discovery and the context of justification in mathematics but also claim a role for experiments beyond that of heuristics.

By way of a single example—that of the PSLQ algorithm—I have shown how the methodology of experimental mathematics features examples of exploratory experimentation similar to the kind recently discussed in the philosophy of the sciences. Thereby, I have pointed out that the practice of experimental mathematics is, indeed, experimental in some of the ways most often associated with physics or chemistry.

On the other hand, even within the community of experimental mathematics, views differ concerning the role of experiment in justifying mathematical claims. They all claim a role for it, but differ on precisely which one. Here, again, I see a role for exploratory experimentation as a framework. It highlights the role of experiments in the process of concept formation that can lead to (formal) theory formation. In these respects, therefore, experimental mathematics *does* pose an example challenging traditional views of mathematical epistemology and bringing forward a suggestion for a more empirically founded philosophy of mathematics that applies, at least, to the domains of mathematics most susceptible to exploratory experiment.

To finally address the question whether (experimental) mathematics is special, I would point out, that even considering exploratory experimentation, "experiment" still means essentially different things in the sciences and in experimental mathematics as was pointed out by, e.g., Baker (2008) in the introduction. However, the emergence of experimental mathematics shares features with exploratory experimentation in the sciences, particularly when it comes to open-ended experimentation, wide instrumentation, and the role of experimentation in concept formation. Thereby, this suggestion raises (again)—and from the novel perspective of comparing with the sciences—the question whether deductive proof is *really* the only permissible mode of justification in mathematics.

# Bibliography

Andrews, G. E. (1994). The death of proof? Semi-rigorous mathematics? You've got to be kidding! *The Mathematical Intelligencer*, 16(4):16–18.

Aste, T. and Weaire, D. (2008). *The Pursuit of Perfect Packing*. CRC Press, Boca Raton FL.

Atiyah, M., Borel, A., Chaitin, G. J., Friedan, D., Glimm, J., Gray, J. J., Hirsch, M. W., MacLane, S., Mandelbrot, B. B., Ruelle, D., Schwarz, A., Uhlenbeck, K., Thom, R., Witten, E., and Zeeman, C. (1994). Responses to A. Jaffe and F. Quinn, "Theoretical mathematics: toward a cultural synthesis of mathematics and theoretical physics". *Bulletin of the American Mathematical Society*, 30(2):178–207.

Avigad, J. (2008). Computers in mathematical inquiry. In Mancosu, P., editor, *The Philosophy of Mathematical Practice*, chapter 11, pages 302–316. Oxford University Press, Oxford.

Bailey, D. H. and Borwein, J. M. (2005). Experimental mathematics: Examples, methods and implications. *Notices of the American Mathematical Society*, 52(5):502–514.

Bailey, D. H., Borwein, J. M., and Girgensohn, R. (1994). Experimental evaluation of Euler sums. *Experimental Mathematics*, 3(1):17–30.

Baker, A. (2007). Is there a problem of induction for mathematics? In Leng, M., Paseau, A., and Potter, M., editors, *Mathematical Knowledge*, pages 59–73. Oxford University Press, Oxford.

Baker, A. (2008). Experimental mathematics. *Erkenntnis*, 68(3):331–344.

Bassler, O. B. (2006). The surveyability of mathematical proof: A historical perspective. *Synthese*, 148:99–133.

Borwein, J. (2008). Implications of experimental mathematics for the philosophy of mathematics. In Gold, B. and Simons, R. A., editors, *Proof & Other Dilemmas: Mathematics and Philosophy*, chapter 2, pages 33–59. Mathematical Association of America, Washington DC.

Borwein, J. and Bailey, D. (2004). *Mathematics by Experiment: Plausible Reasoning in the 21st Century*. A K Peters, Natick MA.

Borwein, J. and Borwein, P. (1992). Strange series and high precision fraud. *The American Mathematical Monthly*, 99(7):622–640.

Borwein, J., Borwein, P., Girgensohn, R., and Parnes, S. (1996). Making sense of experimental mathematics. *The Mathematical Intelligencer*, 18(4):12–18.

Borwein, P., Erdélyi, T., and Littmann, F. (2008). Polynomials with co-efficients from a finite set. *Transactions of the American Mathematical Society*, 360(10):5145–5154.

Borwein, P. and Jörgenson, L. (2001). Visible structures in number theory. *The American Mathematical Monthly*, 108(10):897–910.

Bown, W. (1991). New-wave mathematics. *New Scientist*, 1780:33.

Burkholder, L., editor (1992). *Philosophy and the Computer*. Westview Press, Boulder CO.

Ceruzzi, P. E. (2005). Moore's Law and technological determinism: Reflections on the history of technology. *Technology and Culture*, 46(3):584–593.

Chaitin, G. J. (1993). Randomness in arithmetic and the decline and fall of reductionism in pure mathematics. *Bulletin of the European Association for Theoretical Computer Science*, 50:314–328.

Corfield, D. (2003). *Towards a Philosophy of Real Mathematics*. Cambridge University Press, Cambridge.

Corry, L. (2008). Number crunching vs. number theory: Computers and FLT, from Kummer to SWAC (1850–1960), and beyond. *Archive for History of Exact Sciences*, 62:393–455.

Devlin, K. and Wilson, N. (1995). Six-year index of "Computers and mathematics". *Notices of the American Mathematical Society*, 42(2):248–254.

Echeverría, J. (1992). Observations, problems and conjectures in number theory — the history of the Prime Number Theorem. In Echeverría, J., Ibarra, A., and Mormann, T., editors, *The Space of Mathematics. Philosophical, Epistemological, and Historical Explorations*, pages 230–252. Walter de Gruyter, Berlin.

Echeverría, J. (1996). Empirical methods in mathematics. A case-study: Goldbach's Conjecture. In Munévar, G., editor, *Spanish Studies in the Philosophy of Science*, volume 186 of *Boston Studies in the Philosophy of Science*, pages 19–55. Kluwer Academic Publishers, Dordrecht.

Ekhad, S. B. and Zeilberger, D. (1996). The number of solutions of $X^2 = 0$ in triangular matrices over $GF(q)$. *Electronic Journal of Combinatorics*, 3(R2):1–2.

Epstein, D. and Levy, S. (1992). Message from the editors. *Experimental Mathematics*, 1(3).

Ferguson, H. R. P., Bailey, D. H., and Arno, S. (1999). Analysis of PSLQ: An integer relation finding algorithm. *Mathematics of Computation*, 68(225):351–369.

Franklin, L. (2005). Exploratory experiments. *Philosophy of Science*, 72:888–899.

Gallian, J. and Pearson, M. (2007). An interview with Doron Zeilberger. *MAA Focus*, 27:14–17.

Goldstein, C. (2008). How to generate mathematical experimentation, and does it provide mathematical knowledge? In Feest, U., Hon, G., Rheinberger, H.-J., Schickore, J., and Steinle, F., editors, *Generating Experimental Knowledge*, number 340 in MPIWG Preprints, pages 61–85. Max-Planck-Institut für Wissenschaftsgeschichte, Berlin.

Hales, T. C. (1994). The status of the Kepler Conjecture. *The Mathematical Intelligencer*, 16(3):47–58.

Horgan, J. (1993). The death of proof. *Scientific American*, 269(4):74–82.

Jaffe, A. and Quinn, F. (1993). "Theoretical Mathematics": Toward a cultural synthesis of mathematics and theoretical physics. *Bulletin of the American Mathematical Society*, 29(1):1–13.

Jaffe, A. and Quinn, F. (1994). Response to comments on "Theoretical Mathematics". *Bulletin of the American Mathematical Society*, 30(2):208–211.

Lakatos, I. (1976). *Proofs and Refutations: The Logic of Mathematical Discovery*. Cambridge University Press, Cambridge. Edited by John Worrall and Elie Zahar.

McEvoy, M. (2008). The epistemological status of computer-assisted proofs. *Philosophia Mathematica*, 16(3):374–387.

Medawar, P. (1979). *Advice to a Young Scientist*. Harper & Row, San Francisco CA.

Steinle, F. (1997). Entering new fields: Exploratory uses of experimentation. *Philosophy of Science*, 64(S1):S65–S74.

Steinle, F. (2002). Experiments in history and philosphy of science. *Perspectives on Science*, 10(4):408–432.

Thurston, W. P. (1994). On proof and progress in mathematics. *Bulletin of the American Mathematical Society*, 30(2):161–177.

Tymoczko, T. (1979). The four-color problem and its philosophical significance. *Journal of Philosophy*, 76(2):57–83.

Van Bendegem, J. P. (1998). What, if anything, is an experiment in mathematics? In Anapolitanos, D., Baltas, A., and Tsinorema, S., editors, *Philosophy and the Many Faces of Science*, chapter 14, pages 172–182. Rowan & Littlefield Publishes, Ltd., Lanham MD.

Van Kerkhove, B. (2005). Aspects of informal mathematics. In Sica, G., editor, *Essays on the Foundation of Mathematics and Logic*, pages 267–351. Polimetrica, Monza.

Van Kerkhove, B. and Van Bendegem, J. P. (2008). Pi on Earth, or mathematics in the real world. *Erkenntnis*, 68(3):421–435.

Wilson, R. (2002). *Four Colours Sufficc. How thc Map Problem was Solved.* Penguin/Allen Lane, London.

Zeilberger, D. (1993). Identities in search of identity. *Theoretical Computer Science*, 117:23–38.

Zeilberger, D. (1994). Theorems for a price: Tomorrow's semi-rigorous mathematical culture. *The Mathematical Intelligencer*, 16(4):11–14, 76.

# For a thicker semiotic description of mathematical practice and structure

## Roy Wagner

The Cohn Institute, Tel Aviv University and Computer Science School, Academic College of Tel-Aviv-Jaffa, P.O. Box 39040, Tel Aviv 69978, Israel

E-mail: rwagner@mta.ac.il

## 1 Thick description

As this volume is concerned with sociological aspects and mathematical practice in the philosophy of mathematics, it seems fitting to open with a quotation from an anthropologist. "Once human behavior," explains Clifford Geertz,

> is seen as [...] symbolic action—action which, like phonation in speech, pigment in painting, line in writing, or sonance in music, signifies— the question as to whether culture is patterned conduct or frame of mind, or even the two mixed together, loses sense. The thing to ask about [social practices such as] a burlesqued wink or a mock sheep raid is not what their ontological status is. It is the same as that of rocks on the one hand and dreams on the other—they are things of this world. The thing to ask is what their import is: what it is, ridicule or challenge, irony or anger, snobbery or pride, that, in their occurrence and through their agency, is getting said. (Geertz, 1973, p. 10)

One reason for opening with this quotation is that for many contemporary philosophers of mathematics this quotation explicates why mathematical practice is incompatible with the research framework promoted by Geertz. It's true that mathematical practice is symbolic, and that it is about signifying phonation, lines and gestures. But whether mathematics is a frame of mind, patterned conduct or referenced reality (or the three mixed together)—these questions don't seem to lose their sense, at least not for contemporary philosophers of mathematics.

A key to why questions concerning mathematical ontology retain a sense, which mainstream anthropology has given up with respect to its own objects of study, is provided by Geertz's examples: "a burlesqued wink or a mock sheep raid". In mathematical practice, is there burlesque and mockery? And, if mathematical signs do not risk burlesque or mockery, at least not *inside* mathematics, then perhaps, where mathematical practices are concerned, there's not much point in "asking what is getting said" in the socio-cultural interpretive sense that Geertz promotes (which is not the

Benedikt Löwe, Thomas Müller (*eds.*). *PhiMSAMP. Philosophy of Mathematics: Sociological Aspects and Mathematical Practice*. College Publications, London, 2010. Texts in Philosophy 11; pp. 361–384.

Received by the editors: 26 August 2008; 7 April 2009; 22 September 2009.
Accepted for publication: 23 September 2009.

same as the kind of interpretation offered in a maths classroom)? After all, mathematical practices, unlike Geertz's objects of concern, are not supposed to be about "ridicule or challenge, irony or anger, snobbery or pride"; such practices are never dreamt, but stand firm as rocks; and the rock-solid status of such practices within our cultural world is something that a philosopher should venture to explain.

For a sociologist of mathematics, however, the idea that mathematics might be intractable to the kind of analysis that Geertz promotes would sound absurd; mathematics is, with all things said and done, a practice that is social. But what I'd like to do here is show that philosophers too can follow Geertz's lead and gain a great deal (my engagement with contemporary philosophy of mathematics in this paper will be mostly through structuralist philosophy of mathematics). We'll try to see what philosophical insight into mathematical practice we gain, when we practice what Geertz calls (following Ryle) a "thick description", that is, when we describe mathematical practice as "a stratified hierarchy of meaningful structures in terms of which [all sorts of signifying gestures] are produced, perceived and interpreted, and without which they would not [...] in fact exist [as signifying gestures], no matter what anyone did or didn't do" (Geertz, 1973, p. 7). The purpose of this paper is to experiment with such a thick description of a mathematical case study.

But even at this early stage philosophers may raise an objection to one of the premises of thick description: that without "a stratified hierarchy of meaningful structure", social or mathematical practices "would not [...] in fact exist". This statement should not be taken lightly. Much has been said by philosophers of mathematics about meaningless mathematical signs and about mathematical realities that exist independently of their human, interpreted expression. But a thick description needn't a-priori exclude either. What would not exist without meaningful articulation is mathematical practice. Even a formalist's uninterpreted mathematical sign is distinguishable in its use by mathematical practitioners from a random line in the sand, and is meaningfully articulated as an uninterpreted mathematical mark. And whatever, if anything, lies mathematically beyond practice—such being may safely rest beyond the scope of this essay as well.

Sociologists of science may raise an objection as well. They may justly claim that there is no need to 'experiment' with thick descriptions of mathematical practice, as such descriptions have already been successfully produced. One may bring up Livingston's ethnomethodological work on Gödel's proof (Livingston, 1986) and other proof practices (Livingston, 1999); Rosental's study on the work of logicians (Rosental, 2008b) and on university logic teaching (Rosental, 2008a); Netz's work on Greek geometry (Netz, 1999) and my reaction to his work in Wagner (2009a); this list is obviously just a small sample. These and other texts deal with mathematical writing, institutions, scholars, students, classrooms, publications, errors, competition and more. But it is precisely their ethnomethodological, sociological and cognitive settings that take away the specific edge in which I'm interested here. What's (relatively) special about the approach I take here is the specific slice I carve from the vast range opened by thick descriptions. I will concern myself with the semiotic level of a thick description, which I feel is particularly relevant to the interests of most philosophers of mathematics (but which I don't pretend is more important or more fundamental than other levels of thick descriptions).

## 2   A case study

The mathematical signs that we shall study are 2-by-2 matrices. These are arrays of four numbers ordered in two columns and two rows, such as

$$\begin{pmatrix} 1 & 2 \\ 3 & 4 \end{pmatrix} \text{ or } \begin{pmatrix} -10 & 0.74 \\ 1/2 & 0 \end{pmatrix}.$$

There are many things that a matrix can be interpreted as standing for. One such object is a parallelogram in a Cartesian plain. Another is a linear motion. For example, the matrix

$$A = \begin{pmatrix} 1/\sqrt{2} & -1/\sqrt{2} \\ 1/\sqrt{2} & 1/\sqrt{2} \end{pmatrix}$$

can stand for the square whose vertices are the origin $(0,0)$, the point $(1/\sqrt{2}, 1/\sqrt{2})$, and the point $(-1/\sqrt{2}, 1/\sqrt{2})$ (the fourth vertex is determined by the three given vertices and the postulation that the shape we're describing is a parallelogram). The same matrix $A$ can also stand for a counterclockwise rotation around the origin by an angle of 45 degrees. Another example, the matrix

$$B = \begin{pmatrix} \sqrt{3}/2 & -1/2 \\ 1/2 & \sqrt{3}/2 \end{pmatrix},$$

can stand for the square whose vertices are the origin $(0,0)$, the point $(\sqrt{3}/2, 1/2)$, and the point $(-1/2, \sqrt{3}/2)$, and also for a counterclockwise rotation around the origin by an angle of 30 degrees.

The rule is easy: to interpret a 2-by-2 matrix as a parallelogram, set $(0,0)$ and the two columns of the matrix as vertices. To interpret a 2-by-2 matrix as a linear motion, consider the product of the matrix and vectors in the Cartesian plane (for convenience, I will only use *positive determinant orthonormal matrices* in my examples, so we shall always end up with squares and rotations, rather than general parallelograms and linear motions).

So far we have polysemy—several references for the same sign. This is indeed a prerequisite for a thick description, which is about a "hierarchy of meaningful structures". But in itself polysemy is not terribly interesting, not enough for constituting a thick description, and not what I am after. After all, we can interpret any sign as standing for any object, and there's nothing either thick or thin about describing this fact. That a signifier[1] has several interpretations is less than interesting in this context, unless these interpretations interrelate.

In order to get to the point of interrelation let's consider matrix products. Matrix product is an operation that takes two matrices, and yields another matrix. It's a little more complicated than multiplying the matrices term by term. The product of matrices

$$\begin{pmatrix} a & b \\ c & d \end{pmatrix} \text{ and } \begin{pmatrix} v & w \\ x & y \end{pmatrix}$$

is defined as

$$\begin{pmatrix} av + bx & aw + by \\ cv + dx & cw + dy \end{pmatrix}.$$

Multiplying our above $A$ and $B$, for example, we get

$$A \cdot B = \begin{pmatrix} \frac{\sqrt{3}-1}{2\sqrt{2}} & \frac{-\sqrt{3}-1}{2\sqrt{2}} \\ \frac{\sqrt{3}+1}{2\sqrt{2}} & \frac{\sqrt{3}-1}{2\sqrt{2}} \end{pmatrix}.$$

What's important for us about matrix multiplication is that if $X$ is a matrix that stands for a square, and $Y$ is a matrix that stands for a rotation, then $Y \cdot X$ stands for the square you'd get by applying the rotation represented by $Y$ to the square represented by $X$. So the product $A \cdot B$ above is a matrix that stands for the square you'd get by applying a 45 degrees counterclockwise rotation to the square represented by $B$.

At this point our two interpretations (square and rotation) relate to each other and interact. Mathematical practitioners interpret signs in different ways, and compose these interpretations. One can introduce many such compositions. For example, on top of the above rotation-applied-to-square interpretation of products, one can interpret the product of matrices as a composition of the rotations they represent. The product of a matrix that stands for a 45 degrees rotation and a matrix that stands for a 30 degrees rotation will then stand for the combined 75 degrees rotation. Both these interpretation are useful in practice.

One may of course come up with many different competing interpretations, but so far, the term 'competing' does not appear to be justified. So

---

[1]I am using here the structuralist terminology of the tradition attributed to de Saussure (1966).

far the interpretations that we presented coexist; each is viable in itself. They do not penetrate each other, do not affect each other, and are mutually neutral. Interpretations here do not have the descriptive thickness suggested by Geertz's term "hierarchy". More describing is needed to get there.

So let's look at one more thing that we can do with matrices: raising them to the second power. The formula is:

$$\left(\begin{array}{cc} a & b \\ c & d \end{array}\right)^2 = \left(\begin{array}{cc} a^2 + bc & ab + bd \\ ca + dc & cb + d^2 \end{array}\right).$$

We can think of this operation as a function that takes a matrix standing for a square, and yields another matrix standing for another square. We can check and verify that the angle between the resulting square and the $x$-axis is twice the angle between the original square and the $x$-axis.

So far we're just adding more operations and interpreting them in coherent ways. But what happens when we put things together? For instance, what happens if we note the simple fact that $X^2 = X \cdot X$? Here things start to turn interesting. On the left hand side, our interpretation has only to do with matrices as squares. But when we apply our previous interpretation to the right hand side, we are forced to take the sign $X$, which on the left hand side stood alone with a single interpretation, and impose upon it another interpretation: that of a rotation.

The single left hand side $X$ splits into two at one and the same time. Here we don't just have polysemy. By setting this equality *and* retaining our previous interpretations we force a *shift of meaning*: a matrix designating a square suddenly designates a rotation, because it happened to be raised to the second power and set in a formula. A formal manipulation, combined with inherited interpretations, forced a shift of meaning: from one sign and one interpretation, we turn to a reiterated[2] sign and two interpretations. We can no longer think of the left-hand $X$ as *only* standing for a square, as we could before, because the right hand side and the connecting equality force on $X$ our second interpretation. The left hand interpretation was contaminated by the excess meaning in the right hand interpretation. One interpretation was forced on another following a formal identity.

But we have to qualify in what way this shift of meaning is *forced*. Of course, we need't have acknowledged any of the interpretations I have suggested above. One can do matrix algebra with many other interpretations, including an interpretation that views matrices just as arrays of four numbers. However, here we're dealing with mathematical practice, and *in practice* we interpret. Furthermore, mathematics is useful and interesting

---

[2]For a deep analysis of reiteration as constitutive of signs see (Derrida, 1988); but my use of this term at this point is much more limited in its scope.

because it is interpreted. And in saying that, I am not only referring to interpretation for the purpose of application, but also to interpretation for the purpose of generating mathematical conjectures and proofs. I know no mathematician, who never interprets her or his symbols, when thinking about mathematical problems. I know no mathematician, who sticks to just one interpretation at a time.

Whenever one has an isomorphism, one has (at least) a double interpretation: one in terms of the domain of the isomorphism, and one in terms of its range. My point is that mathematical interpretations can and do bump against each other to force shifts of meaning as above. Going from the left hand side to the right hand side of an equality, a sign may change its interpretation. And this holds not only for specific signs, but for entire domains of knowledge as well. Analytic geometry, for example, is not simply two independent ways of thinking put together—geometric and algebraic. It is a novel geometrico-algebraic way of thinking, which is historically distinct, and not reducible to a disjoint union between classical geometry and algebra based on their intermittent use to interpret signs.[3]

But I want to make the plot even thicker. I want to show how interpretations strike limits. For that purpose, we shall introduce one more matrix operation: transposition, denoted by a superscript T. Its definition is

$$\begin{pmatrix} a & b \\ c & d \end{pmatrix}^{\mathrm{T}} = \begin{pmatrix} a & c \\ b & d \end{pmatrix}.$$

This operation is easy to interpret in terms of both squares and rotations. If a matrix stands for a square, its transpose stands for the square obtained by reflecting the sides of the original square across the main axes. If a matrix stands for a rotation, its transpose stands for the same rotation in the opposite direction.

Now there's an easy theorem stating that $(X \cdot Y)^{\mathrm{T}} = Y^{\mathrm{T}} \cdot X^{\mathrm{T}}$. Let's try to read this theorem in the terms of our interpretations above. If we think of $Y$ as a square and of $X$ as a rotation, then the left hand side makes sense (rotate the square and then reflect its sides), but the right hand side involves multiplying a square to the left of a rotation, which is not something we've considered so far (recall that in general matrix multiplication is not commutative, so we can't just switch the matrices around).[4] If, on the other

---

[3] A modern example for this kind of effect (an arbitrary example among unboundedly many, included here just to give a flavour for how to generalise my claims) is the Gelfand representation of elements of commutative $C^*$-algebras as functions operating on the algebra's maximal ideals. This isomorphism or reinterpretation is much more than two distinct interpretations of the same signs; it is a whole that's greater than the sum of its parts.

[4] Actually, the product of positive determinant orthonormal 2-by-2 matrices is com-

hand, we read $X$ as a square and $Y$ as a rotation, then the left hand side no longer makes sense.

One way to deal with the issue is to offer another interpretation for transposition. For instance, we can stipulate that if $X$ is read as a square, we should read $X^T$ as a rotation, and vice versa. Then, if we decide that $X$ is a rotation and $Y$ is a square, the left-hand side product $XY$ yields a rotated square, while on the right-hand side of the equality, $Y^T$, a rotation, stands to the left of $X^T$, a square, and we can maintain the same interpretation to yield, again, a rotated square. Both sides now make sense. But given this interpretation, the equality between the two sides no longer makes sense. Indeed, on the left hand side, we have a transposed rotated square, which according to our interpretation of transposition must stand for a rotation. But the right hand side is simply the application of a rotation to a square, and is, according to our interpretation, a square. We obtain a situation where a rotation on the left equals a square on the right, which is likely to appear more objectionable than the simple claim that a single matrix can represent either.

Now, when confronted with this kind of interpretive dead-end, we can react in various ways. One reaction is to seek other interpretations for transposition and multiplication that work coherently together, and at the same time allow us to retain a sense of rotating squares for application purposes. This would be a reconstruction of interpretations. Another reaction is to keep using our interpretation *locally*, that is to change our interpretations of multiplication and transposition as we go along, even if it means that $X$ or transposition is interpreted in more than one way across a single line of equality. One can refer to this as superposing interpretations. Another strategy is to set interpretations (in terms of squares or rotations) aside for a while, and bring them up only in specific locations, where they are actually useful. This might be called deferring interpretation. All these approaches have productive roles in contemporary mathematical practice. Mathematics is practiced through and across reconstructions, superpositions and deferrals of interpretations. These processes never come to an end, because formal manipulations never come to an end. Things only get more and more involved, and the example above only traces a few initial steps in a rhyzomatic[5] web of semiosis.

There's yet another strategy, of course, more respectably philosophical: to seek an all consuming ontological grounding or logical reconstruction. But this brings us back to Geertz's initial quotation. Mathematics does not

---

mutative, but if we extend our reading to general parallelograms and linear motions commutativity is lost.

[5]For a discussion of the rhyzome as a model of dynamic structural relations with unstable hierarchies, see (Deleuze and Guattari, 1987).

require a global grounding any more than social phenomena do. Society and mathematics work across, against and in conflict with locally reconstructed, superposed and deferred interpretations. That mathematical marks are there is no more questionable than the fact that winks, either burlesque or 'serious' are there, no more questionable than the existence of rocks and dreams.

Indeed, I would like to emphasise how thoroughly local a local interpretation can be. Here's an example. Suppose we want to decompose the matrix product $(I - A)^{-1}(I - 2A)^{-1}$ as $a(I - A)^{-1} + b(I - 2A)^{-1}$, where $a$ and $b$ are numbers, $I$ is the identity matrix, and the superscript $^{-1}$ stands for matrix inversion. Suppose that we already know that the decomposition is possible, and that the values of $a$ and $b$ are independent of the matrix $A$. One of the most straightforward ways to obtain $a$ and $b$ is to multiply both terms by $(I - A)$ and $(I - 2A)$, and equate them. We obtain

$$a(I - 2A) + b(I - A) = I.$$

Next we substitute $I$ and $I/2$ for $A$. We then get the equations $-aI = I$ and $bI/2 = I$, namely $a = -1$ and $b = 2$.

The point here is to note that the manoeuvre we used to obtain the decomposition is dodgy. In particular, we substituted for $A$ a value, which renders the initial expression (the product of inverses) undefined. Nevertheless, this manipulation is taught and practiced widely. Experienced practitioners will have noted the problem at one time or another, and can easily come up with a rigorous justification (e.g., substituting $(1 + \varepsilon)I$ for $A$ and letting $\varepsilon$ go to zero, or extending the matrix algebra to a consistent framework that sets rules for the acceptability of such a manoeuvre). But this doesn't mean that they will let go of the dodgy manoeuvre or refrain from reproducing it in class for their students, with or without an explicit acknowledgement of its dodgy bits.

This example demonstrates how a local interpretation can involve a local logic, which is at odds with formal rigour. Reducing such local logics to errors or shorthand for rigorous practices is a gross misunderstanding of mathematical practice. The reproduction of such practices depends on authority and local knowledge, rather than on a reterritorialisation into a global logical framework. It is true that such practices can and do lead to inconsistencies, but avoiding such inconsistencies depends on experience, rules of thumb and gut feelings (I witnessed the use of all these expressions in similar contexts by prominent practicing mathematicians) no less than on rigorous reconstructions. And no philosophy of mathematics should repress these thick descriptive facts. Latour instructed us to follow the scientists wherever they go. Since the approach of this essay is semiotic, I suggest here that we obey the slightly different directive of following signs wherever they go. And following signs does not require grounding them in an ontology or a logic—but more on that in the next section.

Before we continue with the argument of this paper, we must acknowledge the shortcomings of the above discussion with respect to our promise to experiment with a thick description of a mathematical case study. In my quotations from Geertz above I suppressed under square brackets, ellipsis and a premature quotation mark Geertz's reference to fraud, parody and rehearsal. This suppression was a strategic device. Suppressing forgery, parody and rehearsal made it easier to graft the anthropologist's notion of thick description onto a realm where it might seem foreign. But once the grafting is performed, it becomes easier to force onto the surface of mathematical practice that which we've been suppressing in our strategic formulation. My purpose is not merely to analyse an existing object (mathematical practice) with a given technique (thick description) but to synthesise them into something that can grow in somewhat less traditional directions.

So let's see if a mathematical sign can express its meaning fraudulently, parodically or in way of a rehearsal. This question has an easy positive answer, if we consider the social framework. Teachers openly 'cheat' in classrooms (misusing signs, practicing invalid manipulations, cutting corners), as do professionals, more or less cautiously, in more or less formal communications. Ideally, all these 'frauds' are correctable, and perhaps merit the title 'white lies' rather than frauds, but cases of scholars making unsubstantiated false claims to gain prestige are not unheard of. As for rehearsals—students rehearse mathematical performances when they do exercises, so that they'll get it right when they're demanded to perform mathematics in exams or in 'real life'. And it is also the case that our academic culture is replete with parodic expressions (algebraic abstractions that are perceived as producing no added value are a target for parody, as are tedious variations on well known themes).

But I stated above that I am interested in the semiotic slice of a thick description, and I should therefore look for fraud, parody or rehearsal at the semiotic level of textual interpretation, not at the level of social context. Indeed, suppose I make algebraic manipulations of matrices pretending that they were real numbers, then check to verify the end result, and finally go back to adapt the original derivation to the non-commutative ring of matrices with its zero divisors. Was my first step some sort of rehearsal? When I follow a geometric argument by thinking of a specific example, or verify it by means of tracing a concrete diagram, am I not fraudulently exchanging the specific for the general? And when Russell introduced his antinomy into Frege's notion of set, did he not follow Frege's own rules and way of working, bringing them to an extreme that exposed an unsubstantiated pretense in Frege's conduct—in other words, performed a parodying mimicry of Frege's practice?

One can come up with good objections to the use of the terms 'fraud', 'rehearsal' and 'parody' for these examples. Indeed, a premature wholesale endorsement of such terms for these examples would underplay the unique features of mathematical practice. A finer reading of the above examples, which will describe them in more accurate terms, may indeed be philosophically interesting, but I will not pursue it here. My point here is simply that there's enough richness to mathematical practice, even at the semiotic level, that justifies grafting the thick description framework onto it. This graft will undoubtedly require some adaptations, but could provide an interesting and challenging analytic framework, the main advantage of which would be its insistence on including aspects that do not fit into prevalent logical or ontological regulative ideas. After all, as I will argue below, a mathematics that's not re-interpreted is a mathematics that's more likely to become irrelevant and outdated.

Finally, inclusion and regulative ideas bring us to the last component of thick descriptions that will concern us here. I am referring here to the "hierarchy" in Geertz's "hierarchy of interpretations". Interpretations are not free floating products of unconstrained human thought. Interpretations are discursive productions, invented, reformed, taught and reproduced. As such, they depend on what Foucault calls a "principle of rarity". This principle reminds us that "on the basis of the grammar and of the wealth of vocabulary available at a given period [that allow unboundedly many combinations], there are, in total, relatively few things that are said", and that the discourse formed by these relatively few actually said statements therefore

> appears as an asset—finite, limited, desirable, useful—that has its own rules of appearance, but also its own conditions of appropriation and operation; an asset that consequently, from the moment of its existence (and not only in its "practical applications"), poses the question of power; an asset that is, by nature, the object of a struggle, a political struggle. (Foucault, 1972, pp. 19–20)

In the philosophical and public discourses about mathematics some interpretations are reproduced more than others. The local, unstable, perpetually deferred aspects of mathematical sign interpretations are often suppressed. They are assigned hardly any philosophical importance or consequence. They are banished to the domain of mathematicians' colloquial practices, which supposedly do not reflect the respectable image of mathematics, Galileo's "language of mathematics", the language of "this grand book—the universe" (Galilei, 1623, pp. 123–123). The grand global narratives (not without grand problems themselves) of set theory, category theory, model and modal reconstructions and foundational ontologies are those that gain the upper hand in the discursive struggle (which, according to Foucault, would be a "political struggle") over philosophical and public representations of mathematics.

I do acknowledge that most foundational theories are interesting and productive, and can, to an extent, serve as interpretations of mathematical practice. As such they are no less worthy than any other local interpretation, each with its various practical roles. But when the debate focuses, as it often does, on which *one* of these theories is the universally fundamental one, rather than on how they interact with each other and with more local interpretations, one ends up with a problematic image of mathematics that has everything to do with protecting its uniquely authoritative place in contemporary politics of knowledge. This leads to a tension that's often felt between the promise of foundational interpretations and the disturbingly incongruous and conflicting results of mathematical statistics and economics, or the spin into 'epicycles upon epicycles' in natural science models. The fault for these tensions is never perceived to lie with mathematics. And indeed it shouldn't. Because mathematics is a web of local interpretations, and cannot support the weight of such a "grand book" as "the universe". What is at fault here are the great expectations projected onto mathematics by holding on to a single grand, and excruciatingly thin, description of mathematics.[6]

## 3   Structure

In order to critically evaluate the argument above concerning local interpretations and shifts of meaning, let us try to present it again with respect to a more elementary case study. Let's consider not matrices, squares and rotations, but integers, rows of oranges and repetitions.

A positive integer $x$ can be interpreted in many ways. For example, a positive integer $x$ can stand for a row of oranges in 1-1 correspondence with the elements of $\{1, 2, \ldots, x\}$. An integer $x$ can also stand for the application of repetitions in 1-1 correspondence with the elements of $\{1, 2, \ldots, x\}$. We

---

[6]One may object that the thick description framework advocated here is just as foundational and universal as the classical approaches, in that it claims that a hierarchy of interpretations along the lines presented above is universally applicable to mathematics. But a thick description is a methodology, which, when applied to different circumstances, draws very different pictures. I do believe that mathematics is always practiced with reconstructions, superpositions or deferrals of interpretations, but this is not a claim I seek to demonstrate. When I find such phenomena in different mathematical case studies and different historical periods (Gödel's proof, contemporary graph theory and algebraic combinatorics, classical Greek geometry and my forthcoming work on Renaissance algebra) I find them expressed so differently that even the theoretical resources I use keep changing (Wagner, 2008, 2009b,c,a). Still different expressions of such phenomena are the *multiplexity* of Dynkin diagrams studied in Lefebvre (2002) and the "useful ambiguity" that Grosholz (2007) finds in the work of Leibniz. There is something universalising in the decree to interpret thickly, but this does not impose a single language or system of interpretation.

can then interpret the product $x \cdot y$ as $x$ repetitions (set side by side) of
the row of $y$ oranges. The $x$ is interpreted as repetitions, the $y$ as a row of
oranges, and the product is an integer that's also interpreted as a row of
oranges. Raising to the second power can also be easily interpreted: if $x$ is
interpreted as a row of oranges, we can interpret $x^2$ as what we'd get by
completing the given row of oranges to form a square.

Now, when we observe an equality of the form $3^2 = 3 \cdot 3$, we find again
two of the phenomena encountered above. First, there's a shift of meaning.
The term 3, a single sign interpreted on the left-hand side strictly as a
row of oranges, is forced, by way of its formal plugging into an equality, to
become two different things on the right-hand side: a row of oranges and
a bunch of repetitions. A formal manipulation, coupled with the retention
of our earlier interpretation, forced a sign to split into two signs with two
different interpretations. Moreover, while on the left hand side we have a
square of oranges, on the right hand side we have a row of oranges, which
our retained interpretations force around the sign of equality, that is, posit
as one and the same.

Now that we're in the philosopher of mathematics' favoured domain—
arithmetic and the integers—the philosopher's response springs up much
more clearly and emphatically. $x$ should not be interpreted either as oranges
or repetitions. $x$ is, and should be interpreted as, number! And if we carry
the same logic back to the previous section, it is obvious that our matrices
should not be interpreted either as squares of as rotations. Matrices should
be interpreted as that mathematical object which matrices are. Of course,
we can apply mathematical knowledge to oranges, repetitions, squares and
rotations—but that's at the level of application. As long as we're doing
mathematics, a number should be a number and a matrix should be a
matrix.

Suppose we allow this marginalisation of what practitioners of mathematics actually do. We are then left with the question: what are numbers
and matrices as mathematical objects? This question catapults us back
into the debate of ontologies and epistemologies of mathematics. In order
not to go again over the already infinite and outmoded debate of realism-
intuitionism-formalism, I suggest we take a detour through the more recent
ontologico-epistemological approaches grouped under the term *structuralism*, where some relatively new interesting remarks can still be made.[7]

The common point for structuralists is that they're interested not in
individual mathematical objects, but in their relational aspects. This, how-

[7]Not that this would be too much of a detour. A quick glance at contemporary forms
of structuralism shows that, like the formalist-intuitionist-realist divide, many of them
carry the traces of mediaeval nominalist-conceptualist-realist positions on the problem
of universals (I do not mean this as a reproach, but as the expression of an important
positive link with the history of philosophy).

ever, can be done in various different ways. One can hold a *methodological structuralist* position, according to which mathematics is concerned with what follows from axiomatically postulated relations, and set aside the entire question of objects. There's the *set theoretical structuralist* position, according to which mathematics deals with what's common to models of axiomatic theories built inside set theory, but defers the question of what sets precisely are, or solves this question non-structurally. There's the *ante-rem structuralist*, for whom structures are abstractions of mathematical models, whose elements are 'objects' that have only relational properties (or at least only relational essential properties, as one can't avoid accidental properties such as 4 being the number of legs that Lassie—or the dogs portraying her in the film – had). And then there are *in re structuralists*, who read mathematical statements not as referring to any specific object, but as quantified over all, or all possible, models of a given structure.

This division above is borrowed from Reck and Price (2000), and is not meant to be exhaustive or authoritative. Variants of these divisions may be found in (Hellman, 2001) and in (Shapiro, 1997), and one should quote at least Resnik (1997), Chihara (2004) and the nice reconstruction of Dedekind's view in Reck (2003)—if not for completeness, then for their interesting highlights. Of course, there's a lively debate concerning the pros and cons of the various structural approaches. These tend to revolve around issues of ontological commitments, uniqueness, individuation, the range of quantifiers and possibilia and the elimination of 'monsters'. Whether or not these problems are essentially shared by all structural approaches (as suggested by Shapiro) or are differently distributed between them (as claimed by Hellman) need not detain us here. Instead, I'd like to look, again, at our case study, in order to point out something that structural approaches tend to leave behind.

So let's go back to our thick description of matrices. As they share the same signs, one could argue, matrices interpreted as squares or as rotations share the same structure.[8] Moreover, shifts of meaning and conflicts of interpretations are set aside if we defer interpretation while doing mathematics, and focus on mathematical structures instead. But the interpretations of matrices as rotations and matrices as squares still leave their trace when abstracting to structures. Our interpretation of matrices as rotations portrayed rotation-matrices as operating on squares, so the structure under hand was conceived of as a set of functions operating on another structure. The interpretation of matrices as squares, on the other hand, did not share this structural feature. The two interpretations are not, after all, fully structurally equivalent.

---

[8]To be precise, one should either restrict to positive determinant orthonormal matrices, or generalise to oriented parallelograms and linear motions, but I'd rather not encumber the argument with this excess terminology.

Now, when considering a given structure, one can always add supplementary relations. For example, compare the integers as a structure satisfying the Peano axioms, and the same integers with the addition of the 'bigger than' relation. Whether we should see these structures as identical, embedded or unrelated is debated among structuralists, and might be undecidable, at least in some versions of mathematical structuralism. If we considered matrices-as-rotations to be the same structure as matrices-as-squares with an additional group operation (matrix product), then we'd be in a similar situation. But that's not the case in the example above.

Our interpretation conceives of rotations as functions over squares. The set-theoretic definition of a function, which involves a set of ordered pairs of domain and range elements, means that when we go from squares to rotations we are not concerned with taking a structure and supplementing it with an extra relation. You can't take just any set and 'add' to its members the 'property' of being functions, as you would add the 'bigger than' relation to a model of the integers.

There's also a difficulty in constructing a single model for both rotations and squares, despite their isomorphism. The canonical set theoretical aversion to sets that are members of themselves prevents the construction of functions that take themselves as arguments. So, if we interpret matrices as rotations operating on squares, we can't simultaneously use the same model both for the rotations and for the squares, even though a single model can model both squares and rotations operating on some other model of squares. What I mean is that we can construct models $A$, $B$ and $C$, where $A$ is a model of matrices as squares, $B$ is a model of matrices as rotations operating on the model $A$ (and therefore itself another model of squares), and $C$ a model of matrices as rotations operating on the model $B$. So the pair $(A, B)$ can model our square-rotations combination, and so can the pair $(B, C)$. But strangely enough, even though the set $B$ can model both squares and rotations, the pair $(B, B)$ can't model rotations operating on squares. From a structural point of view, once a structure starts operating on one of its isomorphs, their previous isomorphism may break into a non-reversible hierarchy. And—here is the crucial point—whether a structure will or will not operate on an isomorph—that is something we needn't decide in advance.

Of course, if we use a set theory that allows functions to operate on themselves, or if our set theory uses ur-elements that can be reconstructed post-hoc as sets of ordered pairs, or if we describe the operation of rotations on squares as a binary operation rather than as a functional relation, the problems above might go away. But I believe that a good philosophy of mathematics should allow us to use (but not confine us to) canonical set theory, and should not deny us thinking of rotations as functions sending squares to squares. And of course, even if the specific problem described above can be avoided by some formal reconstruction, other cases of signs standing for themselves and an excess are likely to surface elsewhere.

My claim is that even the structural interpretation of mathematical signs may be deferred, reconstructed and superposed. Mathematics does not have to have foundations. One can spend a lifetime using matrix algebra for rotating squares without ever having to make decisions concerning splitting the structure of matrices into two structures, of which one operates on the other as functions. This splitting of a structure into two structures, 'active' and 'passive' clones, and the decision whether they are the same or not, can sometimes be a pathology of formal reconstruction, rather than part and parcel of mathematical practice.

Now, I have no intention of denying that some sorts of structuralism in mathematics have contributed to mathematical development (e.g., Dedekind, Noether, Bourbaki). I also do not disqualify structuralism as a philosophical interpretation of mathematics; indeed, structuralism is an important catalyst of interactions between various mathematical interpretations. Furthermore, there are mathematical practices that require the careful articulation of function and domain structures that an analysis of the above example would demand (set theory is, after all, part of mathematics!). But my point is that such articulations need not be made once and for all or in advance. When interpreting matrices, we can consider matrices-as-rotations and matrices-as-squares as the same, as isomorphic, or as essentially hierarchical. And we may also defer or superpose these decisions, or reconstruct them in hindsight (and the same goes for integers, rows of oranges and repetitions). The self-identity of a mathematical object or model is not all that rigid.[9] This is a kind of thickness that structuralism, as well as other universalising ontological-epistemological systems, fail to describe. This is not reason enough to flush structuralism; but this is good reason not to let it monopolise the field.

What underlies the discussion above, and what makes structuralism miss that which a thick description grasps, is the way mathematical structuralists diagnose semiotic problems. To make this point, let's go from contemporary mathematical structuralism to 20th century continental linguistic structuralism. I acknowledge the differences in standards of articulation, which might render even the most rigorous formulations of structural linguists offered by de Saussure, Martinet, Hjelmslev, Jakobson, Trubotskoy or Benveniste less than satisfactory for some contemporary philosophers of

---

[9]The difference and mimesis based philosophies presented in such work as Deleuze (1994) and Derrida (1993), which problematise relations of identity and repetition, enable critical thinking that is not committed to Aristotle's first law of logic ($A = A$). The above argument is, in fact, a restricted and elementary application of Deleuze's interpretation of *eternal return* and/or Derrida's *différance* to mathematical signs.

mathematics. I further suspect that Deleuze's highly philosophical reconstruction of structuralism (Deleuze, 2004), or its post-structural reformu-
lation under the title of "virtual" reality (Deleuze, 1994) would fare even
worse. But I think that mathematical structuralists do have something to
learn from the above scholars of humanities.[10]

A problem posed by mathematical structuralists since Benacerraf (1983)
is that of 'too many' models, each with its own annoying contingencies.
These should either be replaced by an 'ur-model', or by a logical reconstruction of whatever's common to them all. But this was not the problem
of continental linguistic structuralists. Their problem was not that a language such as French had too many models, which they had to unify. Their
problem was that no individuated model of an entire language could ever
be made present. Linguistic structure

> is a fund accumulated by the members of the community through
> the practice of speech, a grammatical system existing potentially in
> every brain, or more exactly in the brains of a group of individuals;
> for the language is never complete in any single individual, but exists
> perfectly only in the collectivity. (de Saussure, 1966, p. 13)

And it is the structural linguist's task to derive the structure from these
local models that are never complete.

But the linguistic structuralists' problem was not simply a problem of
patching together relatively consistent partial models. The different partial models may be in conflict. If one considers all the different ways of
expressing a given linguistic phoneme, one gets an unbounded and fuzzy
realm of vocal manifestations. In strict physicalist terms, no two expressions of a phoneme are ever the same; on the other hand, trying to define a
phoneme in terms of limits on the range of relevant physical properties (the
physical motions of the mouth and larynx, measurements of sound waves)
crashed against grey areas of intersection that were too large to ignore (e.g.,
de Saussure, 1966, p. 106). But not only the distinctions between what's
reconstructed as different phonemes prove problematic; it also made systematic sense to reconstruct phonetic elements that had no vocal expression at
all! (e.g., the presentation of such a manoeuvre by de Saussure in Hjelmslev,
1966, p. 36). As de Saussure put it,

> speech sounds are first and foremost entities which are contrastive,
> relative and negative [...] *In the language itself there are only differ-
> ences* [...] although in general a difference presupposes positive terms

---

[10] An anonymous reviewer suggested that in the Bourbaki group such learning did take
place. Without neglecting the differences between earlier mathematical structuralists
and contemporary structuralist philosophers of mathematics, this would make an interesting topic for a historical survey, which would hopefully be valuable to contemporary
philosophy of mathematics.

> between which the difference holds, in a language there are only differ-
> ences, *and no positive terms* [...] the language includes neither ideas
> nor sounds existing prior to the linguistic system, but only concep-
> tual and phonetic differences arising out of that system. (de Saussure,
> 1966, pp. 117–118)

And nowhere does this force us to hypothesise a consistent and exhaustive
model or modal interpretation of the system, however ideal.

This is not Benacerraf's conundrum of canonizing one model from among
many candidates. This is a problem of extracting aspects that are common
to many incomplete and relatively conflicted forms of expression. My point
here is not to advocate linguistic structuralism as a philosophy of mathematics (it should be quite obvious by now that I am a follower of some of
the reactions to structuralism and its rearticulations subsumed under the
title of post-structuralism). My point is that the problem of mathematical philosophy might not be that of unifying too many models, but that of
coping with the lack thereof.

The position I'm putting here is reflected by Wittgenstein. For Wittgenstein mathematical statements are never simply empirical descriptions of
states of affairs. Indeed, descriptions of states of affairs depend on more or
less accurate measurements. As Wittgenstein puts it,

> "what reality corresponds to the proposition that if you turn a match
> twice through 180° it gets back to its original position?" If this is a
> geometrical proposition, the reality which corresponds is: if we use
> a good protractor, then normally it brings us back, or more nearly
> back the better it is (where "better" is determined by other crite-
> ria). (Wittgenstein, *Lectures on the Foundations of Mathematics*:
> Diamond, 1975, p. 246)

Realities corresponding to mathematical statements are always about "normally", "nearly" and "better". But unlike their corresponding realities or
states of affairs, mathematical statements themselves set standards. That
the measurement and implementation tools which model the standard (the
match and protractor above) are never quite up to the standard—this does
not quite undermine the standard as such. Indeed, Wittgenstein asks

> If [after rotating a match twice 180°] it didn't point in the same direc-
> tion, would you say the protractor was wrong, that it had expanded,
> etc.,—or would you say that in this case twice 180° does not bring
> you back to the same position? (Diamond, 1975, pp. 245–246)

At least there and then, Wittgenstein opts for the former response. The
statement according to which two 180° rotations bring us back to the same
position is a standard according to which we calibrate our instruments. No
practice of measurement or calculation can *model* this standard without

risk of failure. What's most relevant here in this (obviously very sketchy) allusion to Wittgenstein is that, as with continental linguistic structuralism, we do not deal here with what's common to 'too many' good models; mathematics here is a reconstruction related to partial, substandard and interpretation-laden practices.

The breakthrough of the continental linguistic structuralist approach was to understand that a scientific system of rules or differences can be valuable even if it sets an impossible standard, and even if no object, real or ideal, in fact lives up to standard. This holds for mathematical systems as well. They depend for their practice on bits and pieces of substandard formal manipulations and on local, deferred, superposed and reconstructed semantic interpretations. Philosophy of mathematics should affirm this reality and weigh its consequences, rather than restrict itself to attempts at putting together new ideal ur-models or logical interpretations that can stand as referents or senses for mathematical signs. The latter activity can indeed be valuable, but it should be complemented by thick descriptions of mathematical practice, and should not presume to dominate or suppress such descriptions as philosophically inferior.

## 4    Deferring interpretations

I'd like to conclude with some clarification concerning the concept of interpretation that I have been discussing, and how it necessarily thickens the description. I have insisted that in mathematical practice interpretations of signs are subject to local practices of deferral, reconstruction and superposition. A quick look into the term 'superposition' will help us clarify what interpretation is like.

I borrow the term superposition from physics. This borrowing is extremely loose, and its point is not so much to force an analogy as to learn from its limitations. When certain aspects of an object cannot be reduced to a single state (say, a specific location, or a specific spin), we say that the object is in a superposition of states. But to establish superposition it is not enough that we simply don't know the object's state. It is required that we have some testimony to its plurality of states through observed phenomena (such as interference), and that we can learn something about the distribution of those states by measurement. Interference testifies to an object's plurality of states by making it hard to explain the interaction of objects under the hypothesis that each has only one (possibly unknown) state. To explain the interaction one needs (or finds it useful) to assume that the plurality of states of each object influences the results of the interaction. The double-slit experiment is the obvious paradigm.

Can we think of a mathematical sign as being in a superposition of various interpretations, and point out moments of interference? Consider the theorem $(X \cdot Y)^{\mathrm{T}} = Y^{\mathrm{T}} \cdot X^{\mathrm{T}}$ concerning products of matrices, and our interpretation of the product of 2-by-2 matrices as rotations applied to squares. We've already noted that, if the left-hand factor of the product should be interpreted as a rotation and the right hand side as a square, then only one side of the equality can make sense: the left-hand side of the equality if $X$ is the motion, the right hand side if $Y$ is the motion. In order to maintain both the equality and our interpretation, we might want to allow $X$ and $Y$ to be in a superposition of rotation and square interpretations, and consider the equality as an indication of interference.

Of course, one can conclude that this problem rules out the interpretation of matrix product as rotation applied to square. But that would be an overkill, and annoying news for people who use this interpretation in technical applications of geometric linear algebra. But I'm not going to defend this notion of interference of interpretations against such overkill, as my point is not to press the analogy between superposition of physical states and the superposition of interpretations. I am more interested in following the concept of measurement as it goes through the 'superposition of interpretations' metaphor.

In the quantum context, measurement is a process that collapses a superposition of states to a single state. In the context of interpreting signs we may think of committing a sign to a single interpretation as collapse. From a superposition of interpretations we collapse the meaning of the mathematical sign onto a single sense and referent. We may have played with our matrices and their possible meanings, we may have put them through various formal manipulations that effected shifts of meanings, but finally comes the moment of decision. Here, in this application of the bottom line of our reasoning, this or that specific matrix must finally be used in a solid, well-defined particular way.

But is such a collapse necessarily definitive? The obvious objection is our ability to go back to the mathematical text and interpret or use it differently. When a text has been interpreted, it has never been interpreted once and for all; it can always be re-interpreted, and that goes for a mathematical text as well—"a written sign carries with it a force that breaks with its context" (Derrida, 1988, p. 8).

This objection, however, is perhaps too easy. Can't we say at least that at the moment of interpretation or application there is a definitive collapse into a single interpretation? Well, suppose a programmer used a certain matrix to designate a certain square, which will be displayed and rotated on the screen. Has interpretation come to an end? Not quite. The code is to be processed by a certain compiler, then executed on a certain machine, and eventually output onto a certain medium. Anyone experienced with the endless machine specific variations that can ensue is well aware that interpretations have not yet come to an end. And when we finally observe the square rotate on some LCD screen, have interpretation now finally come to an end? They have not, at least as long as someone is there to observe the rotating square and interpret its motions: experience them aesthetically, derive information from the display, act on whatever the rotating square prompts them to do, etc. And things needn't end there. This experience may be remembered, recalled, evaluated, recounted, recontextualised; it may instruct us, reproduce itself in future experiences, enter chains of interpretive expectations and habits; in short, interpretations, like explanations, need never come to an end. They might factually come to an end, but they never need end at any given present moment of time.

But the above example takes our interpretation outside mathematics. Is it not the case that as long as we stay inside mathematics, interpretations must eventually come to an end? In fact, it is never finally decided when an interpretation carries a sign outside mathematics. When I interpret a matrix as a square, is the square no longer mathematical? It depends. A square may well be an empirical object of observation. But it may just as well be a mathematical object.

> Mathematical propositions might quite well be expressed in terms of people, houses, or what not. The word "men" may come in and it may still be mathematics; and the word "lines" may come in and it may not be mathematics. (Diamond, 1975, p. 116)

For Wittgenstein, whether the square (or men, or a line) is inside or outside mathematics depends on how we operate with it. As we observed above, if our dealings with the square set standards (say, if we state that the square's diagonal divides it into two congruent parts, regardless of what empirical measurements suggest), then the square is still mathematical. If our dealings with the square are more empirical or pragmatic, then, according to Wittgenstein, we're no longer inside mathematics. The catch is that "of course the sentence 'The figure I have drawn here [. . .]' may be used either mathematically or non-mathematically" (Diamond, 1975, p. 117). There's nothing in the square itself that forces us to use it either way, that forces us in or out of mathematics. Our interpretation may oscillate in and out of mathematics in ways that might question the topology of these in/out relations.

Unlike the apparently once-and-for-all quantum collapse, mathematical interpretations do not mark a clear boundary of final interpretation or a way out of mathematics. Furthermore, unlike quantum superpositions, which superpose a well defined space of once-and-for-all given states, mathematical interpretation does not take place in a closed and well articulated domain. Mathematical reconstruction of interpretation is an open ended process.

This open-endedness is not the margin of mathematics. It is its decentred centre—a condition of possibility that is not reducible to a stable core. Even if mathematics is grasped as geared not towards practice, but towards some ideality, a mathematics that cannot be reinterpreted is a mathematics that is bound to become outdated, once travelling—historically, culturally, intersubjectively—across our life worlds renders a given interpretation obsolete (a glimpse at the kind of algebraic geometry that dominated $19^{th}$ century professional mathematical literature will provide a fine instance of this claim).[11]

Reducing mathematics from a thick practice of interpretation to a structural ontological or logical core prevents philosophers of mathematics from acknowledging its plurality.[12] By confining mathematics in such manner, scholars actually prevent mathematics from attaining its cross historical and cross cultural ideality—an ideal openness to reinterpretation. Indeed, such confinement makes mathematics inaccessible to the many, who would otherwise access mathematics through the thick of different interpretations. Conceptually unified and confined, mathematics might not survive the historic obsolescence of our fashions and conceptual schemes.

As Cantor stipulated "the *essence* of mathematics lies precisely in its *freedom*" (quoted in Reck, 2003, p. 392). To maintain this freedom is to affirm the thickness of the scientist's reinterpretation of mathematical signs across life worlds and across moments of her own life. To maintain this freedom, even as a freedom to idealise, is to acknowledge the constitutive role of locally superposed, reconstructed and deferred interpretations in the production, transmission, reformation and sustenance of mathematical signs. Without such reinterpretation mathematics might not survive long enough to become ideal.

---

[11]This interpretation relates to Derrida's interpretation of Husserl's work in Derrida (1989). Indeed, even for an ideally oriented thinker such as Husserl, for whom mathematics is a "product arising out of an idealising, spiritual act, one of 'pure' thinking", mathematics must derive from "factual humanity and [the] human surrounding world" (Husserl, 1939, p. 179). Since this is a changing factuality and world, not only historically, but culturally and intersubjectively as well, a scientist, who tries to impose an 'ur-interpretation' on mathematics (structural or other), detaches mathematics from worldliness in general, and ties it down to "something valid for him as a merely factual tradition". His subsequent interpretation therefore "would likewise have a merely time-bound [or community bound] ontic meaning: this meaning would be understandable only by those men who shared the same merely factual presuppositions of understanding" (Husserl, 1939, p. 179).

[12]Greiffenhagen and Sharrock (2006) provide a recent typical example of this manoeuvre: the authors hack off all cultural or historic differences between mathematical practices, interpretations and systems as minor and non essential, avoid the task of articulating what it is—if anything—that survives this hacking, and conclude that mathematics is highly non-relative.

# Bibliography

Benacerraf, P. (1983). What numbers could not be. In Benacerraf, P. and Putnam, H., editors, *Philosophy of Mathematics: Selected Readings*, pages 272–294. Cambridge University Press, Cambridge.

Chihara, C. S. (2004). *A Structural Account of Mathematics*. Clarendon Press, Oxford.

de Saussure, F. (1966). *Course in General Linguistics*. McGraw-Hill, New York NY. Translated by W. Baskin.

Deleuze, G. (1994). *Difference and Repetition*. Columbia University Press, New York NY. Translated by P. Patton.

Deleuze, G. (2004). How do we recognize structuralism. In Lapoujade, D., editor, *Desert Islands and Other Texts*, pages 170–192. Semiotext(e), Los Angeles CA. Translated by Mike Taormina.

Deleuze, G. and Guattari, F. (1987). *A Thousand Plateaus*. University of Minnesota Press, Minneapolis MN.

Derrida, J. (1988). *Limited Inc*. Northwestern University Press, Evanston IL. Edited by Gerald Graff, translated by Jeffrey Mehlman and Samuel Weber.

Derrida, J. (1989). *Edmund Husserl's Origin of Geometry, an Introduction*. University of Nebraska Press, Lincoln NE. Translated by John P. Leavy Jr.

Derrida, J. (1993). *Dissemination*. Athlone Press, London.

Diamond, C., editor (1975). *Wittgenstein's Lectures on the Foundations of Mathematics*. The University of Chicago Press, Chicago IL.

Drake, S. and O'Malley, C. D., editors (1960). *The Controversy on the Comets of 1618: Galileo Galilei, Horatio Grassi, Mario Guiducci, Johann Kepler*. University of Pennsylvania Press, Philadelphia PA.

Foucault, M. (1972). *The Archeology of Knowledge*. Pantheon Books, New York NY. Translated by A. M. S. Smith.

Galilei, G. (1623). *Il saggiatore*. Giacomo Mascardi, Rome. Quoted after the English translation in (Drake and O'Malley, 1960, pp. 151–336).

Geertz, C. (1973). *The Interpretation of Culture*. Basic Books, New York NY.

Greiffenhagen, C. and Sharrock, W. (2006). Mathematical relativism: Logic, grammar, and arithmetic in cultural comparison. *Journal for the Theory of Social Behaviour*, 36(2):97–117.

Grosholz, E. R. (2007). *Representation and Productive Ambiguity in Mathematics and the Sciences*. Oxford University Press, Oxford.

Hellman, G. (2001). Three varieties of mathematical structuralism. *Philosophia Mathematica*, 9(3):184–211.

Hjelmslev, L. (1966). *Le Langage, une Introduction*. Galimard, Paris. Translated by M. Olsen.

Husserl, E. (1939). Die Frage nach dem Ursprung der Geometrie als intentional-historisches Problem. *Revue Internationale de Philosophie*, 1:203–225. Quoted after the English translation in (Derrida, 1989, pp. 155-180).

Lefebvre, M. (2002). Construction et déconstruction des diagrammes de Dynkin. *Actes de la Recherche en Sciences Sociales*, 141–142:121–124.

Livingston, E. (1986). *The Ethnomethodological Foundations of Mathematics*. Routledge & Kegan Paul, London.

Livingston, E. (1999). Cultures of proving. *Social Studies of Science*, 29(6):867–888.

Netz, R. (1999). *The Shaping of Deduction in Greek Mathematics*, volume 51 of *Ideas in Context*. Cambridge University Press, Cambridge.

Reck, E. H. (2003). Dedekind's structrualism: An interpretation and partial defense. *Synthese*, 137:369–419.

Reck, E. H. and Price, M. P. (2000). Structures and structuralism in contemporary philosophy of mathematics. *Synthese*, 125:341–383.

Resnik, M. D. (1997). *Mathematics as a Science of Patterns*. Clarendon Press, Oxford.

Rosental, C. (2008a). Apprendre à voir apparaître des formes, des structures et des symboles: Le cas de l'enseignement de la logique à l'université. In Lahire, B. and Rosental, C., editors, *La cognition au prisme des sciences sociales*, pages 161–189. Éditions des Archives Contemporaines, Paris.

Rosental, C. (2008b). *Weaving Self-Evidence: A Sociology of Logic*. Princeton University Press, Princeton NJ.

Shapiro, S. (1997). *Philosophy of Mathematics: Structure and Ontology.* Oxford University Press, Oxford.

Wagner, R. (2008). Post structural readings of a logico-mathematical text. *Perspectives on Science*, 16(2):196–230.

Wagner, R. (2009a). For some histories of Greek mathematics. *Science in Context*, 22:535–565.

Wagner, R. (2009b). Mathematical marriages: intercourse between mathematics and semiotic choice. *Social Studies of Science*, 32(9):289–309.

Wagner, R. (2009c). Mathematical variables as indigenous concepts. *International Studies in the Philosophy of Science*, 32(1):1–18.